EPIC RETREATS

EPIC RETREATS

From 1776 to the Evacuation of Saigon

By
STEPHEN TANNER

CASTLE BOOKS

This edition published in 2002 by Castle Books,
A division of Book Sales Inc.
114 Northfield Avenue
Edison, NJ 08837

Reprinted by arrangement with Sarpedon
An imprint of Combined Publishing
476 W. Elm Street
Conshohocken, PA 19428

Copyright © 2000 by Stephen Tanner

ISBN 0-7858-1403-5

Contents

LIST OF MAPS

Introduction

Over the past year, upon responding to inquiries with the news that my current project in military history concerned "epic retreats," listeners would as often as not chuckle. Retreats, meaning fleeing for one's life? Bug-outs? The glorious history of armies failing dismally in battle and then trying to get away? I've invariably found it necessary to explain the work in more detail to interested parties. Although the basis for the humor that may exist on the surface of this subject has not escaped me, and, in fact, I have never thought that a work comprised of history's most frantic episodes of cowardice would be uninteresting, such was not my plan for this book.

My idea was rather that during a strategic retreat, armies may need to call upon greater skill and fortitude than they would during an advance. Individual heroism and leadership skill are not necessarily diminished when the strategic situation in a war requires withdrawal; on the contrary, when men have been placed at a disadvantage, particularly due to events beyond their control, they may need to display more courage, discipline and nerve than when the success of a campaign appears to be ensured.

In each of the following chapters, the retreating army was by no means inferior to the pursuing one, at least on a man-to-man basis, but had other problems that compelled it to withdraw. The only exception might be Washington's forces, which were clearly inferior to the British Army in 1776; however, even then the best of the Colonial formations were equal to the best of the British. The intriguing question is what happens when a resolute army is undermined by circumstances out of its hands. In most cases, the force in the dominant position will press its advantage and the fleeing enemy will gradually dwindle into insignificance. Sometimes, however, the disadvantaged army will not go so quietly, and then the retreaters become the most fascinating ones to watch.

This work examines a number of campaigns briefly and thus consists of narrative essays, analytical in nature. The text is sprinkled with quotes from participants—somewhat decreasing with the timeline as the colorful, engaging prose of the 18th century evolves into the bare, factual reports of the 20th—employed primarily to convey a sense of the emotions or urgency felt by participants. After examining thousands of pages of histories of these events, the responsibility to convey each story in shortened form while staying faithful to the larger picture weighed heavily. None of the chapters that follow provide a complete story of the campaigns, all of which have a multitude of military, political and human dimensions, yet I hope to have achieved a certain clarity that will benefit readers who have not yet delved into the topics, while providing a thought-provoking perspective for those who have studied them at length.

In the selection of retreats I have followed my own interests, though with an eye toward providing some historical balance across time. I regret having found no suitable candidate from the American Civil War, and have also wondered why. Strategic withdrawals did occur during that huge conflict, usually to reconnect with sources of supply, but none of them struck me as "epic." At the same time the Civil War is notable for its lack of open field battles of decisive strategic importance such as Napoleon regularly achieved and which decided the Austro-Prussian and Franco-Prussian Wars just a few years after Appomattox. Gettysburg, for example, is considered to be a crucial turning point in the Civil War, but it concluded with the Union holding off a Confederate attack, after which the Rebels withdrew, unbothered by pursuit. As for a connection between the lack of epic retreats and the lack of decisive battles in the Civil War (a large topic), I will here confine myself to the observation that the North and the South were evenly matched adversaries—both without precedent for defeat and possessing more passion than was generally found among European combatants—and that the war's outcome was eventually determined by factors other than prowess on the battlefield.

The most difficult choice for a retreat was which German one to include from World War II. Given the wealth of material, a work larger than this could focus exclusively on German retreats, from North Africa to the Arctic, Italy to France plus several in Russia, not to mention the naval "breakback" in the Channel. The retrieval of the Caucasus offensive, however, struck me as the most interesting, not least because at that point in the war it was only then dawning on the German High Command that they were in far over their heads.

In format, each of the seven chapters consists of a short background to the conflict, followed by an account of an army flush with success. Epic retreats, after all, do not occur unless an army has initially advanced, believing itself to be successful. Then the bell tolls, and the triumphant forces realize that they are in great danger and need to get away—as events unfold, their objective changes from victory to mere survival. There are innumerable cases in military history where an army surrenders, falls apart, or when political leaders simply negotiate an end to the fighting once an immense disadvantage has become clear. It is more exceptional that an army remains determined to stay in the field when the prospect of victory has all but slipped away.

To comment on the chapters in detail, I would be remiss in not stating that the more one learns about George Washington the more one comes to admire him. Having been used to his profoundly uninspiring portrait on the dollar bill, as well as being of a generation subject to his "cherry tree" legend, employed by adults at the time to inspire (intimidate) young children, I had no great fondness for him in my earlier years. Nevertheless he was clearly responsible, far more than any other, for implementing the American Revolution. At least I suspect that if Washington had not maintained command of the American army the Revolution would have degenerated into a bitter unconventional war, and perhaps even have approached the degree of degeneracy experienced by the French a decade later. Though for a while he did not appear to be the most successful of American generals, he was able to survive harsh adversity while never losing sight of an essential "dignity" that had to accompany the cause.

The clash that triggered Washington's series of crises has been called the Battle of Long Island, but I have embraced a recent trend to revive the name, heard sporadically across the decades, the Battle of Brooklyn. General Howe marked his victory dispatch to London "Long Island," no doubt assuming the king would be unimpressed to hear he had conquered Flatbush, Gowanus or the little-known Dutch hamlet of "Breuklynne." In the years since, however, Long Island has become an entirely separate political, not to mention cultural, entity from the locale where the fighting took place. One can traverse modern Long Island's gigantic web of parkways indefinitely and not come within sight of a battlefield; yet a short subway trip under the East River will place one at the scene of the largest battle of the Revolution.

Napoleon's statement to the effect that he could "afford to lose 30,000 men a month" has inhibited many people from sharing the unbounded admiration with which he is viewed by numerous historians. Still, one cannot escape a certain awe when examining his career, after which an entire era in European history took on his name. For the benefit of those who confuse Jena-Auerstadt with Aspern-Essling, or, in other words, all but a handful of the English-speaking population, I have included a synopsis of Napoleon's career prior to 1812 in order to put the Russian campaign into proper context. Such was Napoleon's ability to adjust the course of history to his personal whim that a great multilingual army marched, without knowing why, to oblivion. The confidence these men had in their invincible emperor made all the more poignant the "mother of all retreats" that followed.

The retreat of the Nez Perce was clearly epic, though at best they could summon only 300 fighters. In a way, the enhanced spotlight on the individuals in those battles makes for a more interesting saga; and they did travel 1,500 miles in their attempt to reach safety. I have never subscribed to the view that Native Americans possessed a moral superiority over their white antagonists; although as underdogs, and perhaps also as naturalists, they have received much more retrospective sympathy. No matter how arguments for and against United States expansion are turned, free-roaming tribes did not really have a future by the end of the 19th century except in remote regions of the world like the Amazon rain forest. The Nez Perce, as one of the last tribes willing to take on the overwhelming forces of a modern state in battle, merit a great deal of respect, as do the cavalry troopers who chased them, since the tribe, by then armed with modern guns, was renowned for its marksmanship.

An interesting aspect of the 1940 Battle of France is that the British Expeditionary Force, the most formidable army on the Allied side, exercised its prerogative to retreat from the Continent just two weeks after the campaign had begun. Notwithstanding that other Allied troops were surrendering in droves, Winston Churchill was acutely sensitive to the perception of a British "bug-out." The greatest value of the Dunkirk evacuation was not, at first, the practical benefit of Britain retrieving its army; the RAF was able to forestall German invasion designs by itself later that summer while the British Army did little for the next year but perfect its evacuation skills in southern France, Greece and Crete. The real benefit from Dunkirk was the terrific boost it provided to British morale—alone among the nations targeted by Germany, Britain had ducked the haymaker. The Dunkirk "miracle," combined with the air-waged Battle of Britain, allowed Churchill to enlist a palpable spirit of "destiny" behind his war effort. Even when ultimate victory was not in sight, the British were inspired to fight on with every means at their disposal until, a few years later, they returned to the Continent buttressed by much more formidable allies than the ones they had left behind: the Soviet Union and the United States, instead of France and Belgium.

In 1942 as in 1812, the Russians executed a successful series of retreats, increasing, as much by accident as design, the vulnerability of their invader. In the Soviet Union, I believe that Hitler's abandonment of the territory gained by the German offensive in 1942 marked the end of his hopes to force a favorable decision in the war. Afterward, he could still grasp at straws, hope for the best and cling on as long as he could. Nevertheless, Germany's inability to wrest away Stalin's primary oil fields meant that the Third Reich had a severely limited future. It may be considered ironic that the Germans didn't realize what Rommel was sitting on all that time in North Africa; since they staked their strength in 1942 on seizing the oil fields of the Caucasus, it was there that retreat had the most important effect on the rest of the war.

It is pure coincidence that the final three chapters of this book concern the efforts of Western armies against Communist forces, but such has been the story of the 20th century after the Great War. When the Chinese came into the Korean War, the American Army was forced to re-learn significant lessons from the Russo-German conflict. Communist winter offensives preceded by infiltration and conducted without regard to casualties required the defenders to call upon deep reservoirs of nerve. The Chinese, for their part, were forced to rely on immense courage and self-sacrifice in the face of far greater firepower than they had encountered from the Japanese or Nationalists. During the Chinese Second Phase Offensive, the withdrawal of the Marines from Chosin Reservoir so quickly became renowned that it has always drawn attention from the more chaotic UN retreat that took place in the west; needless to say, my interest has focused primarily on the latter. Neither Eighth Army nor Chinese XIIIth Army Group were specially trained elites; it's difficult to escape an abiding respect for the heroism of each once they came to grips.

The idea of "limited war" was welcomed in Korea because it meant that superpower conflict did not have to result in nuclear holocaust; in Vietnam, however, the flaws in the concept became painfully evident. Did a limited war require only limited support, or limited time and resources? If Vietnamese were clearly not a threat to American security, could they still be considered a "limited" threat due to their alliance with the Soviet Union and China? During the war almost an entire generation of Americans rebelled at the thought of fighting in Vietnam, and the battles of the peace movement to influence U.S. policy appear as a thread throughout the chapter.

In one sense, the U.S. withdrawal from Vietnam, combined with the various South Vietnamese failures, stand apart from the other retreats in this work—the unbowed Americans were not pursued, while the South Vietnamese Army's performace typifies the inability to withstand pressure that we have otherwise, thematically, chosen to avoid. My view, however, is that within the category of great retreats in modern history, America's from South Vietnam holds a prominent place. For four years the retreat was disguised, given cover names such as "Vietnamization," and featured vicious counterattacks against an enemy that for years was unable to prove battlefield superiority. But a huge, agonizing retreat it was. At the very end, in Saigon, the Americans had no further means of concealing the fact that they were trying to bring their people out any way possible. During that evacuation, when U.S. forces returned to Vietnam waters and airspace in strength, it's significant that the Communists held their fire. The Americans, much to the relief of the huge conventional army North Vietnam had by then assembled, had only come to retrieve civilian personnel, completing their total withdrawal.

As it relates to the millions of casualties sustained by North Korea and North Vietnam in fighting for the triumph of their socio-economic

system, the subsequent voluntary dissolution of the Soviet Union 1989-
-91 is almost too ironic for words.

The reader may notice that some chapter titles in the following pages
are not precise, though I am happy that my editors have accepted my
choices. "Washington in New York and New Jersey," for example, seemed
to lack a certain ring. "The Allies Falling Back Through Belgium and
France Toward Dunkirk" seemed overly long, as did "The Nez Perce in
Idaho, Wyoming and Montana." I trust that the gist of the accounts are
not betrayed by my abbreviated title phrases.

I am deeply in debt to historians whose estimable works have examined these subjects before in immeasurably more detail and at far greater
length. The term "Select Bibliography" in this book can be considered
synonymous with "recommended reading." I would also like to thank
Stefan Korshak of the Kiev *Times* for providing translations of several
reports from 1812 from the Russian point of view; Colonel Robert
Hallahan, who fought with Task Force Smith and then to the Chongchon
River and back, for his perspective; and Chris Woods, who has settled the
question of who piloted the last Marine helicopter into Saigon. In addition, thanks go to editors Donn Teal and Rob Weisser for their valuable
input, and to Jay Karamales for his illuminating maps. A special debt is
owed Samuel A. Southworth, editor of Sarpedon Publishers, for his
repeated perusals of the manuscript and learned encouragement through
each stage of development.

Any factual errors that may be found in these pages are, of course,
my own responsibility. I'm certain that the interpretations of the titanic
events that follow are absolutely correct; on the other hand, I would
heartily welcome points of view based on truths that have escaped my
notice, and that will provide further illumination to the events.

Although contained within these "epic retreats" are subdramas of triumph, this book is primarily a study of adversity. At the "hinge of fate,"
to use Churchill's phrase, success can prove fleeting or illusory, and withdrawals in the face of superior enemy forces can be called for. The trick is
to withstand the physical and psychological travail. And, after a retreat
has been successfully executed, who knows what might happen next? A
retreat might indicate an irretrievable defeat; however, it might also provide new options at a later time, and a number of new possibilities.

Chapter 1

WASHINGTON
IN NEW YORK

Beginning with the bloody gauntlet run by a regiment of Redcoats in April 1775 at Lexington and Concord, the Crown's forces had proven unprepared for the size and ferocity of the American rebellion. Independence fever had passed through the merchant class and the rabble-rousing intelligentsia into what the British viewed as the "rabble" itself. In America that meant a predominantly rural population spread across a vast territory, accustomed to struggle and self-reliance. Over a decade had passed since the French and Indian War, but the Americans had retained their grass-roots military organization and could field thousands of experienced fighters. They could also, if decisions were made wisely, field leaders who, if not the "equals" of British officers, possessed as much native talent.

The most important battle of the American Revolution's first year was Bunker Hill on June 17, where General Israel Putnam shouted, even if he didn't invent, the phrase "Don't fire until you see the whites of their eyes." The Americans held their ground behind a steep redoubt against British assaults, repulsing two attacks while inflicting nearly fifty percent casualties on their foes—226 dead and 828 wounded—until Empire bayonets finally achieved a brief orgy of revenge. The British, who had expected the mere sight of their assaulting grenadiers to panic the Colonials, were shocked that their finely trained, regular troops could be mangled by such an ad hoc citizens' army. Nathanael Greene, a young colonel from Rhode Island, commented afterwards, "I only wish we could sell them another hill at the same price." Putnam, once the need for inspirational phrases had passed, remarked, "Cover Americans to their chins and they will fight until doomsday."

Following Bunker Hill, the Americans busied themselves construct-ing more fortifications across the land approaches to Boston, while the Continental Congress in Philadelphia named five men who would thenceforth lead the war effort. Since the initial fighting had been done by New Englanders, Artemas Ward of Massachusetts was named a major general, as was Connecticut's Putnam and New York's Philip Schuyler. These men were uniformly past their prime and would be replaced before the onset of the crucial stages of the war. The final choice for major gen-eral, and the most unique, was a man who had been a colonist for only two years, Charles Lee. A former British officer, Lee had fought through the French and Indian War, beginning with Braddock's expedition. He had been wounded at Ticonderoga and had briefly lived among the Iroquois, marrying the daughter of a Seneca chief. (The Indians called him "Boiling Water.") As a major, Lee fought with the British expedi-tionary force in Portugal, 1761–62 and, when his regiment disbanded, joined the Polish army where he was made a major general. He later returned to the East where he campaigned with distinction in the Russo-Turkish War. It was not until 1773 that he resigned his commission as a lieutenant colonel in His Majesty's forces and emigrated to America, where he promptly became a revolutionary.

For the post of supreme commander, John Hancock of Massachusetts had expected to be chosen, but was denied in favor of an impressive Virginian, George Washington. Aside from Charles Lee, Washington could claim as much executive military experience as any man in the Colonies. He too had served with Braddock and throughout that war against the French, and at age 23 had been commander of Virginia's mili-tia. He was handsome, tall in stature, a superb horseman, and a fervent patriot of clear intelligence. Another of Washington's positive attributes was that he was unusually rich, owning a vast, profitable plantation that employed numerous workers and slaves.

If the British thought they were facing little but a local outbreak from rabid New Englanders, the arrival at Boston of a Virginian commander-in-chief, appointed by the Congress in Philadelphia, challenged that assumption. If they had supposed the American army was an unruly mob of merchants plus what they would term the "lower classes," the aristo-cratic, landowning Washington's arrival informed them otherwise.

Washington and Lee, with their military aides and an ever-increasing retinue, rode together from Philadelphia to Boston, receiving acclaim at stops along the way. One can imagine conversations between the two, whose only common ground was that they were both proven leaders. Washington was an elegant man, modest in his new role as leading citizen turned leading soldier; the rawboned Lee was ambitious, personally unkempt, and always surrounded by a pack of canines. "I must have some object to embrace," he said. "When I can be convinced that men are as worthy objects as dogs, I shall transfer my benevolence . . ." He spoke several languages and in English at least was "wretchedly profane." The British Army has never been accused of failing to produce its share of eccentrics. If Lee resembled a Montgomery to Washington's Eisenhower (or, more drastically, Orde Wingate to Louis Mountbatten), he had still led troops of every ilk, from Mohawks to Cossacks to the King's regulars, with somewhat more success than had Washington with colonial militia. By talking with Lee, Washington familiarized himself as best he could with continental military doctrine and what to expect from the British Army and its personalities in tactics and strategy. At the same time, on that long ride north, Lee no doubt assessed, for his own interest, Washington's potential.

Another decision in summer 1775 was made after the Congress in Philadelphia had grown tired of entreating Canadians to join the rebellion: it instead sanctioned an invasion of the North. An obsession with Canada had been ingrained in American thinking by decades of war with the French, who had used Canada as a base from which to terrorize the Colonies. In American minds, Canada hovered threateningly over the continent and could not be left in hostile hands.

Two dynamic young commanders, Richard Montgomery and Benedict Arnold, attacked Canada from different directions. Montgomery reduced the forts at Chambly and St. Johns and took Montreal. Arnold's army suffered a horrible overland approach march through Maine, but the survivors were joined by Montgomery's men at Quebec to lay siege to that city by December 5. It was believed that the British commander in Canada, Guy Carleton, had only 550 men to defend his territory against invasion; however, Carleton benefited from local Scotsmen who rallied to his banner and otherwise impressed every man he could find into manning the city's defenses. The French population of Quebec province contributed two regiments of volunteers to the Patriots' effort, many of them veterans who had served under Montcalm, but the population as a whole responded far less enthusiastically than the Americans had hoped.

At Boston, meanwhile, the artistocratic Washington spent several months wringing his hands over the collection of ill-disciplined, motley militia he had been given charge of. On August 20, after describing how he had arrested several officers for cowardice and graft, he wrote: "I dare say the men would fight very well (if properly officered) although they are an exceeding dirty and nasty people." The situation had not improved by the end of November, when he singled out Connecticut troops who were about to leave the army: ". . . such a dirty mercenary spirit pervades the whole, that I should not be at all surprised at any disaster that may happen. . . . Could I have foreseen what I have, and am likely to experience, no consideration upon earth should have induced me to accept this command." Still, the Virginian possessed a keen vision of how a proper army should be molded, and he labored prodigiously to that end. A soldier wrote home that "Generals Washington and Lee are upon the lines every day." Washington also aspired to carry battle to the enemy at Boston, especially when the rivers froze over, but each time was voted down in a council of war.

Charles Lee was anxious for action and took a force to Newport, Rhode Island, where a British naval captain had been intimidating the local population. Lee entered the town and arrested eight Tory leaders, forcing them to subscribe to an oath of allegiance he had invented himself. He made them swear they would "neither directly nor indirectly assist the wicked instruments of ministerial tyranny and villainy commonly called the king's troops and navy, by furnishing them with provisions and refreshments." The American Congress objected to what they viewed as his autocratic methods, but Lee believed that the generals of the army were justified in taking independent action. "One must not be trammelled by laws in war-time," he opined. "In a revolution all means are legal." Lee also recommended to Washington that he cease being so deferential to politicians. When the commander-in-chief fretted over whether he had the right to order militia raised from New York and New Jersey, Lee told him, "I have the greatest reason to believe . . . that the best members of Congress expect that you would take much upon yourself, as referring everything to them is, in fact, defeating the project . . ." The key phrase is "the best members of Congress." Lee did not enjoy a high opinion of that deliberative body as a whole.

On December 31, 1775, the entire rebellion was jeopardized due to the expiration of enlistments that would occur on midnight that day. At Boston, Washington was just barely able to recruit enough new men, as well as persuade others to stay on, in order to keep a semblance of strength in his siege lines. At Quebec the situation was more drastic, and Montgomery and Arnold were forced to storm the city on New Year's Eve. Unfortunately Montgomery was shot dead in the assault. On the other side of town, Arnold was wounded in the leg and had to be carried from the battlefield. The leaderless attack then turned into a fiasco and the triumphant British defenders took 400 prisoners, among them a defiant young captain named Daniel Morgan, who was cornered and whose fellow Virginia riflemen barely persuaded not to fight to certain death. (He surrendered his sword to a priest.) The American siege of Quebec continued but its prospects had seriously dimmed.

The siege lines around Boston stayed firm, but the Americans had increasingly become worried about the even more important—and completely undefended—harbor that lay to the south at New York. Charles Lee was requested to cross Connecticut, raising troops along the way, and set up the defenses of what we now call New York City. (At the time, New York town only covered the tip of Manhattan island.) If Boston was a patriotic city with some Tories, New York was more the opposite, and its Tories were emboldened by the presence of a British man-of-war, the *Asia*, that had taken up residence in the harbor, and which could at any moment level the place with burning shot.

While still in Connecticut, Lee received a letter from New York's Committee of Safety that requested he halt at the Connecticut border; the committee feared that their city would become a battleground. Lee had little patience for timidity, even less for Tories, and his reply was not designed to comfort his new correspondents: "If the ships of war are quiet, I shall be quiet," he answered. "But I declare solemnly that if they make a pretext of my presence to fire on the town the first house set in flames by their guns shall be the funeral pile of some of their best friends." A delegation from Congress was sent to mediate the issue of Lee's arrival; thereafter, the general analyzed the defensive problems of the area and assigned places to build forts and redoubts.

On March 4, 1776, Washington threw up works and placed cannon on heights that overlooked both the city of Boston and its harbor. The British now had to view evacuation of the city as an imperative, rather than simply desirable. On March 17, transports took off the army plus over a thousand Tories who no longer had a future there. England's General William Howe traded a promise not to burn the city for an unmolested evacuation and Washington allowed the enemy to leave. The ships sailed to Halifax, Nova Scotia.

On March 1, Lee had been requested to interrupt his work around New York and take over the Southern Department, where a British fleet was headed, precise destination unknown. On June 9, the British mounted an amphibious assault on Charleston, South Carolina, where Lee had arrived in time to supervise the defenses. Fortunately for the Americans, the British troops under General Henry Clinton disembarked on an island in the harbor from which they could go nowhere. Colonel William Moultrie meanwhile commanded batteries on Sullivan's island that blistered the British fleet in a day-long battle on June 28, compelling them to withdraw with severe casualties. Lee paid a visit to Moultrie at the height of the battle, "under intense fire, and personally directed two or three guns." At the end of the day, the British fleet and its troop transports skulked away.

By the beginning of July 1776, the rebellion had proved successful. Notwithstanding the problems of the Canadian venture, which had turned into a full-scale retreat, colonial forces had shown themselves capable of expelling King George III's troops from all but the most remote toeholds in North America. On July 2, an announcement of the formal unification of the thirteen colonies into a new nation was declared. On July 4 it became official. Instead of a grass-roots rebellion against a sovereign, the war would henceforth consist of a self-standing nation, the United Colonies of America, waging a defensive war against an invasion from Great Britain.

Of course, no one thought that the Empire had given up. In retrospect we know that the British frequently begin wars with dismal defeats, only to calmly gather their strength and eventually work through adversity to prevail. In addition, by early summer 1776, all eyes were on the city that had to be the focal point of operations once the enemy renewed its efforts: New York. Nathanael Greene, by now one of Washington's most trusted lieutenants, along with Israel Putnam and General William Heath, was sent to supervise the construction of New York's defenses along the lines recommended by Lee. Washington and most of the New England troops were coming down. The supreme commander asked the Congress to create a "flying camp" of 10,000 men to post itself near the New Jersey shore of the Hudson River. Recruiters in New York were attempting to draw out the local populace for soldiers. Units from Pennsylvania, Delaware and Maryland were arriving from the south.

New York was not only an important city and port but it also sat at the mouth of the Hudson River—an inland highway to the north, navigable almost to Canada. British possession of this harbor and its river could allow them to sever New England from the remainder of the new nation, rendering communications or mutual assistance between the two parts of the country impossible. The "key to the whole continent," as John Adams called it, in accordance with instructions from the Continental Congress, needed to be defended with every resource at Washington's disposal.

On July 9, the Declaration of Independence was read in New York, prompting the violent dismantling of a statue of King George III and wild celebrations. Washington was assembling an army to defend the city that would be the largest he would command throughout the entire war. Men on the shorelines were posted to spot enemy vessels that hoved into view. And, indeed, that summer, the British began arriving in force. An observer stationed on a rooftop gazed at the huge mass of vessels gathering outside the harbor and later exclaimed, "I thought all London was afloat."

In an oddity of history that can only partially be attributed to Britain's reliance on the system of purchased commissions, or nepotism, the commanders of His Majesty's forces in America in 1776 were brothers: General William Howe of the army and Admiral Lord Richard Howe of the navy. Their elder brother, Edward, had been killed at Ticonderoga in 1758, and a monument in his memory had been erected in Massachusetts. Perhaps unwisely, the two commanders of the British expeditionary force had also been declared "peace commissioners." Although everyone hoped for a short war, rather than, for example, a bitter eight-year conflict, the person who most avidly aspired to this result was King

George III. If his commanders on the spot could not only hold the stick but the carrot, perhaps chances that the Colonials would abandon their rebellion would double. In any case, the British expeditionary force forming up on Staten Island was by far the largest in their history. The greater the stick, the more enticing would be the carrot.

Charles Lee had designated the points to be defended in the New York environs, and further had supplied Congress with a strategic analysis of the problem. He didn't believe that the town itself should be a focal point of the defense, but that the surrounding areas could be made "an advantageous field of battle." He recommended that Long Island should be defended by 4–5,000 men at its western end (across from Manhattan) and that the East River and Long Island Sound could be dominated by American batteries against the Royal Navy. The Hudson River could not be closed to British traffic easily, being "so extremely wide and deep," however the terrain along its banks would limit the effectiveness of naval ordnance, while American batteries could inhibit British movement. He recommended that the area around King's Bridge, which connected Manhattan island with Westchester and New England, be heavily fortified. In sum, he concluded: "If our people behave with common spirit, and the commanders are men of discretion, it might cost the enemy many thousands of men to get possession of it." This was an accurate, if not overly optimistic, assessment that pointedly stopped short of recommending the entire Continental Army be staked on the prize. Lee himself would be in the South while the first battles for New York were taking place.

Throughout midsummer, the intimidating British build-up continued, in full view of Washington's fledgling army. On July 12 Admiral Howe arrived with 150 ships to join his brother's 130 that had come from Halifax. Toward the end of July scattered ships came in every day, and then 40 arrived together on August 1—the humbled, perhaps vengeful, expedition returning from Charleston. Between August 12 and 14 still another fleet came in, consisting of 120 ships with a cargo of nearly 10,000 soldiers: Germans.

Since March, the Americans had been aware that King George was seeking to hire foreign troops to help subdue his rebellious subjects. To an Englishman-turned-patriot, nothing demonstrated the malignancy of the sovereign more than such a desperate resort. The monarchy born of the

Glorious Revolution was now so bereft of principle that it had hired mercenaries to fight its battles. The Germans were hired out to King George from half a dozen principalities, but since the greater part of them came from Hesse-Cassel, the entire group became known as Hessians.

The geography of the New York area dictated the British Army's next move. From his staging ground on Staten Island—a broad, pastoral piece of land, largely populated by Tories—General Howe had three options:

He could mount a daring amphibious operation against Manhattan island, thereby cutting off the town of New York and the bulk of American forces from landward retreat. The drawbacks to this plan were that the long approach from Staten Island through the outer harbor, up the Hudson River to a point above the city, would negate any chance of surprise (the East River was not reliable for navigation). Landing troops against such a large concentration of American forces was itself a risk; if cut off on Manhattan, Washington's army could evacuate to Long Island and thence to Westchester or Connecticut.

Another choice was to easily pass over to New Jersey in order to flank Manhattan. The prohibitive drawback was terrain, because New Jersey had huge swamps that ended only when its riverbank became steep cliffs, called the Palisades, that were unsuitable for the large-scale movement of troops.

The third option was to easily pass over to Long Island, which had flat, open beaches for landing, and from which further amphibious jumps could be made against either upper Manhattan or Westchester. The drawback was that the Americans had recognized the importance of Long Island and had thousands of troops already there. Most were deployed along a steep, wooded ridge line that sat forward of their main fortifications. The essential position, which protected the nearest crossing point to Manhattan, was centered on a small hamlet called Brooklyn. Forts on both the Manhattan and Long Island sides guarded the water approaches from New York harbor through the East River. The British could disembark troops along Long Island's sandy beaches; however, once they approached Brooklyn they would have to assault an entrenched enemy that held difficult wooded terrain, as well as an additional position astride the Brooklyn heights.

It is worth noting that, over a year after the event, memories of Bunker Hill still resonated on both sides. This is why the Howes dismissed the idea of an amphibious landing against dug-in American troops on Manhattan. Instead, they would pass over to Long Island, where they could calmly assemble the army and stage a battle of maneuver.

Washington observed the British landings on Long Island, which began on August 22, but properly guarded against the possibility they were a feint. It was the strength he held on Manhattan that prevented the British from assaulting that island directly, and he was careful not to commit the entire army across the East River. The river was only three-quarters of a mile wide and he was well supplied with not only boats but an entire contingent of fishermen from Marblehead, Massachusetts, who knew how to negotiate water. Men could be transferred quickly should reinforcement be required or if, by chance, an evacuation became necessary.

The British meanwhile pondered how best to crack the prepared American positions on the island. Although they possessed a superior number of men—32,000 compared to the 23,000 Washington counted on August 19—the Howes were handicapped by the knowledge that this force was all they would have to work with in the near future, and they could not afford a war of attrition or of indecisive results. Reinforcements would have to be assembled and sailed across the Atlantic Ocean, while the Americans could simply raise more men immediately from their population of over two million. When the British fought, they had to make it count.

Unfortunately for the Americans, Nathanael Greene, who had set up the dispositions before Brooklyn, had taken to bed with a raging fever. Israel Putnam was put in his stead, but had little familiarity with the forward positions and terrain. In the days before the battle, Pennsylvania riflemen camped near the village of Flatbush sallied individually and in small groups against the enemy, seeking primarily to spook the Hessians. Using Indian tactics, the "southern" sharpshooters crept around the enemy emplacements and even forced the Germans to pull back their camp at one point, after they found that cannonfire failed to discourage their nearly invisible antagonists. George Washington, however, disapproved of such scattershot harassment and ordered the old frontier fighter Putnam to rein in his freelancers. Washington described the skirmishing as "a kind of fire which tended to disgrace our men as soldiers and rend our defence contemptible in the eyes of the enemy."

It was General Henry Clinton, an unpleasant man who was Howe's second-in-command, who came upon the solution to Long Island. While the rebels were arrayed formidably along their fortified, wooded heights forward of the main position in Brooklyn, there was an unguarded pass through those heights at Jamaica, just six miles north. Clinton, whose father had once been Royal Governor of New York, had reconnoitered the ground and made a recommendation to Howe for a massive flank attack. Much as Howe hated to follow advice from Clinton, this suggestion was too good to ignore.

On the day of the battle, August 27, 1776, William Alexander of New Jersey, whom the House of Lords had denied title to a Scottish earldom but was nevertheless called by everyone on the Continental side "Lord Stirling," held the American right, near the coast. General John Sullivan held the wooded ridge before Flatbush, where he had heavily fortified the only pass. Samuel Miles was posted on the American left with 800 men. Putnam stationed himself at Brooklyn, in the main American position. Although a direct enemy approach would be from the southeast, Sullivan arranged for five mounted officers to reconnoiter the Jamaica pass to the north, just in case.

The British plan was for General James Grant to move against Stirling's positions near the coast in the early morning. At the same time, General Leopold de Heister's Hessians would demonstrate against Sullivan's troops, opposite Flatbush. The real British stroke, however, consisting of 10,000 men under Clinton, Lords Cornwallis and Percy, and accompanied by Howe himself, consisted of a nocturnal march around the American left, across the Jamaica pass to the north and through the hamlet of Bedford. Once this huge British column had turned the American flank they would fire two cannon shots, signaling to the rest of the army that they were behind the ridge line. Then Grant and de Heister would attack the Colonial line in earnest from the front.

There were moments during the great flank march when Howe feared he was walking into a trap. Dragoons captured the five American horsemen whom Sullivan had dispatched to watch the pass, but the British could still scarcely believe that the entire route was undefended. With daylight, they finally spotted Americans in the woods, but these were a disorganized band that quickly fell apart under attack. Within hours, in fact, Miles' brigade on the American left would cease to exist.

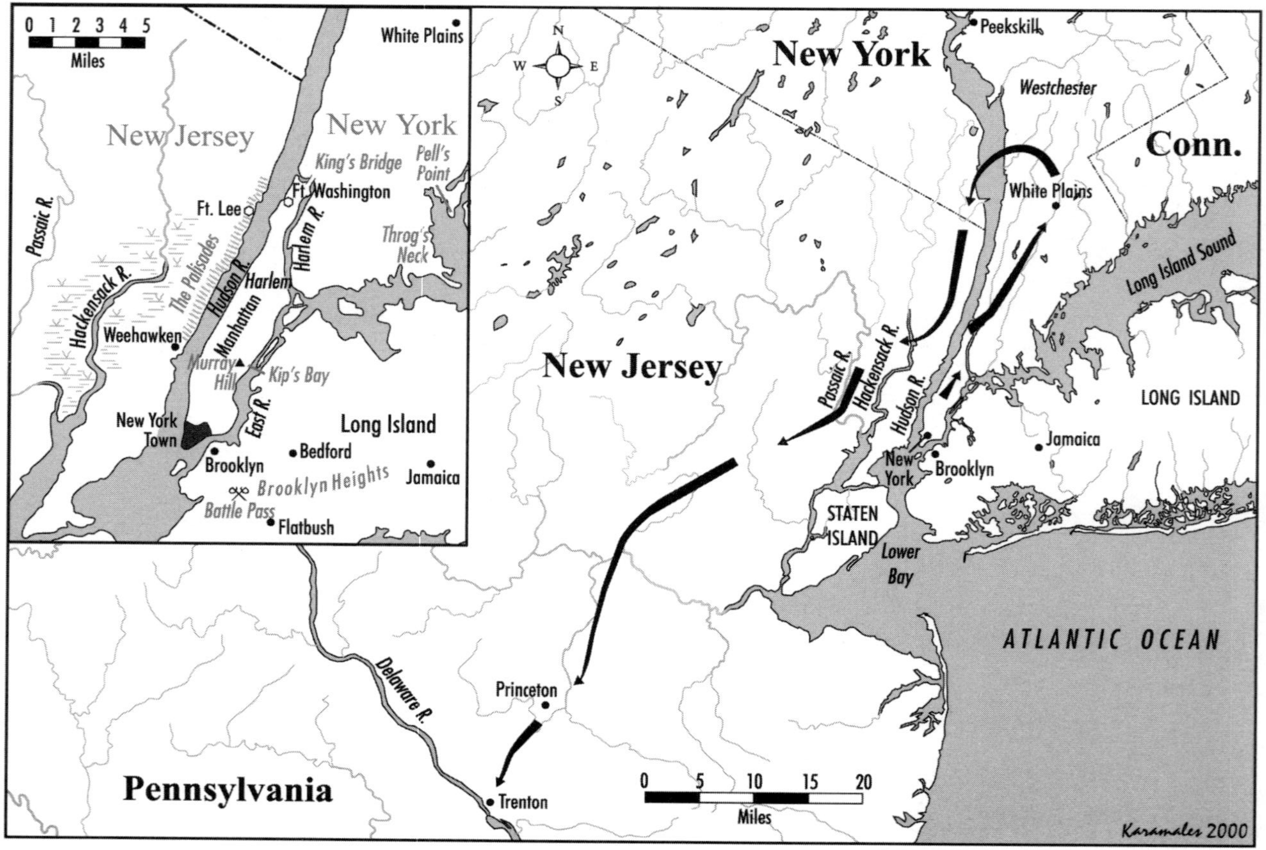

0 1 2 3 4 5
Miles
White Plains
New Jersey
New York
King's Bridge
Pell's Point
Ft. Washington
Ft. Lee
Passaic R.
Throg's Neck
The Palisades
Hackensack R.
Hudson R.
Harlem
Harlem R.
Weehawken
Manhattan
Murray Hill
Kip's Bay
New York Town
East R.
Long Island
Bedford
Brooklyn
Jamaica
Brooklyn Heights
Battle Pass
Flatbush
N
W
E
S
New York
Peekskill
Westchester
Conn.
White Plains
Long Island Sound
New Jersey
Passaic R.
Hackensack R.
Hudson R.
LONG ISLAND
Jamaica
New York
Brooklyn
STATEN ISLAND
Lower Bay
ATLANTIC OCEAN
Princeton
Delaware R.
Pennsylvania
Trenton
0 5 10 15 20
Miles
Karamales 2000

In the early morning on the American right, Lord Stirling was skillfully deploying his troops against the cautious approach of Grant. One of his men recorded that "Lord Stirling gave battle in the true English taste." In the center, the Hessians, alongside the 42nd Highlanders to their left, were lobbing cannon shots at Sullivan but otherwise stayed immobile. Sullivan sent 400 of his men to help Stirling because he assumed the main British effort was taking place on the right. Then, at approximately 9:00, Howe fired off the two signal shots to announce that the flanking movement was in place. The 5,500 Americans in the forward positions immediately realized they were in trouble. In the center, the Hessians stopped their parrying against Sullivan, fired a volley and then charged the wooded heights with fixed bayonets. The Americans fled, only to meet British troops cutting off their escape route. As in many battles, the men who turned their backs first, as well as the fastest runners, had the best chance to survive. Those who hesitated found themselves challenged by British horsemen cutting across their line of retreat or light infantry firing and charging from the woods.

On the British left, Grant had not been able to mount the kind of ferocious assault against Stirling's men that the Hessians had made against Sullivan's. Stirling, besides having some Pennsylvania sharpshooters, commanded troops from Maryland and Delaware, who were the best-trained men Washington had, the latter being the sole contingent with European-style uniforms. They were not easily panicked, and fell back only gradually.

The American center, however, had by now dissolved into a bloody shambles. Colonials fleeing the German and Highlander assault ran straight into British troops charging into the gap between the American forward lines and Brooklyn. For many, there was no escape, and, for many who were not part of large groups, there was no escaping death. An enduring image of the battle is American soldiers being pinned to trees with bayonets. A Scottish officer wrote: "The Hessians and our brave Highlanders gave no quarters; and it was a fine sight to see with what alacrity they dispatched the Rebels with their bayonets after we had surrounded them so that they could not resist. Multitudes were drowned and suffocated in morasses—a proper punishment for all Rebels. . . . Our loss was nothing. We took care to tell the Hessians that the Rebels had resolved to give no quarters to them in particular, which made them fight desperately and put all to death that fell into their hands."

A British officer later stated, "We were greatly shocked by the massacres made by the Hessians and Highlanders after victory was decided." The area surrounding what is today called Battle Pass in Brooklyn's Prospect Park was indeed a landscape of carnage; however, most American casualties were prisoners, including a group of sixty who surrendered with their flag, inscribed with the word "Liberty," to Hessians commanded by Colonel Johann Rall.

George Washington, atop the Brooklyn fortifications, was able to witness the British Army swarming across and behind his forward positions, as well as the many terrified fugitives who were able to make it back to the rear lines. The only American formation still intact was Stirling's on the right, which was still conducting a fighting withdrawal oblivious to the fact it was nearly surrounded. Finally, Stirling, giving way before Grant's assault to his front, encountered Lord Cornwallis' troops in his rear, massed around a stone house that dominated the escape route to Brooklyn. Observers on the heights expected Stirling's men to lay down their arms and surrender as a body to the enemy. Instead, they counterattacked.

Stirling formed up his Maryland brigade and launched it at Cornwallis. The Marylanders were torn apart by British fire, but Stirling repeatedly rallied them and attacked again. The assaults had no hope of success other than to buy time for the rest of the Americans to get back to Brooklyn, through a swamp and a mill pond, in which a number of them drowned. After 256 Marylanders were lying dead on the field, from an original force of just over 400, and the British forces on the scene had been buttressed by Hessians, Highlanders and dragoons come to witness the slaughter, Stirling finally gave up. He refused to hand over his sword to an Englishman and instead sought out the Hessian de Heister in order to surrender.

The Battle of Brooklyn did for the British what Bunker Hill had done for the Americans. The psychological edge between the armies had shifted. British casualties in the battle are reliably reported at 349, including 61 dead. (An additional 21 were made prisoner because they had walked up to the attractively uniformed regiment from Delaware, not realizing they were the enemy.) American casualties are more difficult to determine because of the looser organization in their army. If British estimates on the day of the battle seem high at over 3,000, the confirmed prisoner figure of 1,097 indicates that total losses must have been over 1,500, or at least four times the losses of the assaulting troops. Further, the British had taken all their objectives and had to be restrained by Howe from attacking the Brooklyn heights on the same day.

Howe's decision to hold back his onrushing army from taking the Americans' positions along the East River, after their forward lines collapsed, has remained controversial, in addition to similar decisions he would make in favor of the principle of restraint in the coming months. His stated reason for holding back his troops before the Americans' second fortified line was that the position could be taken at far less cost by "regular approaches"—classic siege techniques. He would have been correct, in fact, if at least one British frigate had been able to make its way up the East River to stand between Brooklyn and Manhattan; instead, contrary winds prevented any sailing ships from gaining that post. Howe subsequently came under criticism from fellow officers for missing the opportunity. In a masterpiece of snide commentary cloaked in flattery, General Clinton wrote:

> I had at the moment but little inclination to check the ardor of our troops when I saw the enemy flying in such a panic before them. I was also not without hopes that His Excellency, who was on a neighboring hill and, of course, saw their confusion, might be tempted to order us to march directly forward down the road to the ferry, by which, if we succeeded, everything on the island must have been ours. I do not mean, however, by this to insinuate that Sir William Howe was in the least wrong in not doing so. . . . had I at the time possessed the same information he was master of, I should possibly, like him, have judged it prudent to wait. . . . Nor can any after-knowledge of the fallacy of that information invalidate the propriety of his conduct under it.

Clinton went on to claim that the rebels had only 800 men at the time, in works that "would have required 6,000 to defend." However, his numbers were off. The battle had been a disaster for the 5,500 Americans holding the outer lines, and had included a number of isolated massacres, but the majority of the men had made it back to Brooklyn. Washington had been ferrying additional troops over from Manhattan all morning, and, when Brooklyn was evacuated less than 48 hours later, 9,500 men were brought off. Still, Clinton was not the only one astonished by Howe's restraint. Captain George Collier of the Royal Navy wrote: "The having to deal with a generous, merciful, *forbearing* enemy, who would take no unfair *advantages*, must surely have been highly satisfactory to General Washington, and he was certainly very deficient in not expressing his gratitude to General Howe for his *kind* behavior towards him." The immensely relieved Putnam remarked, "Howe is either our friend or no general."

The next day there was minor skirmishing as American patrols ventured out to harrass the enemy encampments, and it became clear that Howe was in no hurry. The greatest danger was that the wind would shift, allowing British warships to move up the East River to block the American escape route. On August 29, Washington decided to evacuate Brooklyn. The fishermen from Marblehead rowed the army over to Manhattan after dark, as unit after unit filed quietly down to the river. Aside from the winds that continued to keep British men-of-war at bay, Washington was favored by a dense fog in the early morning hours that concealed his retreat from the enemy. By the time the British realized what was going on, the last American boat was offshore. Washington himself was among the last to leave.

It may be an exaggeration to state that the skillful, and lucky, extrication of Washington's army from Brooklyn to Manhattan "saved" the Revolution. It is granted that Howe, by preserving intact his stupendous victory on the morning of the 27th, may have missed his chance at total victory that afternoon. On the other hand, a second phase of the battle at Brooklyn during the following hours or days, against Americans who were by then "covered to their chins," might have turned into another Bunker Hill. Howe needed to minimize his own casualties, and he disliked attacking fortified heights head-on. There would, indeed, be other chances to destroy Washington's army in the open field.

Admiral and General Howe also kept in mind that they had not been sent merely as avenging angels of the Crown but also as Peace Commissioners. After Washington escaped, General Howe asked one of his prisoners, John Sullivan, commander of the forward lines at Brooklyn, to arrange a parley with the Continental Congress. Of course, the British could not recognize the Congress as a legitimate governing body, but they still hoped influential American leaders would answer the summons. On September 11, Benjamin Franklin, John Adams and Edward Routledge traveled to Staten Island, where they had breakfast with the Admiral. The patriots were in no mood to hear about the Howes' power to "pardon" or to finally negotiate grievances that could have been addressed years before. The Declaration of Independence had altered the playing field. Howe's evocation of his older brother, who had died defending the Colonies against the French in upstate New York, failed to move the American delegation. The Howes were Whigs in British political terms but had nevertheless failed to recognize how implacable the Americans had become. The meeting dissolved to the point of rancor, and at one point Howe said to Franklin, "I suppose you'll endeavor to give us employment in Europe." Franklin, who was about to make a trip to France, would pursue that very goal.

The Howes had failed as Peace Commissioners but they could still succeed as war leaders. Now that the British held Long Island, which paralleled Manhattan all the way to Harlem, it was only a matter of where to apply the knife to sever New York from the rest of the Colonies. Washington had divided his army into a division under Putnam to hold New York town; a division under Heath in upper Manhattan; and a weaker division of militia under Greene, just then recovering from his illness, and Joseph Spencer, to hold the land in between.

On September 15, five British warships assembled in a row, broadside, opposite present-day 34th Street. Eighty-four boats, filled with British and Hessians, began rowing toward American redoubts guarding the steep, rock-strewn landing site. During the initial landings on Long Island, naval covering fire had been unnecessary since the wide beaches hadn't allowed for fortified defensive positions. Now, however, the navy showed how it could cover a landing. Over seventy guns obliterated the American redoubt and forced its defenders to flee for their lives. The Hessians, who had nervously been singing hymns on their slow journey across the river, landed unopposed. Stage two of the British conquest of New York had been accomplished at no human cost to the attackers.

The British invasion of Manhattan at Kip's Bay was less a battle than a case of British cannonfire scattering untrained militia. To George Washington, however, who was on the scene, the landings were traumatic and marked his most pitiful moment as supreme commander of the

Continental Army. "Good God, have I got such troops as those?" he cried as he flailed at fleeing militia with the flat of his sword. Riding back and forth on horseback he tried but failed to rally the men. Three times he threw his hat to the ground. Finally, when advancing Hessians were not seventy yards off, Nathanael Greene grabbed the bridle of the supreme commander's horse and led it, and its rider, away.

General Clinton, who commanded the British advance, paused on present-day Murray Hill to wait for reinforcements from the second assault wave. Then a brigade was dispatched south to occupy the town of New York. Admiral Howe's secretary, Ambrose Serle, reported: "Nothing could equal the expressions of joy shown by the inhabitants upon the arrival of the King's officers among them. They even carried some of them upon their shoulders about the streets, and behaved in all respects, women as well as men, like overjoyed bedlamites."

The British are often criticized for not immediately rushing across the width of Manhattan island to cut off the American troops in New York. Putnam was leading the exodus from that city north toward Harlem, where the new main lines were being formed. Just a day later, however, in front of Harlem Heights, the Americans demonstrated the risk undertaken by unsupported British units advancing too far and too fast.

At Harlem Heights on September 16, a mixed force of 400 British light infantry, Highlanders and Hessians rashly stepped inside the jurisdiction of some 13,000 Americans confined behind the steep cliffs and rocky terrain of upper Manhattan. After they had pushed back a forward patrol of 100 men, the British made the mistake of adding insult to injury, thus bringing upon themselves a violent response. In the words of Washington's Adjutant General, Joseph Reed: ". . . the enemy appeared in open view and in the most insulting manner sounded their bugle horns as is usual after a fox chase. I never felt such a sensation before; it seemed to crown our disgrace."

Reed rushed to Washington, who immediately ordered units of his familiar Virginians into action. A party held the British to their front while three companies under Major Leech and a unit of Connecticut Rangers under Colonel Knowlton were ordered to get behind the enemy to cut them off. The Americans seemed to lack patience for the maneuver, however, and attacked from the flank before getting in the enemy's rear. As the action heated up, Washington threw in two regiments of

Maryland troops, who fired two volleys and then charged. The British broke and ran as the Americans pursued them for over a mile. Lord Percy was coming up with a force of 5,000 to solve the British infantry's dilemma but Washington called back the pursuit just in time. The Americans had suffered about fifty casualties, including Colonel Knowlton, who was killed, and Major Leech, who was wounded, but for the first time they had seen the King's troops run from the battlefield. Reed wrote: ". . . it has given spirits to our men that I hope they will now look the enemy in the face with confidence."

The small scale of the Battle of Harlem Heights allowed Washington to employ only the best units of his army, rather than fill gaps with militia. Lt. Colonel Tench Tilghman wrote, "The prisoners we took told us they expected our men to have run away as they did the day before, but that they were never more surprised than to see us advancing to attack them. The Virginia and Maryland troops bear the palm. They are well officered and behave with as much regularity as possible, while the Eastern people are plundering everything that comes in their way. . . . I like our post here exceedingly; I think if we give it up it is our own fault." British commanders barely even remembered the episode in their memoirs, or in their dispatches to London. General Clinton commented, "On my going the next morning to take the command of the foreposts, I found that the light infantry, having with rather too much impetuosity pursued some parties of rebels toward their works, had got themselves somewhat disadvantageously engaged . . ."

During this early stage of the war, the two armies were a study in contrasts, to the complete disadvantage of George Washington, except to the degree that, unlike Sir William Howe (knighted after his victory at Brooklyn), the American had plenty of space into which to fall back. Howe commanded men who were well equipped, supplied and paid on a regular basis. They were all professionally trained, subject to discipline and served under multiple layers of non-coms and officers whose life's work was the practice of war. Washington, on the other hand, put together his army with whatever bodies he could find, among patriots throughout the colonies, or among local citizens persuaded into temporary service. The militias were a constant source of worry, because not only would they desert back to their homes at first opportunity, or leave en masse after their short terms were up, but their lax discipline prompted consternation among the men whom Washington was endeavoring to count on as "regulars." Furthermore, the local militias were flimsy in battle and not to be relied on. This is not to disparage the courage and patriotism of volunteer militia, many of whom were boys or older men, and in many cases ill-equipped, but only to state that, at best, they were the raw material from which a formidable army might be composed; they were not, in themselves, such a force.

In the implementation of tactics, the British held a great advantage against Washington's militia-heavy army because their men would follow orders. If an attack was commanded against a rebel-held hill, the troops would charge as prescribed by their officers. If holding a position was required, the men would do their best. Washington had a few units in his army that would obey any order, but for the most part he had to ensure his men were always protected by fortifications. The Americans could swarm on the offensive if they sensed vulnerability, and in their ranks were many brave fighters who would sacrifice their lives for the cause. However, in set-piece actions in 1776, the job of the British troops was to attack—in formation and in compliance with the wishes of their officers; the job of the Americans was to hold as best they could, or at least not succumb to panic. During this period when Washington could not undertake offensive operations, and could barely sustain defensive ones, he had as his sole advantage the existence of a vast country into which he could retreat.

As autumn approached Washington did have the consolation that he was no longer beleaguered with requests to detach units to reinforce the American effort in Canada. Since the debacle at Quebec, that invasion had suffered one calamity after another and the remnants of the northern forces were collapsing back into the Colonies. The downside was that these men were being pursued by a 10,000-man British army that was already at Lake George and intent on reaching Albany.

On the evening of September 20, soldiers on both sides of the lines stared with fascination toward the south, where a swirling, bright glow filled the sky: New York was going up in flames. Washington had asked the Congress for permission to burn the town in order to deny Howe the benefit of comfortable winter quarters. Besides, New York was filled with Tories (and would, in fact, become the primary haven for Tories from elsewhere in the Colonies as the war went on). Congress had firmly refused the request, but Patriots in the city took matters into their own hands. Fires started in different places at once, and the British claimed to have caught a number of saboteurs in the act. British soldiers from the lines and sailors from the fleet rushed to help put out the flames, but at least a quarter of the city turned to ashes. Washington commented in a letter to his brother: "Providence, or some good honest fellow, has done more for us than we were disposed to do for ourselves."

In his report on the fire, General Howe stated that "a most horrid attempt was made by a number of wretches to burn the town of New York, in which they succeeded too well, having set it on fire in several places with matches and combustibles that had been prepared with great art and ingenuity." He added: ". . . we have reason to suspect there are villains still lurking there, ready to finish the work they had begun . . ." While the cinders were still warm on September 21, an American spy was captured with sketches of British dispositions and works concealed on his person. He was brought before Howe, who dispensed with a trial and ordered his execution the following morning. The spy, a young officer from Connecticut named Nathan Hale, impressed everyone with his bearing and bravery, and his dedication to the Patriot cause.

Behind American lines, meanwhile, the essential tenuousness of the improvised grand army had already become apparent. Washington's chief of artillery, General Henry Knox, wrote to his brother on September 23: "There is a radical evil in our army—the lack of officers. We ought to have men of merit in the most extensive and unlimited sense of the word. Instead of which, the bulk of the officers of the army are a parcel of ignorant, stupid men, who might make tolerable soldiers, but are bad officers. . . . As the army now stands, it is only a receptacle for ragamuffins."

The next day, George Washington penned a lengthy letter to the president of the Continental Congress, John Hancock, that elaborated on the issue. His thrust was that he could not be expected to face the British with men "just dragged from the tender scenes of domestic life, unaccustomed to the din of arms; totally unacquainted with every kind of military skill, which being followed by a want of confidence in themselves . . . makes them timid, and ready to fly from their own shadows." He was acutely aware that the short-term enlistments favored by Congress meant that entire regiments would soon be walking away. The temporary service of militia not only made it difficult to instill them with proper discipline, but embarrassed efforts to apply such discipline to long-term soldiers.

To the idealists among the Founding Fathers, a "standing army" could only be the tool of an autocrat and the scourge of a true democracy. The new nation, like ancient Athens, would be better served by citizen-soldiers rallying bravely to defend their rights and land. Washington, on the other hand, wanted Congress to provide recruitment incentives that would supply him with professionals: men committed to the field for the duration, so that he could mirror the formidable array of hardened, well-trained British and German units he had to confront on the battlefield.

By packing the American army into the fortified heights around Harlem, Washington had minimized the disparity in the fighting quality of his units, and thus presented the British with a formidable obstacle. Since this position was too strong for a frontal assault, on October 12 Howe made an amphibious landing at Throg's Neck in Westchester, above Manhattan island, to pry the enemy out of its works. It was a mystery to observers on both sides of the lines why Howe had waited so long. Washington's only hope while concentrating the army in upper Manhattan was that the British would choose to impale themselves against his fortified heights—the one kind of battle his army could win. Once Howe finally employed his boats to land troops behind him, Washington had to abandon Manhattan. He left behind only 1,200 troops to man the built-up area centered on Fort Washington (formerly Fort Independence), which, standing opposite Fort Constitution (soon to be renamed Fort Lee) on the Jersey side, guarded the Hudson River against large-scale British naval operations.

While members of Washington's staff usually groaned when they spotted a courier from the north riding into camp (updates on the ill-fated Canadian invasion were invariably depressing), news now arrived of a remarkable battle. Benedict Arnold had built a sort of navy from scratch at the foot of Lake Champlain, and flung it in front of a powerful British flotilla under Guy Carleton coming down from the north. Arnold's "fleet" had been wiped out at the Battle of Valcour Island, but the inexorable British advance had halted. In the face of American willingness to fight back, Carleton decided it was too late in the year to continue his campaign. The British would wait to mount another major effort from that direction in 1777.

Howe's troops on Throg's Neck, meanwhile, found their progress impeded by a swamp, and the flatboats came around once again to move everyone to Pell's Point, a little to the northeast, from where they could march inland. John Glover's brigade of 750 Marblehead men defended the new landing site and gave the British some hot fire. Glover was nervous about his responsibility and later fretted, "I would have given a thousand worlds to have had General Lee or some other experienced officer present." His contingent nevertheless performed well, prudently falling back in stages—from stone wall to stone wall—while inflicting a number of casualties. The British seized control of the coast road that connected New York with Connecticut, and occupied strategically important hills. Washington, increasingly fearful of being cut off, was pulling the Continental Army back to the town of White Plains, 20 miles north of Manhattan, where he planned to make a stand.

On October 15, the good news buzzed through the American camp that General Charles Lee had rejoined the army. His tenure as commander of the Southern Department had climaxed with the successful defense of Charleston, and Congress had asked him to return to where the crucial actions were taking place. With the British forming up before White Plains it was a comfort to Washington to have Lee back to help direct operations. Lee was immediately placed in command of all troops north of King's Bridge. By way of further delineating the contrast between the two men, Washington, upon being named supreme commander, had refused any pay except compensation for his out-of-pocket expenses. Lee, on the other hand, had stopped in Philadelphia on his way up from Charleston and gained a $30,000 "advance" from Congress. Although he could not fault the legislature's generosity, his opinion of that august body remained unflattering. "Not just one or two cattle among them," he put down in a subsequent letter, "but a whole herd."

When Howe, along with Cornwallis and Clinton, came up to the new American defenses they made a good assessment. The American lines stood behind the usual earthworks, as well as the Harlem River; however a hill across that narrow waterway, to the American right, dominated their positions.

Lee had also discerned the importance of Chatterton's Hill and said to Washington, "Yonder is the ground we ought to occupy." Two of the Americans' crack formations, John Haslet's Delaware brigade and the remnants of Smallwood's Marylanders, were ordered to man the height, along with McDougall's brigade of Massachusetts militia under Brooks and Webb. When McDougall's men came up they stationed themselves directly behind Haslet's troops, who promptly demanded they move to the left. The Delawares had no wish to be caught in a crossfire between British cannon to their front and edgy militia to their rear. Smallwood's Marylanders similarly jockeyed with Brooks' regiment on the right before finally settling in somewhat in advance and toward the center. As it turned out, when the British opened fire, their second roundshot wounded one of Brooks' men in the thigh and the whole regiment ran away, only to be forced back with some difficulty by Haslet and other American officers.

At the Battle of White Plains on October 28, the British demonstrated across the American front with cannonfire and long-range musketry, but their major objective was to take the hill. Line abreast, British troops stormed the height, but after meeting strong resistance formed into columns in order to present a smaller target. Clinton, who along with Howe was observing the action, thought this maneuver would win the day, but then the lead column stalled. The officer in charge had fired his musket and then, halfway up the hill, stopped to reload. The men behind him began dropping dead or wounded while the officer painstakingly rearmed his musket. "If the battle should be lost, that officer was the occasion of it," Clinton exclaimed. Cornwallis made the same observation. The British column broke and had to reform.

The hill was being plastered by over a dozen British guns while Alexander Hamilton directed his battery against the attackers from the main American line. There was only one fieldpiece on the hill, but it was positioned poorly and Haslet had to help drag it to a spot where it could be brought to bear. On the way, a cannonball shattered its carriage and the artillerymen ran. Eventually a few returned to fire one or two shots before pulling the piece to safety.

British on the left had meanwhile attacked up the hill against the Americans' far right flank, where Brooks' regiment had been reformed. These men were unable to stand against the onslaught and ran for the main American line, accompanied by the other militia regiment on the hill's left. Most of Haslet's Delawares and the Marylanders in the center kept firing for a few more minutes but finally had no choice but to retreat.

Haslet recounted afterwards: "The left of the regiment took post behind a fence on the top of the hill with most of the officers, and twice repulsed the Light Troops and Horse of the enemy; but seeing ourselves deserted on all hands, the continued column of the enemy advancing, we also retired." The Americans were able to pull back the right flank of their main line so it was out of range of small arms on the hill.

Howe spent October 29th reassessing the situation, and resolved to attack again on the 30th. But a gigantic rainstorm, followed by sudden cold, intervened. When the British geared up once more for attack on the 31st they found the American lines deserted. Washington had fallen back north of White Plains, to a position flanked by hills and a valley escape route to its rear.

On November 5th it was the Americans' turn to be surprised when they discovered that the bulk of the British army, too, had vacated its positions. With Washington barely holding together his army in the face of desertions and expired enlistments, he was in no shape to intervene against British designs. He could only worry where they would attack next, and hope it would not be across the Hudson toward Philadelphia.

Washington made the decision at this time to split the American army. A council of war determined that the New York Highlands—a vital stretch of territory between the city and Albany—could be defended by 3,000 men. These were placed under General William Heath. To remain in Westchester and defend the approaches to New England, while also being available should the main theater of operations switch to New Jersey, 5,000 men were placed under the direct command of Charles Lee. Washington himself crossed the Hudson River to New Jersey, which the British would have to traverse to get to Philadelphia. Aside from some 2,000 southern troops from the main army whom he transferred over the Hudson, he would have the "flying camp" at his disposal, the New Jersey militia, and the men stationed at Fort Washington in Manhattan.

It is difficult to imagine the woeful letters the commander-in-chief wrote in Westchester becoming any sadder, but when he arrived in New Jersey he found that the "flying camp," which he thought still had 5,000 men, barely consisted of 1,000, most of whose enlistments were nearly up. The New Jersey militia, in addition, was refusing to turn out. The worst development, however, was that on November 16, Fort Washington ceased to exist. That afternoon, the enemy renamed it Fort Knyphausen.

When the main American army had vacated Manhattan for White Plains, Washington had doubted the wisdom of leaving valuable soldiers and enormous quantities of munitions behind in the fort that bore his name. Charles Lee took a dim view of the decision. The rising star Nathanael Greene, however, had made a number of good arguments for retaining a strong position in Manhattan. The importance of New York lay not primarily in its town, or harbor, but in the fact it was the door to the Hudson River, by which the British could bisect the Colonies and join hands with their forces in Canada. Such an accomplishment would doom New England or, alternatively, doom the southern Colonies. At this stage, the strategic thinking of both sides was still keyed on the British presence in the north.

The Americans had stretched a boom of sunken vessels across the Hudson between Forts Washington and Lee to prevent enemy ships from making use of the great inland waterway. Though the British proved early on that they could get by the boom, even while under fire from the forts, this was not a pleasant undertaking and not feasible for more than a couple of fast frigates at a time. On October 27, the *Pearl* had run the gauntlet and lost ten dead and as many wounded in the process. The forts and the boom were not impenetrable barriers, but they were important obstacles that prevented large-scale enemy movement.

In addition, what history calls Fort Washington was really a three-mile-long stretch of high ground protected by rivers on two sides and steep ascents on all sides. The only land approaches were narrow passages against cliffs buttressed by man-made redoubts (the latter constructed with typical American enthusiasm). Assault from the Hudson River side was impossible; a crossing of the Harlem River in the east was easier, but those attacking troops would then have to mount a steep hill. The "fort" was an earthwork more or less in the center of the lines that was set up so its cannon could bombard passing British ships. On the day the *Pearl* barely got past the boom, British troops under Lord Percy had been badly bloodied in a skirmish to the south. On November 2 an attempt by Hessians on the Harlem River side was turned back.

Greene's most persuasive argument for not abandoning this strong position was that its garrison could always be evacuated should the worst occur. In fact, he had reinforced it with additional battalions until the position now bristled with 3,000 men. Meanwhile the British, should they attempt to assault this nearly impregnable bastion, would need to expend enormous amounts of time and men to do so. Since the British could never take the position quickly, Washington could at any time order the defenders to evacuate, crossing the Hudson River to New Jersey under the guns of Fort Lee.

Greene, however, was disastrously wrong. On November 16, Fort Washington fell in four hours, including the half-hour the Americans were given to consider surrender terms.

It was the kind of set-piece battle the British could execute in 1776, but which Washington could only dream of with the army then under his command. Howe completely surrounded the American position, then launched a four-prong attack—five, if one counts the frigate that stationed itself in the Hudson to fire on the main American redoubt. From the north, two columns of Hessians crossed Spuyten Devil Creek and attacked the precipitous heights. From the east, British Guards regiments crossed the Harlem River and charged uphill. To their left, the 42nd Highlanders made a feint to draw off American strength. From the south, Lord Percy had easier terrain to negotiate but the enemy had thrown up three full lines of works. Percy easily took the first, but the Americans fell back to the second. At this point the Highlanders were ordered to charge and they unhinged the second line of redoubts. The Americans, being hit in the flank as well as from the front, skipped their third defense line and ran back to the fort.

The Hessians in the north meanwhile took terrible punishment from Virginia and Pennsylvania riflemen in their climb up the cliffs. A German in the regiment of Johann Rall recounted that they had to "creep along up the rocks, one falling down alive, another being shot dead. . . . we got about on the top of the hill where there were trees and great stones. We had a hard time of it there together. Because they had no idea of yielding, Col. Rall gave the word of command, thus: 'All that are my grenadiers, march forwards!' All the drummers struck up the march, the hautboy-players blew. At once all that were yet alive shouted, 'Hurrah!' Immediately all were mingled together, Americans and Hessians. There was no more firing, but all ran forward pell-mell upon the fortress." The defending riflemen had become unnerved by the sound of gunfire to the south approaching their backs and this line, too, collapsed back on the fort.

As mentioned, however, Fort Washington itself was only a big earthwork designed to overlook the river. It had no ditch, barracks, or even a well. When nearly 3,000 Americans crowded within, the British had only to bring up some artillery in order to turn the enclosure into a slaughter pen. The Hessians were first on the scene, and furthermore had suffered the most in the assault. General William von Knyphausen sent an officer under a white flag to offer surrender. The Pennsylvanian Colonel Robert Magaw asked for four hours to consider the offer, but the Hessian responded that a half-hour would suffice. General Washington had meanwhile dispatched a brave young officer named John Gooch to cross the Hudson from New Jersey, run up to the fort and request Magaw to hold on until dark, when an evacuation might be effected. Incredibly, Gooch got Magaw's reply and ran back down to the river, dodging Hessian bayonet thrusts along the way, rowed back to New Jersey and informed the supreme commander that Magaw couldn't delay any longer. Washington watched somberly from across the river while the flag over his namesake fort came down.

Fifty-nine defenders were killed in the assault, 96 wounded, and, according to the official British count, 2,837 marched into captivity. The attackers suffered 452 killed and wounded, most of them Hessians. Many among the American garrison might have cheered at the sight of British-occupied New York burning on the night of September 20; however, now that they were prisoners billeted in the town they endured dreadful hardship, packed densely together in the few buildings left available for their shelter during the winter. Many died because of their cramped, dirty conditions and from neglect.

One result of the disaster was that among those vying for influence in the American chain of command, it became open season on Nathanael Greene. Since the siege of Boston, Washington had come to view Greene—a tall, energetic and analytical Patriot not unlike himself—as the most promising figure among his senior commanders. On the occasion of Washington's previous worst disaster, Brooklyn, Greene had been ill with fever. The tremendous losses suffered at Fort Washington, which nearly gutted the American army west of the Hudson, however, could clearly be lain at Greene's doorstep—and, by inference, Washington's. Charles Lee, commenting on the debacle that he claimed to have predicted, wrote to his commander-in-chief, "Oh General, why would you be overpersuaded by men of inferior judgment to your own?"

The fall of Fort Washington was followed by the solitary quick strategic move the British made during the entire campaign for New York: the prompt reduction of its counterpart on the New Jersey side, Fort Lee. Cornwallis, with 4,500 men, crossed the Hudson and scaled the nearly sheer cliffs north of Weehawken, then marched on the fort. Washington had decided that once the Manhattan bastion that guarded the Hudson was gone the Jersey one no longer served a purpose, so he had ordered Fort Lee's abandonment. Still, Cornwallis found twelve drunken soldiers inside, as well as cannon and supplies. The Americans had evacuated so hastily that cooking pots were still simmering. The British also captured over a hundred Colonials wandering the nearby woods.

George Washington in New Jersey now faced a dilemma different from those he had suffered before. He had perhaps 3,000 ragged, beaten soldiers with whom to defeat the British juggernaut that now seemed intent on pouring through New Jersey to Philadelphia. While at Boston, Brooklyn, Harlem and Westchester, Washington's letters described a man bending under the task of challenging one of the world's great powers on equal terms with the improvised, polyglot forces he had been given charge of. In New Jersey, however, Washington no longer had such a formidable job. If he could hold the army together at all, it would be a miracle. If he could escape destruction, it would be an accomplishment. He and his small army were now fugitives.

William Howe, who has since been criticized for being slow, was at this time quite pleased with himself. He was already looking forward to "winter quarters," where he and the army could rest, keep warm and discuss their recently won laurels. The weather in the fall of 1776, however, was unusually good, prompting more operations. Howe thought that the reduction of the American Hudson forts was already icing on the cake of a successful campaign. Now Cornwallis was spearheading a drive across New Jersey in pursuit of Washington, who hardly even seemed to remain a worthy opponent. Part of Howe's problem was that every time the British faced up against the Americans they won. It was difficult for him to restrain his commanders given that kind of success. At least he had gotten rid of his annoying second-in-command, Clinton, who was ordered to put to sea on December 1, along with 6,000 men, to attack Newport, Rhode Island.

Howe's own view was that one more sweeping campaign would be necessary in the spring of 1777, during which Philadelphia would be conquered and Washington's tattered forces would finally suffer the consequences of a large, Brooklyn-type battle that would render them irrelevant once and for all. Still, he couldn't relax in comfortable quarters in New York while so many of his generals—who had the king's ear—were convinced that the rebellion could be crushed immediately. Howe resigned himself to join the pursuit of Washington in New Jersey, even if the weather was very soon to turn abominable and his true wish was to rest and bask in the victories his men had already earned.

Washington's troops, meanwhile, trudged through town after town in New Jersey, presenting a poor picture to the inhabitants with their bleeding feet and shivering shoulders. Immediately after the Continental Army's departure, Cornwallis' regiments would march in, providing a stark contrast between victor and vanquished. An American officer wrote home: "Our people, instead of behaving like brave men, behave like rascals, and to add to that, it seems that the British troops had gone into the Jersies, only to receive the submission of the whole country. People join them almost in captains' companies to take the oath of allegiance. . . . A Hell itself could not furnish worse beings than subsist in the world where our army are now posted." Samuel Adams gloomily remarked, "Nothing can exceed the lethargy that has seized the people." The Howes had proclaimed to the residents of New Jersey that all was forgiven if they would only reaffirm their loyalty to the king.

Washington may not have doubted the inherent wisdom in his decision to leave Charles Lee with half the army in Westchester, north of New York. After all, who could have foreseen that the British would not have set out to conquer New England during this period of American weakness? However, now it had become clear the British were intent on Philadelphia, and part of the plan was that Lee would come to Washington's aid should the British move south. But Charles Lee, for some reason, was refusing to move.

Washington first issued his order on November 20 in polite, respectful terms, having an aide write: "His Excellency thinks it would be advisable in you to remove the troops under your command on this side of the [Hudson] River . . ." Washington personally followed up with: "I am of opinion and the gentlemen about me concur in it, that the public interest requires your coming over to this side . . ." On the 21st, Joseph Reed, Washington's Adjutant General, wrote to Lee privately, and more frankly:

> I do not mean to flatter, nor praise you at the expense of any other, but I confess I do think that it is entirely owing to you that this army and the liberties of America . . . are not totally cut off. You have decision, a quality often wanting in minds otherwise valuable. . . . Nor am I singular in my opinion—every gentleman of the family, the officers and soldiers generally have a confidence in you—the enemy constantly inquire where you are, and seem to me to be less confident when you are present. . . . Oh General—an indecisive mind is one of the greatest misfortunes that can befall an army—how often have I lamented it this campaign.

Lee, buttressed by such votes of confidence in his judgment, ignored what he termed "His Excellency's recommendation" and instead tried to order General Heath to reinforce Washington with 2,000 of his own men. Washington quickly corrected Lee's mistaken impression that the defense of the New York Highlands should be denuded of troops, and reaffirmed that he wanted Lee's army to cross the Hudson. Lee responded that he was anxious to come to Washington's aid; however, there were matters he preferred first to attend to in Westchester.

If Washington sensed that something was going on, his apprehension heightened after opening a letter from Lee that was addressed not to him, but to Joseph Reed. The talented, energetic Reed had been by Washington's side since Boston and was privy to the supreme commander's worst problems and most intimate thoughts. Reed was not in camp when the letter arrived, so Washington opened it to discover that his closest aide and his most highly regarded lieutenant had turned on him. Lee's missive read:

> I received your most obliging, flattering letter—lament with you that fatal indecision of mind which in war is a much greater disqualification than stupidity or even want of personal courage. Accident may put a decisive blunderer in the right—but eternal defeat and miscarriage must attend the man of the best parts if cursed with indecision.

Washington promptly wrote Reed a note explaining that he had read the private letter only because he had assumed it concerned the army's business. It was fortunate that Washington had not also read the letters Lee was sending to other American leaders, in which the collapse of the Continental Army's power—with the exception of that part commanded by Lee himself—was similarly lamented. Lee was known to complain to those around him, "Have I come from gathering laurels in many other parts of the world to lose them in America?"

From the point of view of Charles Lee, Washington had proven that he was not the best man to lead the American war effort in the field. He had taken an army of over 20,000 men in July 1776, and in the process of losing New York had been reduced to commanding a ragged division of 3,000 intent on scrambling across New Jersey to avoid British pursuit. One of the "great man's" primary attributes was his habit of constantly deferring to the Congress in Philadelphia for instructions, keeping the politicians apprised of his actions. Lee possessed no such obsequiousness and had even written to James Bowdoin in Massachusetts on November 24: "Affairs appear in so important a crisis that I think the resolves of the Congress must no longer too nicely weigh with us. We must save the community in spite of the ordinances of the legislature. There are times when we must commit treason against the laws of the state, for the salvation of the state." Lee was more interested in results than consultation. Further, after Brooklyn, Washington had claimed American casualties were surpassed by those of the enemy, which was nonsense. He had attempted to paint the minor skirmish at Harlem Heights as a stupendous victory; likewise he had depicted White Plains as a successful action. He had even tried to minimize the catastrophe at Fort Washington by claiming only 2,000 American troops were taken, when 3,000 were lost.

Now that the supreme commander had been divested of regulars, deserted by militia, and no longer commanded a significant military force, naturally Lee was being begged to join and reinforce his army. Washington had panicked himself into thinking the British would extend themselves in midwinter all the way to Philadelphia. If they did so, nothing would better serve America's cause than an intact army under an experienced general threatening the northern flank of the enemy supply route to New York. While an independent army under Lee on the British flank might win a victory that would reverse the terrible effects on the country of Washington's ineptitude, Lee as second-in-command of a concentrated army, contributing his brilliance to nothing except the public credit of "His Excellency," would not ultimately serve the cause. All Washington had to do, in Lee's view, was get behind the Delaware River and, once there, leave the soldiering in the field to professionals. The aristocratic Virginian could then write letters to the Congress all day long, while Lee considered how best to fight the war.

Still, Lee had to follow orders, or at least appear to do so. In the meantime, if he tarried long enough, Washington might finally self-destruct, if not fall prey to Cornwallis. Once Lee finally crossed the Hudson on December 2, the British became worried by his presence, as he had anticipated, inhibiting any late-year ambitions they might harbor against Philadelphia. Before the march Lee had weeded his army of all those who were frail or unable to see the campaign through. By December 4, Lee's army, now comprised of 3,000 veteran regulars and 1,000 militia, was sliding ever so slowly toward a rendezvous with the supreme commander.

Washington had meanwhile gained breathing room by crossing the Delaware and by seizing the boats along a 70-mile stretch of the river so the British couldn't follow. The Continental Congress was preparing to abandon Philadelphia for Baltimore. Some 2,000 militia had walked away on December 1 after their enlistments had expired; the bulk of Washington's men were due to go home on January 1. Horatio Gates was on his way with eight regiments from the Northern Department, Heath was coming with a detachment from Peekskill. Some Pennsylvania militia were turning out, including a cavalry detachment from Philadelphia. Still, it was Lee's army, once rejoined with Washington's own, that would propel the Americans into a position of strength. Once the army was together again, it might even attempt an offensive operation.

On December 8, Lee wrote Washington: "If I was not taught to think that your army was considerably reinforced, I should immediately join you; but as I am assured you are very strong, I should imagine we can make a better impression by hanging on [the British] rear. . . . I shall look about me tomorrow, and inform you further." To General Heath, Lee wrote: "I am in hopes here to reconquer the Jerseys. It was really in the hands of the enemy before my arrival."

Washington no longer had time for politeness or persuasion: he wrote back to Lee, confessing how much he needed him. "The force I have is weak and entirely incompetent [to defend Philadelphia]," he wrote. "I must therefore entreat you to push on with every possible succor you can bring."

Lee answered with a suggestion that, instead of joining Washington directly, he cut across the rear of Cornwallis' army and cross the Delaware at Burlington, below Trenton. He also inquired about places to cross above Trenton. He also mentioned, "The militia in this part of the province seems sanguine. If they could be sure of an Army remaining amongst 'em, I believe they would raise a very considerable number." Fury and helplessness must have vied for precedence in Washington's reaction to this letter. He immediately replied: "I am much surprised that you should be in any doubt respecting the route you should take." By now Washington had provided his subordinate with orders containing precise route instructions many times, only to be ignored. Much as the supreme commander could sense a rising tide of blame about his shoulders for the perilous military situation of the republic, he had still not lost confidence that by level-headed coordination between the army, the Congress and American Patriots, the cause could be retrieved. By now Washington had seen through Lee as a military adventurer, albeit one who instinctively resisted authority in whatever form, while surrounded by his ever-present hounds. A valuable general, yes; the man upon whom American destiny should rely, no. Washington had already learned what Joseph Reed thought: he was for Lee. Greene was loyal, but what did Gates think? Knox, Arnold, Heath, Putnam? Despite recent events, it was Washington himself who had the responsibility to defend the new nation against the forces of the Crown, even if the Congress was just then packing its bags to abandon the capital.

On December 13, a young British officer named Banastre Tarleton stepped into the pages of American history. His exploit that day was the first of many that would earn him a reputation as the most notorious cavalry commander on either side of the Revolution. By the end of the war he had inspired a term, "Tarleton's quarter," which meant "No mercy," and he claimed to have "butchered more men and lain with more women than anybody else in the Army." On that Friday the 13th in December 1776, he was only involved in a small unit action, but it resounded throughout the colonies. He captured General Charles Lee.

Tarleton was then a cornet under Colonel William Harcourt, a former subordinate of Lee's, whose regiment of dragoons was scouring the north New Jersey countryside for the American army en route from Westchester. He came upon two American soldiers and, as Tarleton put it, "The dread of instant death obliged these fellows to inform me . . . of the situation of General Lee." The British learned that he was in a tavern not a mile away, well to the rear of his army.

In that tavern, James Wilkinson, a courier from General Gates, was looking out the window when he saw Tarleton and his dragoons come charging down the lane. He said to Lee, "Here, sir, are the British cavalry." Lee took a look and replied, "Where is the guard? Damn the guard, why don't they fire? Please, sir, see to the guard." Wilkinson grabbed his pistols and ran downstairs, only to dodge British fire in the doorway and glimpse American sentries surrendering to the British horsemen. In a letter to his mother Tarleton described the capture.

> The sentries were struck with a panic, dropped their arms and fled. I ordered my men to fire into the house thro' every window and door, and cut up as many of the guard as they could. An old woman upon her knees begged for life and told me General Lee was in the house.
>
> This assurance gave me pleasure. I carried on my attack with all possible spirit and surrounded the house, tho' fired upon in front, flank and rear. General Lee's aide de camp, 2 French colonels and some of the guard kept up a fire for about 8 minutes, which we silenced. I fired twice through the door of the house and then addressed myself to this effect: "I knew General Lee was in the house, that if he would surrender himself, he and his attendants should be safe, but if my summons was not complied with immediately, the house should be burnt and every person without exception should be put to the sword."

Lee icily rejected a woman's suggestion that he hide under a bed; Wilkinson meanwhile had placed himself in a spot where he "could not be approached by more than one man at a time"—in a chimney. When it became clear his guards were no longer resisting, Lee, after pacing back and forth for a few minutes, gave himself up, dressed in slippers, a housecoat and a filthy shirt. The British tossed him atop Wilkinson's horse and spirited him back to their own lines, racing along different backroads from which they had come. Once he had been turned over to Cornwallis, then Howe, there was some question of treating the American general as a deserter from the British Army, but Lee pointed out that he had properly resigned his commission prior to emigrating to America. He was subsequently treated in captivity with all the respect due his rank. Colonel Harcourt exulted in a letter to his brother that he had seized the "most active and most enterprising of the enemy's generals."

Lee's reason for tarrying in a house far removed from his troops is a matter of speculation, but the fact that women were present may provide a clue. Tarleton described an "old woman" who begged for mercy, however to a twenty-one-year-old of his ilk that description might have applied to any female over thirty. In the moment prior to his capture, Lee had signed another anti-Washington letter, this one to General Gates, that began: "The ingenious maneuver of Fort Washington has unhinged the goodly fabric we had been building. There never was so damned a stroke. *Entre nous*, a certain great man is most damnably deficient."

Although to many American Patriots, news of the capture of Charles Lee came as a disaster, George Washington could be forgiven for seeing the event as less than tragic. The cabal of whisperers who had been busy promoting Lee as supreme commander had suffered the major loss. Washington gained because now Lee's army would promptly join with his own, without further machinations or unsanctioned strategic schemes on the part of its commander. The real danger to the American cause— already rippling through the camps—was the perception that Lee had turned traitor and had voluntarily surrendered to the enemy. If one can picture the effect on American morale if Hancock, Franklin, or Washington himself had suddenly approached Cornwallis with the conviction that the patriotic cause was bankrupt, such would be the impact if Lee, who many considered to be America's only professional general, had deserted to the enemy. In America's greatest hour of peril it was essential to inform the public that Lee had been captured; he hadn't given up.

The day after Lee's capture, General William Howe announced that the 1776 campaign was over. He returned to New York, where his mistress was waiting to provide him with more pleasurable activity than slogging endlessly through snow after a ragtag band of amateur soldiers. He left behind a string of outposts along the Delaware River to protect the acquiescent King's Colony of New Jersey against last-ditch American banditry. Like Washington, Howe benefited from intelligence supplied by sympathizers in enemy-held territory, and was aware that the American army was due to dissolve on December 31 because of expired enlistments. He might not have considered, however, that for exactly that reason Washington would need to mount an operation before the end of the year. By mid-December, American strength had grown to over 7,000 men, encamped on the Pennsylvania side of the Delaware. And George Washington needed a victory, both to save the cause and to retrieve his reputation.

On Christmas Day, 1776, Colonel Johann Rall's Hessians in the southern New Jersey town of Trenton on the Delaware, celebrated with feasts and drinking. The Germans had a tradition of setting up and decorating small pine trees in their quarters in recognition of the holiday. As they gave themselves over in that foreign outpost to serious religious contemplation, the consumption of rum, or simply sad nostalgia for the families they had not seen for over a year, they were unaware that the Continental Army was on its way for a visit. George Washington was no longer retreating.

Chapter 2

NAPOLEON
IN RUSSIA

After 1776 the French government could feel amused at British failure to contain the anti-monarchical rebellion in America, and Louis XVI heartily assisted the Patriots' cause. A decade after the end of that long, costly conflict, however, the last laugh belonged to King George III. The Bourbons suffered their own revolution, closer to home, and in 1793 King Louis, his wife and hundreds of other aristocrats were beheaded before cheering crowds in Paris. That the French Revolution subsequently unleashed a wave of ideological fervor across Europe might have been predicted; unforeseen was that one man would rise from the chaos and transform "liberty, equality, fraternity" into dominating military power. That man was Napoleon Bonaparte, and as French historian Louis Madelin remarked, "The French people watched in amazement how . . . a star began to rise that would change the image of the earth."

During the two centuries since his rise to power, Napoleon Bonaparte's career has continued to inspire admiration, even awe, on the one hand, and hatred or contempt on the other. The opposing camps are represented by the terms "Napoleonic age," which respectfully recognizes his impact on history, and "Napoleonic complex," a modern term describing an urge for overachievement on the part of a less-than-average individual. Indisputably, Napoleon was one of history's greatest achievers. That he is loathed in some quarters to this day, however, is because he plunged Europe into two decades of warfare, ostensibly on behalf of France but in fact for his personal gain.

What George Washington had fought in America, from Brandywine to Yorktown, resembled a Cabinet War compared to the huge, winner-take-all battles that Napoleon initiated in Europe. The scale of battle, the carnage and the involvement of entire civilian populations increased throughout his rule of France. The logical extension of the kind of warfare of which Napoleon was modern history's leading practitioner did not finally run its course until 1918. The term "War to End All Wars" has been ridiculed, but in one sense it was accurate. Massive bloodletting inspired by competitive jealousies between European monarchies, or nation-states steeped in that tradition, no longer took place after the last of the Great War's ten million dead had been put in the ground. The Nazi phenomenon that followed, together with Japanese aggressions, represented a 20th-century leap into new motives for battle. Warfare had simply become too destructive to take place on behalf of an individual's ego or because of petty quarrels between rulers; henceforth, far larger causes would be necessary to inspire populations to sacrifice themselves.

Of course, even in the post–Cold War world, conflicts can be instigated by individual statesmen. In 1991, Iraq's Saddam Hussein sought to cover himself with glory by undertaking an "old-fashioned" conquest of the nation of Kuwait. He was promptly confronted by a gigantic global coalition determined to roll back his aggression. A few years later, Serbia's Slobodan Milosevic encountered a similarly powerful alliance intent on quashing his designs in the Balkans. Bloodletting in Africa's Rwanda and Congo, meanwhile, was allowed to take place. Small or regional wars can thus continue to be fought—depending on the interests of great nations and their allies—but today the horrific consequences of modern warfare mitigate against entire continents going up in flames because of a single individual's lust for power.

Napoleon Bonaparte was not only an energetic, creative head of state but also the greatest general of his time, and so was able, during his unique era, to take control of history. It is significant that he lived at a time when the arts were at a peak of sophistication, and when armies, providing quick ascension to social status, attracted ambitious individuals who today might be lawyers, doctors or business leaders. His battles took place when military science was a fertile field for intellectual thought, yet before the repeating rifle, machine gun, and high explosive appeared to inflict mass slaughter. The casualty rate caused by low-velocity, round balls was still acceptable enough then to permit open field maneuvers by rigid formations of gaily dressed infantry and heroic charges by sabre-wielding cavalry against cannon.

While the low-tech firepower of Napoleon's time abetted his inclination to pursue constant warfare, the horse-and-cart-based logistics of the era eventually ensured his downfall. Bonaparte became so skillful at conquest that he eventually marched his armies farther than the points to which they could be supplied. A single rail line into Russia would have allowed him to fulfill all his ambitions; but then, had Napoleon lived to see a railroad he would already have found warfare becoming too ghastly to be provoked year after year on his personal whim.

The modern figure most commonly compared to Napoleon is Adolf Hitler. Because Bonaparte's primary genius was as a general and Hitler never led armies in the field, the two men's careers are fundamentally different; but to the degree that they are the only two post-medieval leaders who conquered nearly all of Europe, they do have something in common. When Napoleon began his role as First Consul he rode the tumultuous wave of ideological fervor prompted by the French Revolution, and in that sense was modern. He dropped the façade in 1804 and declared himself Emperor (making his brothers kings), and from that point it became clear that his own ambition had always taken precedence over the state's. In retrospect, Napoleon can be compared more easily to Alexander the Great or Julius Caesar than to Hitler.

The German dictator, unconcerned with establishing a personal dynasty, was obsessed with a nationalist ideology that he initially sold to the public through electioneering, a thoroughly modern technique. When Hitler's mask came off, however, it was seen that he was also motivated by an impulse toward mass murder of civilians and thus, too, had put his private ambition ahead of the state's. Although in view of Nazi horrors it is a serious insult to Napoleon to compare him with Hitler, in military terms the Nazi leader can be considered the French emperor's closest replica. Prior to World War II, generals and statesmen were invariably flattered by suggestions that they might be "another Napoleon," but after Hitler's career the phrase became an ominous slur. While Napoleon represented the epitome of "personal" warfare as it could still take place prior to the technological age, Hitler unveiled the depths of depravity even leaders of great nations could sink to, and how horrific the consequences could become in the 20th century. There is little doubt that while Hitler could feed upon Napoleon's achievements for inspiration, Bonaparte would have recoiled had he ever received a premonition of his successor.

The question of absolute rule by self-made men aside, however, the most visible bond between Napoleon and Hitler is that they both ensured their own downfalls by making a decision to invade Russia.

Due to the fact that Napoleon's attack into Russia resulted in a gigantic calamity and history's most poignant retreat, there has always been an effort to match the causes of the war with the consequences that followed. It is impossible to do so because no one expected the scale of suffering that would follow. What Napoleon sought, and anticipated in 1812, was one great summer battle between the Russians and his French-led, pan-European host. His differences with Tsar Alexander I only motivated him to teach the Tsar a lesson, after which the Russians would resume their compliance with French policy, and perhaps lose a slice of territory or two, which was customary after a monarch had dared to resist the Grand Army. Napoleon had already marched east several times in order to inflict decisive defeats on the Prussians, Austrians and the Russians themselves. This time he was having to enter Russian territory in order to project his might. He had no intention to go all the way to Moscow; he only wanted to meet the Russian army during the summer months and give it a thrashing.

One occasionally comes across the theory, invariably put forth by historians from a certain English-speaking isle off Europe, that Nazi Germany invaded the Soviet Union only as a backdoor way of subduing Britain. Although the theory is far-fetched as applied to Hitler in 1941, it is central when applied to Napoleon in 1812. Britain had been France's primary rival for power and influence for hundreds of years; with the onset of the colonial age that rivalry had continued on a global scale in the Americas, Africa and the East. Unfortunately for Napoleon, the British Navy was far superior to the French one and was able to confine his ambitions to the European mainland. Just before Napoleon's stellar victory at Austerlitz, Lord Horatio Nelson had utterly destroyed the French fleet at Trafalgar, rendering any prospect of a cross-Channel invasion, and thus a confrontation on British soil, remote.

Since Napoleon had no hope of enticing Britain into a decisive land battle, he responded with the "Continental System" that barred British trade from European ports. The system was designed to bring Britain to its knees through economic hardship and at the same time enhance the French economy by making it the fulcrum of continental industry. The

British responded with a blockade of Napoleon's Europe and a high-handed policy toward neutral ships bound for European ports. (British determination to control the sea lanes to France's detriment would eventually earn them another war with the United States in 1812.)

The Continental System was sound in theory, and in fact created prosperity in France, but for the other European nations that were forced to join it only created hardship. Cut off from global trade by Napoleon's feud with Britain, entire factories were forced to shut down; stocks of food and raw materials went to waste in rotting piles, their owners ruined and the workers destitute. It was a fantastic time to be a smuggler; however, the spread of that profession did nothing to enhance state coffers. Britain only seemed to be growing richer, and British gold was available to finance any sovereign willing to put armies in the field against the French.

Napoleon was able to close off the northern European coast to the British only because he was able to defeat any opposing land army on the battlefield. In 1811, when Russia, despite its earlier treaties with the French, suddenly withdrew from the Continental System, it invited an invasion. There were also geopolitical matters to be settled in the east, and there may, too, have been the fact that Napoleon, once he had passed forty years of age, simply didn't know what to do with himself if he didn't have a war. In any event, before or since Napoleon's invasion, the French had had no particular axe to grind with Russia. Napoleon's major goal was simply to force Russia to be a solid ally in France's historic struggle against the British.

To understand the dilemma encountered by Russia in 1812, when the greatest army in world history began trampling across its borders, it may be helpful to step back a few years to provide a recap of Napoleon's achievements to that time.

The Terror in Paris during the French Revolution, culminating in the execution of Louis XVI in 1793, alarmed the remaining monarchies of Europe. Appalled at the mob rule and executions, they converged on France to quell the bloody tide. The French people responded with "Revolutionary armies" that sallied forth to their borders to defend their nation, and to export their principles of freedom.

Twenty-six-year-old Napoleon Bonaparte, who was born in 1769 on the island of Corsica, was one of the more successful French Revolutionary soldiers. Trained as an artillery officer, he first gained notice by directing his cannon against the British after their seizure of the port of Toulon in southern France. In Paris he won favor by directing his guns against a counter-revolutionary mob. Operating in northern Italy as a general, he won fame through a string of victories against the Austrians. By 1796 the French Revolutionary armies had won their battles on the border against Europe's vengeful monarchies, but Napoleon had approached the center of power in Paris and the question for the civilians became what to do with him next. They countenanced his plans to invade the Mideast, to disrupt the British trade route to India. After the spectacular failure of his expedition to Egypt and Palestine (Lord Nelson had destroyed the French fleet behind him), Napoleon returned to Paris in 1801. The government, as usual, was in flux. Napoleon, for his rare combination of dynamism and common sense, was named one of the triumvirate to henceforth rule France. He quickly surpassed his fellow consuls and became the sole leader of the country.

Napoleon's unique ability as head of state was that he could not only administer the country, with often innovative domestic policies, but could leave his desk to take command of the army in the field whenever he chose, winning spectacular victories against foreign powers. Aside from his title of First Consul, he was also the best general in Europe, and his triumphs enhanced French power as well as his own. Frederick the Great had been the last European ruler who possessed such skill, but Frederick did not rule such a large and powerful nation.

Napoleon's initial campaign as head of state was crucial, because it would serve to consolidate his position. While he had been in Egypt, a Russian army under Aleksandr Suvorov had reversed all the gains he had achieved in northern Italy. (The Russians then dropped out of the anti-French coalition, due to squabbles with their allies.) As his first venture onto the battlefield as First Consul, Napoleon crossed the Alps into northern Italy and conducted a brilliant campaign, winning a decisive victory against the Austrians at the Battle of Marengo.

French revolutionary ideals were at this time still in play. A number of citizens without privilege living under the aristocracies of central Europe welcomed the somewhat anarchic French tide. Just as important, the French had stumbled upon a new innovation in warfare that caught the monarchical armies by surprise. Because the Terror had beheaded much of the French officer corps, and compelled many survivors into exile, the Revolutionary armies were incapable of disciplined linear tactics that were considered state-of-the-art among armies more strictly controlled. Instead, the French stressed column tactics, which presented their opponents with dense masses of individuals aimed at narrow sectors of the front. These columns were officially comprised of successive platoons or battalions, but in reality, after a few steps into the face of fire, represented a mass charge. One effect of the Revolution had been to free up more men from the countryside, regardless of class and with the benefit of enthusiasm, to toss into battle. This huge influx of recruits could not quickly be molded along traditional lines. More often they were best utilized by simply propelling them at the enemy.

At the same time, enough traditional military expertise remained in the French army so that it could feature both linear and column techniques. An enemy anticipating a set-piece battle would not know what to expect. Unfortunately for his opponents, Bonaparte himself had been trained in artillery. French élan also lent itself handsomely to the cavalry, where a young man named Joachim Murat emerged as an incredibly brave commander. The new generals of France were not hidebound to theory and instead were promoted on the basis of their success, often improvised. And France, being the largest country in Europe, with the most aroused population, confronted the traditional armies of the monarchs with an especially formidable challenge.

Napoleon Bonaparte did not create the armies, with their tactics and their energy, which he would command; but he was an exceptional strategist with an eye for opportunity. He had a rare talent for assessing a battlefield and was always cool, always aggressive. His lieutenants, promoted strictly through merit, became household names in their own right— Murat, Massena, Ney, Junot, Bernadotte, Davout, Bessieres—and the French soldiers came to see themselves as invincible. Napoleon was idolized by his troops, and in 1804 the French people cheered his ascension to the title of emperor. At his inauguration he took the crown from the Pope's hands and placed it on his head himself.

In 1805, what is known as the Third Coalition took the field against Napoleon, led by the Austrians, backed by the Russians, the Prussians promising to join and the British helping wherever they could, primarily at sea. Napoleon marched east and with a well-orchestrated series of maneuvers was able to reduce the main Austrian forces at Ulm. His subsequent masterpiece was Austerlitz, where he was outnumbered by a predominantly Russian army, yet assessed the field and anticipated the moves of his enemies so precisely that he ended the war in one day's battle. Both Tsar Alexander I of Russia and Emperor Francis II of Austria were present, somewhat inhibiting the initiative of the allied general in charge, the Russian Alexander Kutusov. The allied left was totally destroyed, and their right, led by Russian Prince Peter Bagration, was forced to withdraw. After the French victory, Austria sued for peace while the Russians limped home, leaving some 20,000 casualties behind.

The Prussians, still clinging to the glory won by Frederick the Great and chafing under French domination, took the field the following year, with a promise of a renewed Russian commitment. Napoleon met and wrecked a Prussian army at Jena. On the same day, fifteen miles north, Marshal Louis Davout took on the other half of the Prussian Army at Auerstadt with success. The new problem for the French was that the Russians were coming, along with winter. At Eylau the two armies met in a raging blizzard. With snow blowing into their eyes, French infantry waded into a Russian artillery concentration, and were then assailed from the rear by their own guns firing into the whiteness. As a massive Russian counterattack threatened to overwhelm the Grand Army, Joachim Murat placed himself at the head of the entire French cavalry reserve of 10,000 horsemen and charged into the Russian center. In two columns the cavalry broke through the enemy infantry, overran a combined battery of 70 guns, then reformed into a single column and charged back through the lines. The spectacular attack bought time for Davout's corps to arrive from the south and begin hammering in the Russian left. Only darkness brought a halt to the gruesome battle, under the cover of which the Russian Army withdrew.

At Eylau the outnumbered French, despite appalling conditions, had bested the Russians, but Napoleon had been forced to realize that when operating in the east, in bad weather, his brilliance and willpower needed to be buttressed by better logistics. The army remained in winter quarters until summer, when Napoleon trapped the Russians against a river at Friedland. A third of the enemy army was destroyed and the rest fled in panic from the field. After this victory, Napoleon met with Tsar Alexander on an island in the Niemen River called Tilsit, where they signed an amicable treaty. Alexander promised to join the Continental System, barring British trade from his country.

In 1808, Napoleon decided to dispose of the monarchies of Spain and Portugal and sent troops into the peninsula to enforce French rule. The Spaniards resisted with a guerrilla war while Portugal opened its ports to a British expeditionary force. In 1809 Napoleon had to enter the peninsula with elements of the Grand Army to correct the situation. He defeated the Spanish and chased back the British, under John Moore, forcing them to evacuate the peninsula through the port of Corunna. By that time Austria had once again taken the field, so Napoleon returned to Paris to prepare another continental campaign. (Four months later the British would return to Portugal under Wellington.)

In 1809 Napoleon found the Austrians, under Archduke Charles, to be more formidable than in prior years. The French took Vienna but then Charles caught the Grand Army in the middle of crossing the Danube near the villages of Aspern and Essling. Rather than back off from a clearly disadvantageous position, Napoleon attempted to win the battle anyway, but eventually had to withdraw his bridgehead. Two weeks later he forded the entire army downstream and confronted the Austrians at the village of Wagram. After a huge battle resulting in over 70,000 combined casualties, Charles sued for peace. Napoleon lopped off some more slices of Austrian territory as the price of accommodation and returned, with the Grand Army, to France.

After Wagram, Austria and Prussia did little for two years but attempt to heal their wounds. Having succumbed to the French Emperor, with considerable loss of territory, the monarchies of Francis I and Wilhelm III sought to reestablish support among their people and regather their strength, if not self-esteem. The new map of Europe, as Napoleon had redrawn it, consisted of a huge France that incorporated all or parts of modern-day Belgium, the Netherlands, Germany, Switzerland and Italy. The British had swooped down on Copenhagen in 1807 to destroy the Danish fleet, so Denmark was now a French ally. Napoleon had taken over the Vatican and all of Italy; Joachim Murat was made King of Naples. Outside Austria and Prussia, the Germans of Europe were formed into the Confederation of the Rhine, a collection of monarchical states subject to French rule. The poor, primitive Spanish and Portuguese continued to resist proper government (Napoleon had decided to make his brother Jerome king of Spain), but ten years of warfare had essentially quieted the continent under French rule.

Alone among the great powers of mainland Europe, Russia continued to act as if it were uncowed by the Emperor of the French. In 1810, Napoleon had inquired about marrying the Tsar's younger sister but was given a subtle brush-off. Beginning on the first day of 1811, Tsar Alexander renounced the Continental System and opened his ports to British vessels. The Treaty of Tilsit had been signed between equals, Russian Tsar and French Emperor, notwithstanding the serious Russian setback at Friedland. If the Tsar were to continue guiding his country according to French policy, however, the French were bound to ensure economic benefit. Instead, the embargo on British trade had only caused poverty.

The other great issue that divided the French and Russians was Napoleon's resurrection of Poland under the name Grand Duchy of Warsaw. Poland had disappeared in 1795, divided up between the Russians, Prussians and Austrians, but Napoleon, after defeating all three of those powers, had demanded back enough territory to reconstitute the Polish state. It was not a move taken lightly by the Russians. Poland had once been a great nation that had occupied Moscow as well as most of Ukraine and the Baltic coast. A reconstituted Poland under Napoleon's auspices could only be perceived as a threat, if not a staging ground.

In St. Petersburg, there was little sentiment toward acting as a client state of France. On the contrary, the great Suvorov had proven early on that a Russian army was to be feared. And Napoleon had learned at Eylau that the Russians were tougher than his other opponents. The little faux Frenchman had been charming at Tilsit, but young Tsar Alexander had not been intimidated. An alliance with the French could be accepted for practical purposes, but if the alliance did not live up to its promises, providing benefits to the Russian empire, it would be discarded. If the Corsican wished to contest the issue, Russia sat huge and immobile on the edge of the continent, willing to fight any invader. To Tsar Alexander and his coterie of young, dashing officers, the vote was unanimous: Let him come.

In the spring of 1812, Napoleon assembled the largest army in history to that time, including soldiers from nearly every nation in Europe. Aside from the French Grand Army, comprising a third of the force that would invade Russia, he had Spaniards, Italians, Croatians, Dutch, Swiss, and an amalgamation of forces from the Confederation of the Rhine: Saxons, Bavarians, Westphalians, Wurtembergers, Hanoverians, Hessians and others. In the north he would have 20,000 Prussians advancing on his flank and in the south 30,000 Austrians. The largest contingent of foreign troops, naturally, were the Poles, amounting to 70,000 under Prince Joszef Poniatowski. If few of Napoleon's other soldiers knew exactly why this war was being fought, the Poles recognized that a Napoleonic victory would result in the revival of their independent nation. Napoleon could have benefited from as many as 200,000 Polish volunteers if he had enthusiastically proclaimed the resurrection of their country; however he also had the Prussians and Austrians to consider, who still held Polish territory and who could field more disciplined troops.

In all, Napoleon mobilized 660,000 soldiers for his invasion of Russia, 449,000 of whom would cross the Niemen River in the initial wave. Enormous care had been taken to provide supply for the troops, but, as in other respects during his career, Napoleon had proven to be a man ahead of his times. His skill at assembling such a vast, pan-European host was superior to his ability to supply it with the horse-and-cart system that was then the apex in technology.

Tsar Alexander, like everyone else, believed that the war would be won by a brilliant victory or two in the summer. The gory revolution in France had inspired a number of noble French emigrés to enlist in the Tsar's service. Napoleon's subsequent conquests had compelled many Germans, Prussians and Austrians to head east, toward the one state that could still defy the French. The Tsar's closest military advisor was Ernst von Phull, a Prussian; his northern army would be commanded by another Prussian, Prince Ludwig Wittgenstein; his First Army by a Scotsman with a French name, Barclay de Tolly (born in Livonia on the Baltic coast). Count Levin Bennigsen, from Hanover, was a close adviser, as was the Swede Count Gustaf Armfeld. Another officer in Russian service was a young Prussian named Carl von Clausewitz, who would go on to compose a number of universal military principles based on his observations in the coming campaign. The rank and file might have been Russian, but most of the ranking officers were foreign. In fact, the language of the Russian court at this time was French. Alexander's most prominent native commanders were Peter Bagration of the Second

Army, born in Georgia, and Matvei Platov, who commanded the Cossacks.

If the Tsar of Russia benefited from a stable of modern military commanders from throughout Europe, albeit from countries already defeated by France, the Emperor of the French could call upon an even more impressive array of subordinates—marshals who had all won their batons under the gaze of Napoleon, and who had become thoroughly accustomed to victory. As it turned out, the problem for the French during their advance into Russia was not defeating the enemy armies, but negotiating the factors of space and time given their logistical foundation.

On June 23, 1812, the invasion began when three companies of French skirmishers crossed the Niemen River in small boats. Pontoon bridges were quickly erected and the Grand Army began to cross in force. Napoleon took position on a hilltop, accepting cheers from the men as they filed to the river and spread out across the vast expanse of empty territory on the other side. On a reconaissance just prior to the crossing, Napoleon had been thrown to the ground after his horse had shied from a rabbit. When word of the mishap got around, a number of officers at headquarters took it to be a bad omen; they would have been cheered had they known that on the previous day Tsar Alexander's staff had erected a huge outdoor canopy under which to hold a moonlight ball, but it was shaky and collapsed. The staff hurriedly set up another before the Tsar and his glittering guests arrived to dance.

The Russians were split into two main armies when the French invasion began. The largest force, some 100,000 men, was led by Minister of War Mikhail Barclay de Tolly, with the Tsar and his advisers also in attendance. Prince Peter Bagration led the Second Army of some 40,000 on the left, to the south. A cossack force of 15,000 under Platov connected the two Russian wings, though Platov was more inclined to lean toward Bagration, a native hero as opposed to a foreigner. There was some idea at the beginning that Bagration would counterattack into the French rear, but those plans were abandoned when the Russians realized the magnitude of the offensive.

Napoleon's first intention was to keep the Russian armies separated so he could easily destroy either one. It was a sound idea, validated some hundred years later by the Germans at Tannenberg in World War I, but in 1812 it backfired on the French. An irony of Napoleon's offensive is that he desired, above all, a decisive confrontation with the Russian Army, but by maneuvering to keep its elements separated he made it impractical for them to risk battle. Bagration in the south, as well as Barclay in the north, could only retreat, drawing the French further into Russia until which point they could make a stand. Given the vast space of Russia, French maneuvers aimed at "divide and conquer" turned out to be self-defeating.

A further irony is that the Russian's own master plan at the start of the war was to unite all their forces for a decisive battle to take place at Drissa on the Dvina River, 70 miles inside Russian territory. The brain-child of Tsar Alexander's chief military advisor, the Prussian Phull, the plan called for Barclay's army to man an extensive network of fortifications against which the French would impale themselves, while the Second Army attacked the enemy from behind. Thus stymied and bloodied, the French would be further weakened by the onset of winter and would abandon their invasion, leaving Russia's major cities intact. Thousands of serfs and peasants spent months preparing the works at Drissa for the climactic event.

On June 28 the French entered Vilna, the capital of Lithuania, while Barclay fell back to Drissa. Since the Lithuanians had more in common with the Poles than the Russians, Napoleon expected cheering crowds, but was disappointed. In contrast to the well-fed Russian army which had just vacated the place, the French descended on the city like ravenous wolves, looting houses and stealing food wherever it could be found. Even though the invasion was only four days old, most of the French army had marched hundreds of miles over a period of weeks or months just to reach their start line on the Niemen. At the bridge crossings supplies were already backed up; torrential rains at the end of June worsened the problem, and thousands of horses were falling dead along the routes of advance. Napoleon stayed for over two weeks in Vilna, setting up a sort of Paris East and a major supply depot that could serve the army during the operations to come.

To the south, meanwhile, Prince Bagration was in trouble. Napoleon's brother Jerome's army on the right wing was slow to get started, but the French divined that the Russian Second Army of the West was retreating toward Minsk. Napoleon dispatched Marshal Davout with two divisions to head for that city to cut Bagration off and hold him until Jerome, commanding the V, VII and VIII Corps, could arrive to deliver an annihilating blow. On July 4 Napoleon had admonished his younger brother to hurry, saying, "You are compromising the success of the whole campaign on the right flank."

In earlier years, Andre Massena was probably Napoleon's best subsidiary commander, but by 1812 that status belonged to Louis Davout. A cold personality who did not even attempt to cultivate charisma, Davout was nevertheless one of the few men Napoleon could trust to take the strategic initiative, and his I Corps was considered the best in the army. Napoleon issued an order that in joint operations between Davout and Jerome, Davout was to hold overall command.

Davout got to Minsk on July 8, ahead of Bagration, capturing piles of Russian supplies in the city. Bagration veered away. Jerome continued to be laggard in his advance and on July 23 Bagration tried to break through Davout's lines at Mogilev on the Dnieper River. The first significant battle of the campaign consisted of less than 20,000 men on a side: Davout had left three of his five divisions behind with the main army; Bagration, on the other hand, was unsure of French strength and hesitated to commit his entire force. After a fierce series of attacks and nearly 7,000 combined casualties, Davout's men held, forcing Bagration to retreat to find another way to link up with the First Army. By now, Napoleon was enraged at Jerome. When, in addition, Jerome learned that he had been made subordinate to Davout, he decided to leave the army and return to his most recent appointment, King of Westphalia. Bagration continued east, his cossacks under Platov in constant skirmishes with Davout's cavalry and Poniatowski's Polish lancers, who had finally arrived in the vicinity.

To the north, the Russian First Army had meanwhile reached the great fortified position at Drissa and Barclay's entire staff was appalled. The site was half finished, ill-conceived, and could only serve as a death trap for the army if it were to sit there and wait for Napoleon to arrive. Besides, Bagration was still 180 miles away and could not execute the flank attack as called for in the master plan. On a positive note, the Tsar succumbed to gentle arguments at this time that he would be of far better use to the war effort if he left the army and concentrated instead on mobilizing support in the hinterland. He left a large coterie of advisers behind, but Barclay insisted that the imperial headquarters stay a day's march ahead of the army. This, as Carl von Clausewitz commented, "placed it in the same category as heavy baggage."

Barclay moved to Vitebsk, the first great Russian city in the path of Napoleon's advance, to await Bagration and then give battle. Joachim Murat skirmished daily with the Russian rearguard, his flamboyant uniforms and dashing conduct on the field making him the most visible figure on either side of the lines. On July 25 Murat found elements of the Russian IV Corps near the village of Ostrovno. The battle grew as French units arrived on the scene until 35,000 men were engaged. The climax came when a regiment of Italian infantry was attacked by guns and then beset by a charge of Russian cavalry from out of a stand of trees. Murat put himself at the head of a regiment of Polish lancers and attacked into the melee, deciding the day and forcing the Russians to withdraw. From prisoners he learned that the Russian First Army was in place before Vitebsk. Napoleon, who had seemed somewhat lethargic until then, became stimulated and it was said that the old gleam returned to his eye. He sent urgent missives to all the elements of the Grand Army to converge on Vitebsk; on the 26th, aggressive Russian cavalry parried with French horsemen. The next day Napoleon eagerly consolidated his forces for the battle that would win the war, and end the campaign.

On the evening of the 26th, however, Barclay received word of Bagration's failure to break through the intervening French lines. Barclay declined to fight a major battle against Napoleon by himself and ordered the First Army to retreat, giving up Vitebsk without a fight. On the morning of July 28 the French discovered that the Russians had vacated the field, leaving not so much as a discarded musket behind. This was in marked contrast to French march routes, which were littered with broken wagons, ill soldiers and dead horses. A depressed Napoleon entered Vitebsk, snapping at his staff, "Do you think I have come so far to conquer these huts?" At this point, Napoleon called a council of war at which he seriously considered aborting the campaign right there. He had been able to mount the largest, hardest punch in the history of European warfare only to deliver it into a pillow. The Russian armies simply ran, abandoning their cities. At the same time, Napoleon's unprecedented army was melting away because of difficulties with supply.

One reason that French armies had been so dynamic, ranging far and wide following the Revolution, was that they broke away from traditional

methods of supply. Instead of relying on massive baggage trains or a pon-
derously constructed series of depots like monarchical armies of previous
eras, they had become expert at forage, living off the land or local
resources. The fact that they were originally buttressed by ideological fervor, and then the thrilling sense of victorious destiny that Bonaparte was
able to impart, made this technique less oppressive that it would have been
if, say, the Prussians had tried it. Historian George Nafziger has described
how as the years of Napoleonic warfare went on, foraging techniques had
become institutionalized in French units; companies routinely designated
ten or so men and an officer to gather food on behalf of the others.

While it is commonly recognized that the combat skill of Napoleon's
army had been diluted by 1812, due to casualties and retirements, combined with the need for ever-larger forces, it may have been the parallel
loss of foraging skill that truly doomed the French in Russia. The
Wurtembergers, Italians, Hanoverians and others who had been enlisted
as part of Napoleon's great host simply had no idea how the original
French armies used to sustain themselves. They needed food and water
and fodder for their horses and had no notion how to get it for themselves. Napoleon could arrest a noticeable decline in combat efficiency by
varying his tactics—for example, by depending more and more on concentrations of artillery—but foraging skill on a small-unit level was a
more subtle problem. It was already apparent that discipline in the army
was breaking down and that the invading troops, to Napoleon's chagrin,
were ruthlessly looting everything in their path.

According to historian Richard K. Riehn, after the first week of
Napoleon's invasion 50,000 of his troops were freebooting in Lithuania,
raiding villages and estates, sometimes in organized bands. Uncounted
others had just headed back west. Davout's corps held together the best of
Napoleon's formations, but during the first month of the invasion over
150,000 men had been lost, despite no major battles. The Russians tried
to multiply the French hardships by executing a scorched-earth policy in
front of the advance, even as the war grew increasingly bitter on the
flanks. Russian officer Boris Uxkull wrote:

> The saddest thing of all is that our own soldiers spare nothing. They
> burn, pillage, loot and devastate everything that comes to hand.
> All around, for a circle of 100 versts, you can see immense fires

which indicate the road taken by the enemy troops and our own, for we have vowed to leave nothing to the enemy. Bread, fruit, supplies, animals—everything is wiped out. The wells and streams are ruined, and the foragers don't dare show themselves for fear of being clubbed down by the partisans. . . . I saw a French prisoner sold to the peasants for 20 rubles; they baptized him with boiling tar and impaled him alive on a piece of pointed iron.

Since history provides many warnings not to underestimate Napoleon, we might also suppose that the gradual disintegration of his invasion force may not have come as a surprise. He had desired that his invasion would prompt the Russians to offer him a major battle, which would have energized the army and brought all his foreign contingents into the spirit of victory. His dilemma during the advance was that if he pushed the troops too hard they would outrun their supply wagons, leaving the march routes littered with dead and sick and flanked with deserters. If he did not press the offensive energetically, however, the Russians would continue to escape, drawing the French ever farther into the abyss.

On the Russian side, after six weeks of campaigning, their failure to stand and fight the invaders was becoming a national disgrace. Bagration was beside himself with fury at Barclay de Tolly for continuing to lurch back in continued retreats. Bagration and Platov thought they were facing the entire offensive by themselves, while Barclay and his Prussian, German, Swedish and French advisors played theoretical wargames, abandoning mile after mile of Mother Russia because they lacked the spine to resist the onslaught.

The French were now advancing on Smolensk, which was more important than either Vilna or Vitebsk; it was a holy city of Russia and the subject of a proverb: "He who has Smolensk also has Moscow." Napoleon paused at Vitebsk in order to collect his forces and bring up supplies, allowing time for Bagration and Barclay to finally unite. The Russians attempted a counteroffensive against the approaching French but it only succeeded in stringing out their units. On August 14, Murat found a Russian division on the march near Krasnoye and destroyed its rear elements. Before Smolensk Napoleon paused, reviewing his troops on his birthday, August 15, and consolidated his forces for the battle that would finally come. The center of Smolensk was surrounded by an ancient wall, though suburbs had spilled out to the north and south beyond its protection. Barclay, however, decided to give Napoleon only half a battle; manning the wall and defending the suburbs, he simultaneously decided not to risk his entire army. The Russians were already retreating, even as a few divisions were left back to defend the city.

Napoleon arrived at Smolensk on August 17 and surveyed the site with a telescope from a nearby hill. Davout had come up and by now the main army consisted of 220,000 men. The French attacked that morning from the south and by afternoon had blasted their way into the city. Unfortunately, it began to catch fire during the battle. By the time the last Russians had vacated, the prize was a smoldering ruin, hundreds of wounded caught and killed by the flames. Even as the fighting raged, the French noticed columns of Russians withdrawing east on the Moscow highway. Napoleon subsequently entered another useless Russian metropolis, in full knowledge that his army was steadily growing weaker, and the Russians had still not given him the great battle that would convince Tsar Alexander to accede to his will.

Joachim Murat set out in pursuit of the Russians. If cooler heads in Napoleon's councils urged that at this point the invasion be declared a success, and thus terminated, Murat persisted in the belief that the Russians could be brought to bear. On August 19 Murat, along with Marshal Michel Ney, in fact, was able to pin down several divisions of Barclay's First Army, which had become lost during their retreat and accidentally circled back near the French. Prince Bagration, disgusted with the course of operations, had withdrawn along the Moscow highway, leaving Barclay to fend for himself. Marshal Andoche Junot's Westphalian corps, inherited from Jerome, was in perfect position to get behind Barclay's men and trap them in a pincer. When Murat came riding from the heat of battle to urge Junot onward, he was cheered by the Westphalian cavalry, drawn up and eager to join the fighting. Junot, however, declined to advance because Napoleon had not given him any orders. Some of the Westphalian units followed Murat back into the battle but they were too few to block the Russians from escaping. The battle of Valutina-Gora, in which 41,000 French had been available against 22,000 Russians, resulted in another French failure to win a decisive victory. They lost 9,000 men in the attempt, while Russian losses were considerably less.

One of the mysteries of the Russian campaign is how Joachim Murat, who insisted on dressing ostentatiously while personally leading the advance of the Grand Army, avoided getting shot. A clue may be provided in the following account from Captain Victor Dupuy of the 7th Hussars:

> This prince, herculean in strength, excessively gallant, admirably cool in the midst of danger, had inspired an extraordinary veneration among the Cossacks. . . . If, as frequently occurred, he put himself in view of any Cossack skirmishers, these halted suddenly and ceased fire. One day I witnessed this almost magical respect. I received orders from my general to go and reconnoiter a path through a swamp which we had to cross, and I rode ahead with three men. At the far end of a small wood . . . I saw the King of Naples, all alone, while in front of him on the other side of the flat ground some forty mounted Cossacks were leaning on their lances and looking at him. . . . Thinking it was an attack, I shouted to the men with me: "Hussars! Cover the King!" We had no sooner taken up positions in front of him than the Cossacks fired several shots, one of which struck my horse in the leg. The King noticed this and said to me with a laugh: "Just as well! What have you come for? They never say anything to me!"

On the other hand, the King of Naples was persistently attacked by Marshal Davout, who claimed Murat was running the French cavalry into the ground. In fact, not only were the horses of the cavalry expiring at an alarming rate but Murat was also wearing out his supporting infantry and artillery by making them constantly rush to support often pointless skirmishes. When the argument was presented to Napoleon he backed Murat—better someone who was too aggressive than not aggressive enough—however, before long Davout's point would come to be sorely felt. One of the cavalry's problems was that each night the Russian rearguard would stop at the best site they could find, near water and forage, forcing the French to set up a camp nearby, regardless of its desirability. In addition, as a French officer wrote, "The King of Naples, and the generals copying him, were much more concerned with themselves than with their troops. . . . The method, which had become a mania, of forming large corps of cavalry, so as to give large commands to ambitious generals, is one of the causes of the cavalry's ruin."

With the end of summer approaching, Napoleon was halfway between Vilna and Moscow. If he were to halt the advance, all of Europe would assume he had been defeated by the Russians. It was also not clear that he could hold his multinational army together during the winter at Smolensk if they could not be inspired by the hope of an imminent victory. He might have considered, too, that only the weakest elements of his army had thus far been cast off; like the branch of a tree thrust into a windstorm, all the leaves would fly away, but the sturdy wood that remained represented the limb's true strength. He commented at this time that "the French were not made for defensive war but only for the advance." In any event, the burning ruin of Smolensk was no sort of capital in which Napoleon could rest. The army continued to push east in pursuit of the Russian forces.

By now the pressure on Barclay de Tolly to stand and offer a major battle to the invaders had become irresistible. Although an objective observer might suppose that a Scottish Minister of War, with French, Prussian and German advisors, was not necessarily the worst combination of military minds to oppose Napoleon, to native Russians who saw nothing but city after city going up in flames, then looted by the French, there was something alarmingly suspicious about the course of operations. Two months into the war, Russian arms had fought at least 50 small engagements and had yet to be proven inferior in combat. Was Barclay planning to fall back all the way to Moscow? Was there any point at which the army would stop retreating and confront the invaders with all its forces?

The Russians had a tradition of murdering Tsars (including Alexander's father, Paul) if they did not reflect the national will, much less German-speaking Lutheran generals with French names. Barclay de Tolly finally decided to make a stand. He chose a site 100 miles short of Moscow and halted the army to prepare works. However Tsar Alexander, under pressure of his own, announced an even greater event at this time. He named a new general to supercede Barclay and take over both Russian armies: the new supreme commander would be a native son, Alexander Kutusov. In his younger years Kutusov had been a student of the legendary Suvorov, and he had just completed a successful campaign against the Turks on the Danube. As Clausewitz wrote with some irony,

"Kutusov's arrival awakened new confidence in the army; the evil genius of the foreigners was exorcised by a true Russian, a Suvorov on a somewhat smaller scale, and no one doubted that a battle would follow at once, which would mark the culminating point of the French offensive."

There has been much speculation that 43-year-old Napoleon Bonaparte, beset with various health discomforts, as opposed to his 29- or 35-year-old self, would have wrapped up the Russian campaign at an early stage. However his new opponent, whom Bonaparte had already destroyed once at Austerlitz, struck an even lesser figure of vitality. In 1812 Kutusov was 67 years old, with one blind eye, and was so obese that he could barely get on a horse. The new Russian champion had lost more than a few steps over the years, but his instinct for survival seemed to have remained undiminished. His first decision was not to commit the Russian Army on the ground that Barclay had chosen—after all, it was Barclay's— but retreat a bit more until a new site was found. His chief of staff, Karl von Toll, a Prussian borrowed from Barclay, found a new position around the village of Borodino, 70 miles west of Moscow.

On September 5, the French Army arrived and rejoiced that, instead of endless marching into the unknown on barely sustained supplies, they would have a chance to win the war. The combined Russian Army was drawn up before them. The Russians had erected a redoubt some 2,000 feet ahead of their lines and the French immediately assaulted it, forcing the Russians to withdraw after dark. It is difficult to explain what purpose the redoubt at Shevardino had been meant to serve, since it was too far forward to be supported by the main Russian lines. Then again, before diving into a pool, people invariably test the water first with their toe so they will know what to expect. The next day the various French columns deployed before the main enemy positions while the Russians frantically built up more works, anticipating the fight that would decide the fate of their homeland.

Bagration's Second Army manned the Russian left while Barclay's First Army held the right. Kutusov set up well behind the lines in a farmhouse well stocked with food and champagne. Bagration had the weaker terrain on the left so he constructed three flèches—V-shaped redoubts— as obstacles against French attack. In the center of their lines, in front of Raievsky's division, the Russians erected a breastwork over 100 yards long, afterwards called the "Great Redoubt." There was a ditch in front of it and thirty yards of wall on either side.

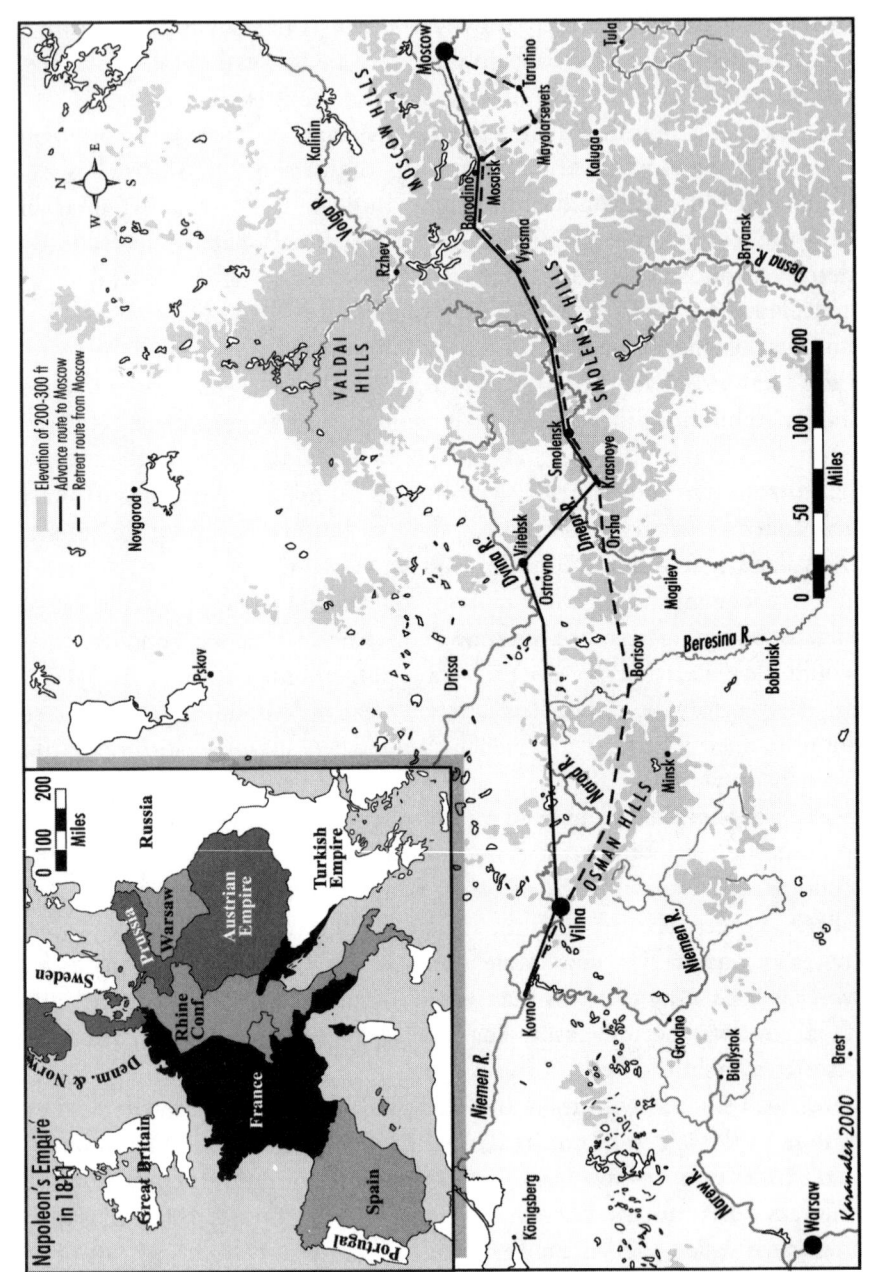

Napoleon's Empire in 1811

Great Britain
Denm. & Norw.
Sweden
Prussia
Warsaw
Russia
Rhine Conf
France
Austrian Empire
Turkish Empire
Portugal
Spain

Miles
0 100 200

Elevation of 200-300 ft
Advance route to Moscow
Retreat route from Moscow

Novgorod
Pskov
Drissa
VALDAI HILLS
Rzhev
Kalinin
Volga R.
MOSCOW HILLS
Moscow
Tarutino
Mayolarsewets
Tula
Kaluga
Borodino
Mosaisk
Vyasma
Dvina R.
Vitiebsk
Ostrovno
SMOLENSK HILLS
Smolensk
Krasnoye
Dnepr R.
Orsha
Bryansk
Desna R.
Mogilev
Beresina R.
Borisov
Bobruisk
Minsk
Narod R.
OSMAN HILLS
Vilna
Niemen R.
Kovno
Grodno
Bialystok
Königsberg
Brest
Warsaw
Norew R.

Miles
0 50 100 200

Kevember 2000

On the day before the battle Kutusov paraded through the Russian Army with the Black Madonna of Smolensk, a sacred icon that inspired the troops to commit their greatest effort in the fight to come. On the French side, Napoleon received a courier from Paris who brought a portrait of his 18-month-old son, Napoleon II, aka the King of Rome. The Emperor placed the portrait outside his tent so that veterans of his Imperial Guard could pass by to see it.

On September 7, the Battle of Borodino took place. Napoleon's main army had by now been reduced to 130,000 men determined to win a brilliant victory, pitted against 120,000 Russians determined simply to survive the day. The Russians had 640 guns opposed to 590 French, but many of the Russian cannon were deployed on their far right where no major fighting took place; the French, as instigators of the battle, were able to concentrate all their artillery on the crucial spots. There were no great maneuvers on the field because the French simply tried to break through the Russian positions on the left and center in frontal assaults. Napoleon had dismissed Davout's suggestion of a flanking movement against the Russian left as "too risky." After the first half-hour the battle developed into a more or less equal number of French attacks and Russian counterattacks.

By eight o'clock in the morning, the combined corps of Ney and Davout succeeded in taking Bagration's first two flèches, but they had been unaware of the third one that sat beyond on higher ground and that raked the French attackers with grapeshot. Due to the arrowhead shape of the flèches they were difficult to hold against Russian assaults and the fighting raged back and forth. Early in the morning Davout's horse was killed and the Marshal was thrown to the ground, stunned. A far worse calamity happened to the Russians when Prince Bagration was felled by a wound to his leg that turned out to be fatal. On the French far right, Poniatowski's Poles attempted to turn the Russian flank but were not strong enough to do so.

In the center Napoleon's stepson, Prince Eugene, took the village of Borodino, but then Russian cavalry charged and nearly wiped out the attacking regiments. Eugene threw more men into the fight and secured the village; the main effort then switched to the Great Redoubt. Just as with Bagration's flèches, every successful attack was met with an equally powerful counterattack. At one point, when the French were being pushed out of the Great Redoubt, General Bonammy, about to be skewered by a Russian bayonet, shouted, "I am the king!" For a while the Russians thought they had captured Murat and word spread like wildfire through the army. They were disillusioned, however when the real King of Naples remained in evidence all over the field, leading attacks and urging the French on against vulnerable spots in the Russian lines. Anatole de Montesquiou wrote, "The King of Naples charged forward impetuously in the midst of his squadrons like the boiling Achilles . . ." If Murat was Achilles on this day, Hector was personified by the figure of Barclay de Tolly, who had put on his dress uniform with all his medals, almost as if he were inviting a French bullet to put an end to his ordeal. The despised commander who had guided the Russian retreat prior to this day turned out to be the bravest of fighters once battle was joined.

Boris Uxkull and the Chevalier Guards were poised to execute one of the Russian counterattacks:

At 8:30 we were put on a hill, where we were quite exposed but where we could see everything. The spectacle was most majestic, most impressive. Before us, 300,000 men were squeezed into a narrow space, fighting with fury and tenacity. The cannon fire increased from one minute to the next. The defensive firing could hardly be heard. The whole plain was covered in smoke. For three hours we stood there motionless, looking, trembling and losing patience. The shells hit and killed a great many horses, fewer men. At 11:00 the massacre grew general; the earth trembled, the air darkened.

Suddenly the noise stopped! The enemy cavalry, 25,000 strong, charged our columns and batteries. The reserves started moving. The earth shook and groaned beneath the weight of the cuirassiers. There was a collision, but the dust prevented us from distinguishing our adversaries. It was the carabiniers and lancers. The carnage lasted five minutes. The horses trampled the dying and the wounded. We repulsed the enemy, but we had scarcely formed our squadrons when the fire from a masked battery nearby decimated our ranks. The trumpet assembled us. There was a second attack, with less success. The enemy meanwhile had forced the left wing and we were dispatched across the battlefield.

Regiments of cavalry, divisions of infantry and batteries of artillery were flung back and forth as commanders on both sides sought to plug holes and exploit weaknesses. A major blow to the Russians occurred when the commander of their reserve artillery, Kuitasev, was killed while rashly putting himself at the head of a cavalry counterattack on the Great Redoubt. From then on the French were able to bring superior numbers of guns to bear on the most important Russian defenses.

At one point General Jean Rapp was hit four times, making a total of 22 wounds he had suffered under Napoleon, and was forced to leave the battle. While his wounds were being dressed the Emperor walked over and said, "So it's your turn again. How are things?" Rapp said he thought that it was about time to send in the Imperial Guard, to which Napoleon replied, "I shall take care not to. I do not want it destroyed. I am certain to win the battle without the Guard becoming involved."

The climax of the day occurred when the Saxon heavy cavalry finally decided the issue at the Great Redoubt. They had been standing in formation for over two hours under Russian artillery fire when the order came to charge. Their ardor was unsurpassed. Some of the horsemen got beyond the redoubt and entered it from behind. Others simply jumped over the earthworks into the Russian defenders. The Saxons would all have been killed, and no more than a dozen were still alive, swinging about with swords, when following French infantry began climbing over the ramparts to seize the redoubt once and for all.

By mid-afternoon the battle was dying down, though the French artillery continued to fire against remaining Russian targets. Clausewitz, who had been posted on the largely inactive Russian right, found it interesting how the battle "gradually took on the character of weariness and exhaustion." The Russians had abandoned Borodino, Bagration's flèches and the Great Redoubt, and their entire line pulled back over a mile. The French had achieved their objectives and inflicted superior casualties on their foe—over 40,000 as opposed to just under 30,000 of their own. If the Russian Army had still not collapsed, it was swaying in the breeze, gutted and rickety. It would not be able to withstand another push.

It was at this point that the Russian supreme commander, Kutusov, earned his pay. Once the battle had died down, Barclay de Tolly dispatched an aide, the Prussian Ludwig Wolzogen, to Kutusov's headquarters to report on the situation and request further instructions. According to Wolzogen,

> When I said that all important posts had been lost on the right wing and to the left of the high road and that the regiments were all extremely tired and shattered, Kutuzov shouted: "With which low bitch of a sutler have you been getting drunk, that you are giving me such an absurd report? I am in the best position to know how the battle went! The French attacks have been successfully repulsed everywhere, and tomorrow I shall put myself at the head of the army to drive the enemy without further ado from the sacred soil of Russia!" At this he looked challengingly at his entourage, and they applauded with enthusiasm.

After a moment of anger, Wolzogen realized what the sly old general was doing for the benefit of the nobles and the representatives of the Tsar's court who surrounded him. Kutusov, despite having done nothing in the battle, controlled communications to the rest of Russia, including the Tsar, and shamelessly caused the country to believe that he had defeated Napoleon. The effect on Russian morale was fantastic; the Tsar became more adamant than ever not to negotiate for peace. Of course, when Kutusov subsequently ceded Moscow to the French, the Tsar began to catch on to the lie; by that time, however, the entire situation had changed. The thin shred of truth behind Kutusov's claim was the fact that his army had not been utterly destroyed—somewhat to the surprise of the Russian soldiers themselves. For this, Napoleon has been blamed ever since.

On the day of the battle Napoleon was personally weak, due to a bad cold and a bladder problem. His marshals had once again proven their skill and bravery in combat, but throughout the day Napoleon hung on to his 20,000-man Imperial Guard, repeatedly refusing to commit it to the battle. He was beset by requests, first by Murat and Ney once they had pushed in Bagration's left; then by Murat and Eugene once they had weakened Barclay's center. During the morning Napoleon had said, "I must be able to see more clearly on this chessboard." In the final afternoon debate, not even Murat could recommend that the Grand Army's last reserve be committed. The obstinate Russians could be killed but they could not be routed. Given the great density of their front and the French decision to hit it head-on, the enemy had been pushed back but there were still no wide gaps for exploitation. There had not been a maneuver so great as to open the enemy lines for a decisive thrust; why commit the Guard into a meat grinder?

Countless commentators have accused Napoleon of failing to think of, or risk, a great maneuver that would have earned a decisive victory. On the other hand, if one looks at Napoleon's victories in prior years, more often than not he was able to orchestrate a counterstroke against an enemy offensive movement. A fluid battlefield, on which he could not only attack, but also react instantaneously to enemy attack, gave him twice as much opportunity to wield his forces to decide the outcome. By 1812, his reputation for winning battles was such that from the first day of the invasion it was clear the Russians were afraid of him. Napoleon's own greatest fear was that if he undertook an ambitious flanking maneuver they would only retreat again. The French were not ecstatic about the outcome of Borodino, but from Napoleon's point of view the battle had opened the gates to Moscow. He would later term it "the most terrible of all my battles," a statement that is usually interpreted as a reference to casualties. The true meaning is that at Borodino he was compelled to forego any creativity and simply take advantage of the opportunity to finally fight the enemy. Napoleon's intimidating reputation had worked against him during his advance into Russia, but that same reputation would turn out to be invaluable once the French themselves had begun to retreat.

Murat's cavalry fought one more engagement on the succeeding days, against his flamboyant Russian counterpart Mikhail Miloradovich— according to Prince Eugene, "a knight in the strictest sense of the word, unsurpassed in bravery"—but a Russian council of war had already decided to abandon Moscow. An elementary fact that may have robbed Napoleon of victory in the war was that Russia had two capitals. Moscow was still the traditional heart of Russia and its most important city, but the Tsar and government resided in St. Petersburg. That, along with the fact that the Tsar had not witnessed the carnage at Borodino, allowed him to retain the moral and pragmatic resolve not to consider making peace. Miloradovich made a deal with the French advance guard to allow several hours for the Russians to retreat through the city, in exchange for an unmolested French entry.

On September 15 Napoleon entered Moscow, accompanied by about 100,000 soldiers, all that was left to him of the huge army with which he had first crossed the Niemen. To be clear, thousands of troops had been left behind to garrison occupied towns; thousands of others were employed on the supply route back to Poland. Schwarzenberg to the south, and Macdonald and St. Cyr to the north still had intact corps that had never been part of the central thrust. Nevertheless, desertion, sickness and battle casualties had weakened Napoleon to a shocking degree. Clausewitz, who was privy to Kutusov's councils, commented afterward that if the Russians had known how weak the French had become they would have retreated in a different direction rather than lead Napoleon down the Moscow highway. For his part, Napoleon had never planned to march all the way to Moscow, but now that he had done so he expected the prize to win him the war. It came as a horror when, starting on the first day of the French arrival, the city began going up in flames.

At first no one thought that a few isolated fires were unusual. The French imagined it was their own soldiers being careless in the pursuit of loot or food. Napoleon went to sleep early in the Tsar's rooms of the Kremlin, but in the early morning hours of the 16th was awakened and asked to look out the window at a roaring sea of crimson. By that time a number of Russian arsonists had been caught and had confessed. Napoleon said, "A demon inspires these people. They are Scythians! This is a war of extermination. What a people! What a people!"

The Mayor of Moscow, Fyodor Rostopchin, had removed all of the city's firefighting equipment and instructed his police to stay behind, in disguise, to set the fires. He also freed hundreds of criminals from the city's jails on the promise that they would win redemption for their sins by contributing to the conflagration. Ironically, in order to spare the city from being ravaged, the French had posted the bulk of their troops outside town. Now they had no way to stop the flames and, instead, French troops from the outskirts dashed into the city to grab what loot they could before racing back to their safe camps. The Imperial Guard was meanwhile fighting a losing battle to keep the fire away from the Kremlin. At four o'clock in the afternoon, Napoleon was persuaded to leave. He was already isolated from his troops, and who could say that the Russians had not planted a cache of powder beneath the Kremlin that would explode at any moment?

If the center of Moscow had subsequently become the funeral pyre of the French Emperor and his entourage, one could almost say that the city's destruction served a purpose by ending the war while wreaking a poetic revenge on Russia's antagonist. Napoleon, however, was able to escape through veritable tunnels of flames, and set up his new headquarters at the Petrovsky Palace to the north. Three-quarters of the city was destroyed, and the rest of it thoroughly ransacked amid the disorder, for no practical reason other than to intensify the passion of the war. To the south, the Russian Army watched the immense pillar of smoke during the day and the glowing sky at night, and assumed the French had cruelly decided to destroy the heart of Russia. There would henceforth be little mercy extended to the invaders who would soon be falling into Russian hands in droves.

After the fires burned out, Napoleon returned to the Kremlin and sent off peace feelers to Tsar Alexander, explaining, too, that the destruction of Moscow had not been his fault. Alexander refused to reply. One can sympathize with the Emperor of the French, who had undertaken enormous difficulty and expense, and had caused over 100,000 deaths to reach this point. What more was he supposed to do? If he had had one-tenth the inkling of what this campaign would entail, he would never have begun it. He also sent a representative to Kutusov to explore the possibility of ending the war.

Joachim Murat stood in the field to the south of Moscow, and while Napoleon's emissary waited to be received by the Russian commander, he rode up to chat with his opposite number, Miloradovich. "How much longer is this war going to last?" Murat asked. "This is no climate for a King of Naples." "It is not we who started it," answered the Russian cavalryman.

At headquarters, Kutusov was too crafty not to listen to Napoleon's ambassador, but he couldn't let the Tsar think he was carrying on negotiations. He was already in enough trouble with his sovereign for giving up Moscow after having announced he had won a spectacular victory. Kutusov's army, unlike Napoleon's, was growing stronger every day, militia and new recruits filling the gaps made at Borodino. Further, those Cossacks who had not been eager to oppose the French advance were now proving more enthusiastic. English General Robert Wilson, an observer of the campaign, wrote, "The Don regiments continue to pour in. Such a reinforcement of cavalry was perhaps never equalled . . ."

In Moscow, Napoleon continued to dither on his next move, partly lulled by an unusually warm string of October days. On October 18, the Russian Army made up his mind for him by springing a surprise offensive at Murat's lines. Tarutino was the first battle of the war in which the French had become the prey and the Russians the aggressors. Murat was outnumbered two-to-one and lost his entire left wing, but due to Russian failure to execute their original plan he was able to fight his way back to Moscow. Napoleon realized that he couldn't procrastinate any longer; the Russians were not in the least inclined to make peace.

On October 19 the great retreat from Moscow began. The procession resembled a hybrid between a Western army and an Asian caravan, with a touch of Mardi Gras. All the best carriages and buggies of Moscow had been commandeered, laden with paintings, candelabras, carpets, silverware and furniture, as well as food and clothing. Many of the soldiers were wrapped in furs and other finery. A French officer wrote: "Supply officials and actors, women and children, cripples, wounded men and the sick drove in and out of the throng . . . countless servants and maids, sutlers and people of that sort accompanied this march."

Napoleon decided to take with him the giant cross that adorned the Kremlin, as well as 200 years' worth of flags and standards that the Russians had won in battle from the Turks. He also tried to blow up the Kremlin. A glimpse of Moscow after the French had gone is provided by Andrei Norov, who had lost a leg at Borodino and was one of the first Russians to reenter the city:

> Everywhere you looked, wherever the eye fell, was black. The tall chimneys of the buildings were either tottering or fallen in, houses were burnt completely, the churches from top to bottom seemed to be covered in crepe . . . inside the golden inscriptions were torn away and replaced with what looked like charcoal etchings. In some of the churches the doors had been torn down and broken open, with the various holy articles and furniture carried out into the street and partially burned.
>
> But the worst sight of all was the Kremlin! The Ivan the Great cathedral had lost its cross, it looked like he was wearing the remains of a golden bowl on his head. It looked less like a memo-

rial than a post, as its magnificent wing, which had had two cupolas and massive bells, had been blown up and lay in rubble. When we got closer and looked from the river bank, we saw . . . that a huge crack ran the entire length of the main building, from the top to the bottom. . . . We were barely able to make our way through the rubble.

Napoleon's three options were: to retreat to the north and then west, which would pass the army through poor but unravaged territory; or he could move south and then west, providing richer, untouched resources for the men and horses, but which would mean having to fight the main Russian army en route; or, he could go back down the center, the way he came. Clausewitz disdains the notion that Napoleon had any other choice but to fall back along the line of his original approach.

We have never understood how people could so stubbornly stick to the idea that Bonaparte should have returned by a different route than on the one on which he had come. What could he have lived on, except his depots? What good was an area that still had foodstuffs to an army that had no time to lose, that always had to bivouac in concentrated masses? What officer would have been willing to ride ahead of the army to organize the collecting of food, and what Russian officials would have obeyed his orders? The army would have been starved out in the first eight days.

Clausewitz may have been correct, but then Napoleon's logistical foundation had by now become just as weak as the army itself. While the distance from Smolensk to Moscow should have comprised the most heavily garrisoned and supplied stretch of territory in order to allow for an orderly retreat, instead the French depots were only strong from Smolensk back to Poland. Deep inside Russia supply columns had come under attack from Cossacks and partisans. There had been no large contingents of troops to drop off at Mosaisk, Vyasma and all the other towns after Smolensk. When the French left Moscow they were taking a big leap into the unknown.

Their initial advance was toward the south. On October 24 at the village of Mayolarsevets, Davout and Prince Eugene with 24,000 men met an equal number of Russians under Docturov, who had replaced Prince Bagration. Kutusov could have poured more forces into the battle but he was not yet ready for another Borodino. He hesitated, fearful of Napoleon's aims, and the French held the village, inflicting 8,000 casualties while suffering 6,000 of their own. In the early darkness of the next morning Napoleon rode up to reconnoiter the ground, accompanied by his chief of staff, Louis Berthier, former ambassador to Russia Armand de Caulincourt and about a dozen guards. Suddenly they were in the middle of a swarm of Cossacks who had slipped behind the French pickets. Napoleon and his companions dismounted and drew their swords amidst gunfire and the clang of pikes and sabres. Fortunately, four squadrons of Imperial Guard cavalry came rushing onto the scene and in the semi-darkness were able to fight off the intruders.

Historians have debated whether this incident, along with the tough battle his army had fought the previous day, influenced Napoleon's decision to change the direction of his retreat. Clausewitz, as mentioned, thought the French had no other choice in the first place but to fall back along their original route. Napoleon's initial drive south may have been a ploy to scare off Kutusov, backing him away from the Grand Army. Then the French could steal a few days' march and head directly back via the central route. The Russians cooperated after Mayolarsevets, in fact, by falling back to Kaluga, and began to disperse the supplies there because they were afraid these would fall into Napoleon's hands. It took the Russians, or at least Kutusov, a few days to realize that the French had abandoned Moscow, their entire force now flowing in a long, unwieldy column north to Mosaisk and from there to the west on the Moscow–Smolensk highway. The great Russian retreat had ended once and for all; the beast had turned tail and, from the Russian point of view, the pursuit was on.

The weather held up during late October but due to encumbrances of loot, wounded and civilians the French Army became strung out across sixty miles. Thousands of stragglers were falling behind; many others were without food. The wagons of wounded were the first to be abandoned, once their horses had given out because of lack of forage. Russian prisoners were shot, because it was not considered wise to let them go back to the enemy with full knowledge of the dire straits of the Grand Army. Many French soldiers dropped out of formation because they couldn't keep up day after day without eating. On October 29–30, the French endured the depressing ordeal of retraipsing the Borodino battlefield, where 30,000 human remains lay more or less exposed, half-eaten by wolves.

Cossacks were already harrying the column at its weakest points, but on November 3 Miloradovich's regular Russian cavalry caught up from the south and severed Davout's I Corps from the rest of the army near Vyasma. Eugene's Italians and Poniatowski's Poles had to go back and fight their way through to rescue Davout while Ney held a blocking line against further Russian advances. To some, the near annihilation of the once-proud I Corps marked a turning point in the retreat, toward desperation.

On November 6 the first heavy snowstorm occurred. By that time, after two and a half weeks of marching, the Grand Army had begun to fall apart. Food was the major problem, because after a few days marching without sustenance, people became frantic and deserted the columns. The English observer Robert Wilson roamed the area of Vyasma and wrote:

> The naked masses of dead and dying men; the mangled carcasses of ten thousand horses, which had, in some cases, been cut for food before life had ceased, the craving of famine at other points forming groups of cannibals; the air enveloped in flame and smoke; the prayers of hundreds of naked wretches, flying from the peasantry whose shouts of vengeance echoed incessantly through the woods; the wrecks of cannon, powder-wagons, military stores of all descriptions . . .

At one village Wilson found that fifty French prisoners had been burned alive; at another place fifty had been buried alive by peasants. At the latter spot he noted a dog belonging to one of the victims which came every day to "sit and moan over the newly-turned earth." Elsewhere, he encountered sixty naked prisoners with their necks lined up along a felled tree while male and female peasants "singing in chorus and hopping round, with repeated blows struck out their brains in succession." Sergeant François Bourgogne of the Imperial Guard recalled the night of November 8 as especially disastrous amid months of additional hardship. He and his men were about to camp in the open when they were "roused by an extraordinary noise." It was a sudden onset of the north wind, bringing heavy snows and crippling temperature. The men rushed to escape the blanketing cold but for many there was no refuge. Bourgogne wrote,

> We heard them shouting as they ran about towards any fire they saw; but the heavy snowstorms caught them, and they could soon run no more, or if they tried to do so, they fell and never rose again. In this way many hundreds perished, and thousands died of those who had stayed where they were camped. . . .
> At daylight, to regain the road, we were obliged to go down to the ravine, where the evening before the artillerymen had made their bivouac. Not one was left alive; men and horses were all covered with snow—the men still round the fires, the horses harnessed to the guns, which we were forced to leave there.

On November 9, the French reached Smolensk, which had been turned into a major supply depot. All hopes of an orderly distribution of food were lost when the first arrivals turned out to be a horde of famished stragglers. These were held at bay until the Imperial Guard had passed, but then the mob pushed its way into the city and began plundering the stores. By the time Eugene, Davout and others reached the city all was chaos. Ney's corps, which was now bringing up the rear, had the worst pickings.

Napoleon and the Guard departed Smolensk on the 14th, with Eugene, Davout and Ney to follow on succeeding days. The Russian Army, however, by advancing along parallel roads to the south, had gotten in front of the French, converging near the highway at Krasnoye, 40 miles southwest of Smolensk. It was Napoleon's turn to save the rest of the Grand Army. On the 16th, the Imperial Guard, down to about 15,000 men with 50 guns, lined up against 35,000 Russians with 500 guns. Napoleon attacked the Russian center and drove it back. Kutusov had thought he had come upon a weak flank of the French, but when he discovered the Emperor himself was present he refused to commit more men to the battle. Even Miloradovich, commanding the Russian cavalry, hesitated to move. Still, Russian cannon swept the highway and local firefights burst out on all sides. On the main road the Cossack swarms were everywhere, scooping up stragglers and raiding the baggage.

On the 17th Kutusov asked a peasant who had just come from Krasnoye if the French commander on the scene was a short plump fellow. The peasant said no, he was tall (probably meaning Marshal Mortier of the Young Guard). The Russians geared up for an attack when suddenly Kutusov received word that French soldiers in the village were wearing busbys: Napoleon's personal troops. The attack was called off. Meanwhile, Eugene had gotten through the gauntlet of Russian artillery and Cossacks on the main road. The next day Napoleon continued to hold off the main Russian army while Davout's corps escaped. By now it could only be a matter of minutes before Miloradovich realized that Napoleon's holding force was so small. There was not a moment left to wait for Ney. The Imperial Guard rejoined the retreat as the Russian Army poured across the highway behind it, taking thousands of starved, frozen stragglers as prisoners. When Ney came out of Smolensk he found the Grand Army gone and thousands of Russians deployed to greet him. Miloradovich sent him an offer to lay down his arms but Ney replied, "A Marshal of France does not surrender." Ney attacked against overwhelming odds but had no hope of breaking through.

On November 19 Napoleon had supper in a peasant hut, on what must have been the most depressing evening of his life. Ney was gone, cut off, surrounded. The army had simply not been strong enough to extend its hand to pull him out. Also, by leading the retreat during its initial stage Napoleon had been able to avoid witnessing the disintegration of his forces, but now that he had done a stint as rearguard, the pitiful human wreckage along the march route had become all too apparent. As for the future of his remaining men, new dispatches revealed that their prospects were dimmer than ever. In the northwest, Marshals Victor and Oudinot were supposed to have held Vitebsk but instead they had been pushed back by a superior Russian army under Wittgenstein. In the southwest, Schwarzenberg's Austrians had retreated to Poland before a superior Russian army under Admiral Tchichagov. The city of Minsk, with its immense wealth of supplies, had been retaken by the Russians. Even if Napoleon could stave off the main Russian force under Kutusov, two more Russian armies, each of 60,000 men, were now converging across his line of retreat. At this point the French retained around 20,000 effectives with a greater number of stragglers, many of them civilians.

On the very next evening, Napoleon's headquarters became a picture of happiness. As Armand de Caulincourt wrote, "Never did a battle won cause such a sensation. The joy was general . . ." Marshal Ney had broken through the Russians and rejoined the army. After his bloody repulse outside Smolensk, Ney had fallen back to the east. After dark his III Corps came to a stream, and, breaking the ice, they determined the direction of its flow, which had to be toward the Dnieper. The next day their march was accompanied by Cossacks, while regular Russian divisions followed. At night Ney lit numerous campfires to indicate he was bedding down, but instead marched off again, losing the Russians behind him. Constantly harried by Cossacks, after two more days Ney finally reached the Dnieper and passed his remaining men across, though the ice was too thin to support his guns and wagons. A courier got through to the town of Orsha where Eugene sent out a division to receive the prodigal marshal. Ney had gotten back with 900 men under arms as well as several hundred stragglers. It was at this point that Napoleon pronounced him "The bravest of the brave."

The feat of Marshal Ney provided a terrific boost to morale, at least at headquarters, but the army continued to disintegrate. Dr. Heinrich von Roos wrote, "On this day we were made particularly aware of the fact that soldiers were becoming weak-sighted and, indeed, many had gone totally blind. One saw men dragging their comrades along by sticks, like beggars. Massage with snow improved the condition for a while." Roos thought that the problem was caused by men holding their faces too close to the smoke of campfires, as well as by staring at snow. Another observation during the horrendous ordeal came from Russian General Andrault de Langeron, who commented, "One cannot imagine the number of women the French army dragged along, and I noticed that they stood the cold better than the men did." Napoleon's chief surgeon, Baron Dominique Larrey, who might have set a record at Borodino for most amputations performed in one day, opined:

> I noticed that people with dark hair and an emotional, labile temperament, mostly from the countries of southern Europe, stood up better to the severe cold than did fair-headed men of phleg-

matic temperament and coming for the most part from northern countries. This is contrary to the view usually held. The circulation of the first group is no doubt more active; their vital forces have more energy. . . . From the same cause their morale remains higher; they do not lose heart; and thanks, of course, to a care for their self-preservation, they know better how to avoid dangers than do the usually apathetic inhabitants of cold, damp climates.

A less debatable discovery was made by French soldiers who found that when live horses became numb from cold they could be cut for strips of flesh, or for blood to make warm soup. Jakob Walter, a Westphalian who spent most of the retreat as a straggler, recalled, "In every bivouac soldiers who looked like specters crept around at night. The color of their faces, their husky breathing and their dull muttering were horribly evident; for wherever they went they remained hopeless; and no one allowed these shades of death to drag themselves to the fire."

Unfortunately the Grand Army's greatest tribulation was still to come. The Beresina River had been no great obstacle to the French during their unopposed summer advance, but it now threatened to present an insurmountable barrier to their retreat.

Ironically, if the weather had been colder the Beresina would have frozen over at this time of year and everyone could have walked across. Instead, due to a number of mild days the river was only half frozen, ice chunks churning downstream. The real problem was that Tchichagov's army was waiting on the opposite side; Wittgenstein's army was coming down from the north on the near side and Kutusov's main army was slowly but surely catching up. All the Russian forces were converging on—surrounding—the remnants of the French just as they got to the river. Even with Oudinot and Victor arriving from the north to reinforce the Grand Army—imagine their shock at seeing the state it was in—French combat strength would be outnumbered by at least three to one, while their stragglers and noncombatants would be equally trapped with nowhere to escape. It was an occasion for Napoleon's luck, which seemed to have deserted him throughout the campaign, to reappear one last time.

The town of Borisov, which possessed the only bridge in the vicinity, had already fallen to Admiral Tchichagov's army, coming up from the south. Napoleon ordered Oudinot to retake the town, and he did so with

élan. His 4th Cuirassiers achieved surprise and the Russians began streaming back over the bridge, shouting, "French! French!" Tchichagov, who
had just started to eat dinner, fled on foot. Oudinot took a thousand prisoners and 300 supply wagons, plus the remains of the Admiral's hot meal.
But the Russians destroyed the bridge behind them and congregated on
the far side. It now became a guessing game as to where the French would
try to cross, with Napoleon holding the advantage that the Russians still
didn't seem to realize how weak he was. The 70,000-man column that the
Grand Army presented was in reality now 30,000 armed men who stayed
with their depleted corps, and 40,000 unarmed stragglers or civilians.
Many of the vehicles that the Russians thought were guns or caissons were
civilian carriages with women inside.

The French found a ford north of Borisov, near the town of
Studyenka, where the river was shallow, not more than eight feet deep at
its center, but over a hundred yards wide. General Jean-Baptiste Eble's
engineers dismantled the wood of the village for bridge materials. At
Orsha, Napoleon had considered the choice of retaining either his pontoons or a number of cannon and had, typically, chosen the guns.
Fortunately Eble had held on to his nails and clamps, as well as some
portable forges. His men began assembling the bridges behind a hill near
the ford. Meanwhile, Napoleon made every effort to feint toward the
south in order to mislead the Russians about his true crossing site.

On the Russian side, Oudinot's vicious counterattack at Borisov had
resonated throughout the various commands. The French Army still had
teeth. Clausewitz noted that Napoleon was lucky to have found a suitable
ford, but that "his reputation as a soldier had made the real difference,
and he now benefited from capital built up long before. Both
Wittgenstein and Tchichagov were afraid of him, of his army, of his
Guards—just as Kutusov feared him at Krasnoye. No one wanted to be
defeated by him." Tchichagov became convinced that the French were
going to cross the river south of Borisov, and marched his army away,
leaving just one division to guard Studyenka. French troops were making
loud demonstrations toward the south and Kutusov too fell for the feints,
relaying his thoughts to Wittgenstein, who he hoped would reinforce the
main army should Napoleon have to be confronted.

On November 26, only four Russian soldiers guarded the Grand
Army's crossing site. The day before, an entire division had been posted

there, but had prudently followed Tchichagov's orders to move to the south, even though they alone among the Russian combatants must have known where the French Army was. At eight in the morning, 80 Polish cavalrymen, two-to-a-horse, forded the river and cleared the opposite bank. Several hundred men of Oudinot's corps followed on rafts. Forty French guns were lined up along a ridge on the east side, aimed at the far bank.

Under the eyes of Napoleon, Murat and other French leaders, Eble's engineers waded neck-deep into the river to set up the two bridges, one for infantry and another for artillery and wagons. The bridges themselves were a string of large sawhorses tied together in parallel rows, with planks or sticks connecting them to provide a platform. The planks weren't tied together so it was treacherous footing, especially for horses. The bridges sagged and swayed and were not far above water level, sometimes beneath it in the center of the river.

The footbridge was completed at one in the afternoon and Oudinot's corps of 7,000 men went across. Napoleon and the Guard stayed behind in anticipation of an attack by either Wittgenstein or Miloradovich, at the head of Kutusov's advance guard. That evening Ney's corps of 4,000 was passed over. Davout and Eugene were still approaching; Victor's corps marched south toward Borisov to maintain the feint and hold off any approaching Russians.

At eight in the evening, part of the large bridge gave way and Eble sent half his engineers into the bone-chilling water to fix it. At two in the morning it gave way again and the other half of his engineers went in, repairing it by six o'clock. A few stragglers had been permitted on the footbridge during the night, but when Davout and Eugene arrived, Napoleon crossed with his Imperial Guard. Admiral Tchichagov had finally realized where the French were crossing and the far side was under attack. On the near side, Cossacks were skirmishing with the French rearguard. Wittgenstein had insisted on marching straight for Borisov, despite the commotion at Studyenka, but he was now on his way north.

The afternoon of the 27th saw the first ugly panic on the bridges as the disorderly crowd tried to force its way across, a procedure not unlike pouring pudding through a thimble. In the evening Davout and Eugene's corps cleared their way through, leaving Victor alone to hold off the Russians approaching from the east. One of Victor's divisions became surrounded near Borisov and laid down its arms. On the night of the 27th, the army of stragglers settled down by campfires, even though the bridges were clear and French officers moved among them urging them to cross while they could. On the morning of the 28th, the battles on both sides of the river grew more intense as the main Russian armies came into the battle.

On the wooded west side, where Tchichagov's divisions were swarming against the bridgehead, Oudinot was wounded and Ney took over the advance lines. On the Russian left, the 3rd Cuirassier Division caught the Russian 18th Division forming up in a meadow and took 1,500 prisoners, sabreing hundreds of others in a frantic charge. On the right, the Young Guard stormed into a formation of 4,000 Russians near the river and reduced them to 700 survivors. By darkness Napoleon and the stub of his Grand Army were in control.

On the east bank, on the other hand, Victor's rearguard was being steadily pushed in by enemy pressure. On the riverbank it was the Götterdämmerung of the stragglers as 30,000 humans and horses, in a solid mass, fought to reach two thin, wooden lines to safety. One survivor wrote: "Every moment I found myself stumbling over corpses. . . . I know no more horrible sensation one can feel than treading on living beings who cling to your legs and paralyze your movements as they try to get up again." Another wrote, "One had to beware of neighbors who, finding themselves about to fall, hung on and pulled one down with them."

The Russians got close enough so that two cannon began shooting into the panicked crowd, which was over a mile wide and 400 yards deep. As Victor's men fell back, more Russian guns, including small arms, were able to fire into the dense mass. Dr. Roos witnessed:

> . . . a beautiful lady of 25, wife of a French colonel who'd been killed a few days ago. Indifferent to everything that was going on around her, she seemed to devote all her attention to her daughter, a very beautiful child of four. . . . Several times she tried to reach the bridge, and each time she was repulsed. A grim despair seemed to overome her. . . . Almost instantly her horse was hit by a bullet and another shattered her left thigh above the knee. With the calm of silent despair she took her crying child, kissed her several times and then with her bloodstained skirt, which she'd taken off her broken leg, she strangled the poor little girl; and then,

hugging her in her arms and pressing her to herself, sat down beside the fallen horse. Thus she reached her end without uttering a single word and was soon crushed by the horses of those pressing forward on the bridge.

Jacob Walter, who was in the crowd on horseback, thought that "To be on foot was to lose all hope of rescue." His technique for pushing through the crowd was to repeatedly cause his horse to rear up, "whereby he came down again about one step further forward." He noted that "everyone was screaming under the feet of the horses, and everywhere was the cry, 'Shoot me or stab me to death.'"

At ten o'clock on the evening of the 28th, Victor's IX Corps marched across the bridges. Incredibly, the horde of stragglers and civilians settled down once again for the night, warming themselves by campfires. General Eble, who had orders to burn the bridges at first light the next day, moved among the crowd, imploring people to cross while there was still time. The next morning the panic began again and Eble waited as long as he dared. At 8:30 the bridges were set afire and Napoleon continued his retreat to the west. Thousands of French who were stranded on the east bank could only stare at the calmly approaching Cossacks.

After the Beresina, the Grand Army fought its hardest battles against "General Winter," a cruel opponent who did not distinguish between the opposing sides and who was gaining strength by the day. Aleksei Karpov, an officer of the pursuing Russian army, wrote:

One night we bivouacked on a hill and were unable to find any wood to make fires. In our company three men froze to death. There were many nights like that, during which sleeping soldiers froze, especially among the infantry. . . . The winter was extremely cold, and there were many sick on our side. In our army during the follow-up to the French retreat, perhaps half the soldiers were sick, people said. It would be fair to say that in our company no one was healthy at all. . . . There were less than 100 men in our unit, equipped poorly, attired in light shirts and almost at the end of our strength with exhaustion and sickness. But we were ready to attack the French even from our hospital beds, which we actually did on many occasions.

On December 5, Napoleon announced to his marshals that he need-ed to leave the army and race back to Paris, to assemble a new force of 300,000 men that would re-stabilize the entire situation. His decision was greeted with dismay, but its logic was difficult to dispute. When news of the French debacle reached Europe, Austria and Prussia would be tempt-ed to join the Russians in a retributive war to reverse all the Empire's gains. At this stage in the war, Napoleon's place was clearly in Paris, not on a snow-covered road surrounded by weakened troops in Lithuania.

Murat was given command of the Grand Army as it inched excruci-atingly toward safe haven at Vilna. Sergeant Bourgogne recalled, "The roads were like battlefields, there were so many dead bodies; but as the snow fell all the time, the horror of the sight was softened." Those few who survived the weather owed their further salvation to Marshal Ney, who led the army's rearguard against the ubiquitous Cossacks. Dis-charging his task with a Crusader's zeal, Ney fought day after day in the line, musket in hand, inspiring the troops with his example. Constantly outnumbered, the rearguard occasionally benefited from healthy garrison troops who joined the fight as the retreat proceeded past French depots. The French column was raided and harried at every step, but its hard crust of fighting men in the rear—sometimes numbering no more than 300 against thousands of attackers—staved off total annihilation by Cossack lances.

When the French arrived at the city, Vilna's governor noted: "Murat was huddled in huge and superb furs. A very tall fur hat added to his already large stature, making him rather like a walking colossus." The French troops who poured into the streets, however, were more a plague of freezing, starving beggars. Many of them had survived the Russian countryside only to die in the town. A woman who ventured outside wrote: "I saw terrifying corpses in the streets, seated on the ground, lean-ing against walls, preserved by the cold, their limbs shrunken and stiff in the position in which Death had overtaken them. They had died of hunger, of pain, and without physical or spiritual help."

Murat departed Vilna on December 10 for Kovno, the invasion's starting point on the Niemen River. He had slightly over 4,000 armed men left, mostly the Imperial Guard. The Russians said they found 15,000 Grand Army dead, sick and wounded in Vilna. On an icy hill out-side the town, the last of the French artillery, plus the Imperial treasure chest, were abandoned because the horses couldn't make the height. It is said that Cossacks mingled happily with French stragglers to divide the gold handed out by Napoleon's paymaster, who didn't want to see the treasure go to waste.

After Kovno, the Grand Army that had marched into Russia ceased to exist, though tiny elements were still alive to make their way back to their homes. In the Prussian town of Gumbinnen, French General Mathieu Dumas found warm lodging, but that night he was surprised by a visitor. He recalled,

> We had just been served with some excellent coffee when I saw a man wearing a brown coat come in. He had a long beard. His face was black and seemed to be burnt. His eyes were red and glistening. "Here I am at last!" he said. "What, General Dumas! Don't you recognize me?"
>
> "No. Who are you?"
>
> "I am the rearguard of the Grand Army, Marshal Ney. I fired the last shot on the bridge at Kovno. I threw the last of our weapons into the Niemen, and I have come as far as this through the woods."

Including stragglers, around 25,000 men of Napoleon's central army group got back to safety. Counting the forces on the northern and southern flanks and garrison troops, some 83,000 were still available for operations in 1813. Half a million men had been lost. As a testimony to the difficulty of the winter march to Poland, Kutusov's Russian army, which had begun with 100,000 men near Moscow, reached Vilna with less than 40,000 troops remaining under arms. Of course, the Russian sick and stragglers did not have to deal with Cossacks or the wrath of peasants when they had fallen behind.

Napoleon Bonaparte, who had reached Paris on December 19, made good on his promise to assemble a new army to restore the French position in Europe. The following fall he fought the largest battle of his career, at Leipzig in eastern Germany. He was able to field over 200,000 men against an Allied force of 320,000. Subsequent to that French defeat, Russian, Prussian and Austrian armies overran northern France, and in April 1814 Napoleon was exiled to an island in the Mediterranean. He returned a year later, however, for one hundred days, to fight the most famous battle of his career, at Waterloo in Belgium. In the very last battle of his life, Napoleon was finally able to face the British, under their best general, in a decisive engagement on land. Unfortunately, by that time, the French Grand Army had become an improvised shadow of its former self. The original had been destroyed years before in Russia.

Chapter 3

THE NEZ PERCE
IN MONTANA

Between the final defeat of Napoleon and the rise of a united Germany, Western Europe enjoyed the longest period of peace in its history, recently matched by the current stretch that has followed World War II. The largest war during that period was fought by the young immigrant nation, the United States, which tore itself apart along geopolitical lines, a rift that required 600,000 lives to mend. Given America's isolation, bordering oceans rather than other great powers, the Civil War remains the greatest conflict in that nation's history, fought between its own citizens. Once the irresistible weight of the North had forced the South to succumb, however, smaller battles continued to flare on the American continent as they had ever since European immigrants arrived on those shores. These battles, on the western frontier of the rapidly expanding nation, were fought year after year between Americans of European descent and a wide variety of culturally and linguistically distinct peoples of Asiatic origin, collectively known as Native Americans or Indian tribes.

In the years following the Civil War, a state of perpetual, low-level warfare existed in the West that resembled the almost numbing stream of engagements the American public observed a century later during the Vietnam War. To readers of East Coast newspapers, the ultimate outcome was seldom in doubt but there could still be flashes of drama. In 1867, the Sioux Chief Red Cloud was able to wield superior force against the U.S. Cavalry on the northern Great Plains. The United States conceded by closing down the Bozeman Trail, and abandoned several army outposts in order to quell the violence. Elsewhere, however, there was no stopping the unlimited resources and persistence of the whites.

In 1872 the Apache Cochise was finally defeated in the Southwest. The following year in northern California, Captain Jack, with about fifty armed Modocs, tied down 1,000 U.S. troops until he was betrayed by one of his own followers and caught. Captain Jack (birthname Kientpoos) would no doubt hold a higher place in the pantheon of Indian heroes if he hadn't murdered U.S. General Edward Canby, who had come to speak to him under a flag of truce. The State of Texas, meanwhile, had gained enough population to deal with the fearsome Comanche tribe, and in 1874 Quanah Parker was defeated by volunteers and cavalry, his Kiowa allies forced to succumb the next year.

In the spring of 1876 the large Sioux tribe once more took the field. Men like Crazy Horse and Sitting Bull, who had never been beaten, were joined by others who chafed under white policy, and by Cheyenne and Arapaho fighters from the south. In what could be compared to the Indian version of the Tet Offensive, though without its practical effect, they won the most spectacular victory in Native American history against the whites' "greatest Indian fighter," George Custer, wiping out his command at the Battle of the Little Bighorn. Afterward the tribes were methodically rounded up or dispersed by the U.S. Army. Sitting Bull was able to escape with 2,000 followers to set up an impoverished camp across the Canadian border.

By the late 1870s, Indians no longer harbored the illusion that the white men could be defeated in war, but this is not to say they lacked good reasons for trying. Their ancestral land as well as their way of life and culture were at stake. Further, the whites, who loved to hold elaborate conferences at which grandiose contracts were signed, often broke their promises, leaving greedy and corrupt individuals to implement arbitrary terms. The U.S. government's intent was to place all Indians on limited parcels of land—reservations—confining them away from white settlers who desired freedom to live or work where they chose. To Indian warriors, on a man-to-man basis perhaps the best fighters in the world, submission to the whites had become a severely painful option, but increasingly, in view of the fact that they had families to protect, their only one.

In Washington, DC, the necessity to deal with dozens of independent, albeit small and primitive, "nations," presented a constant challenge with no shortage of controversy. Time and again, the government would designate a large reservation for a tribe, only to find that the social Darwinism represented by U.S. citizens heading west made its plans moot. The worst cases occurred when gold was discovered on Indian land. The great equalizer between the haves and have-nots of nineteenth-century white society, each gold discovery in the West was accompanied by a horde of immigrants. First came the prospectors, followed by mining companies, then merchants, blacksmiths, families, teachers, undertakers and all the other supports of a permanent community. Even where gold was not evident, immigrants from the East were pouring gradually onto Indian land as cowboys or farmers. The population of the United States was over forty million by 1870 and rapidly growing; the number of Indians in North America by then had fallen to about half a million, a number that was shrinking. With or without the best intentions of the government in Washington, Manifest Destiny had taken on an irrefutable logic of its own.

The Nez Perce Indians lived in a part of the West—at the junction of the present-day states of Idaho, Oregon and Washington—that provided temporary isolation from both the white tidal wave that had spilled onto the Pacific Coast and from the vicious battles that raged on the Great Plains. They were far north and inland from the California gold fields, protected by rough country from large-scale white immigration from the Pacific. On the other side, the high Bitterroot Range of the Rockies comprised a sort of wall between Nez Perce territory and the open plains of Montana. The Nez Perce had ties with both the Pacific Coast Indians and the Plains tribes, but more similarities to the latter in their dress, style of battle and affinity for horses. A constant stream of Nez Perce hunting parties crossed the mountains to the Plains to kill buffalo, bringing back the valuable hides.

Their name—"pierced nose"—was bestowed on them by French Canadian trappers working for the British Hudson Bay Company in the early 1800s. The best guess is that a few members of the tribe might have pierced their noses with fish bones in those early years; some suspect the name comes from an error in translation—among other Plains tribes the sign language for Nez Perce was a finger swiped beneath the nostrils. In any case, by the 1870s they were known as the Nez Perce, though nose piercing was not evident; in their own language they were Niekamp, or "the people."

Although since the war in 1877 the Nez Perce have been overly idealized by the public, to the point where references to the tribe's pre-war days bring to mind one of those innocent races on Eden-like planets recurrent in science fiction, contemporary accounts nevertheless described them favorably. Lewis and Clark, who stumbled out of the Bitterroots half-famished in 1805, were the first to leave behind a detailed account. Clark said they were "much more cleanly in their persons and habitations than any nation we have seen since we left the Illinois." A member of their expedition called them "the most friendly, honest and ingenuous" of the Indians he had seen.

The Nez Perce, divided into a number of autonomous bands, were the largest tribe in the Northwest. They feuded occasionally with the Spokanes and Couer d'Alenes to their north, but were more of a formidable guardian for the Columbia River tribes against the Paiutes, Bannocks and Shoshones to the south, as well as against the Sioux to the east. The Nez Perce were rich in horses, and may have been the only Western tribe that practiced selective breeding. Along with their Palouse neighbors, they are thought to have developed the Appaloosa. Their primary weakness was a lack of agriculture, which resulted in periodic famines when the weather wouldn't cooperate with hunting, fishing or the gathering of other sustenance from the ground.

In 1831, four Nez Perce caused a sensation among church groups in the East when they traveled to St. Louis in search of the white man's "medicine." In response, the Presbyterian minister Henry Spalding and his wife set up a mission and school in Nez Perce territory. Some members of the tribe converted to Christianity and began farming. Spalding left in 1847 due to an uprising, but his influence continued to be felt. When war came three decades later, the division between those Nez Perce who fought and those who stayed on the reservation could be roughly marked by which Indians had been influenced by Spalding's mission.

By 1855 there was still no great pressure on the Nez Perce from settlers; nevertheless, U.S. policy dictated that the tribe should be assigned exact boundaries. The government also desired that the Nez Perce should elect a single leader to represent the tribe, in addition to the traditional large council representing all the bands. A minor chief who spoke English took on the leadership role and the whites called him Lawyer. With a solemn treaty the United States and Nez Perce agreed that tribal land would henceforth consist of some 100,000 square miles bordered by the Snake, Salmon and Clearwater rivers. It spread across northeastern Oregon, southeastern Washington and western Idaho, basically the tribe's traditional homeground, and, at twenty-five square miles for every man, woman and child, was not a bad deal for the tribe. Then, in 1860, gold was discovered on the land.

By the summer of 1861, 5,000 white prospectors were on the reservation and the town of Lewiston had been created, where "drunkenness, gambling, crime and murder were in full blast." The local Indian agent, seeking to bring some order to the chaos, drew up an agreement with Lawyer and the tribe's council to designate an area in which the miners could operate. By way of easing the tension, long-promised government supplies, including cattle, began to arrive. All the best efforts of the Nez Perce chiefs and responsible whites, however, could not control the gold seekers. Miners continued to pour in and set up shop wherever they chose. By mid-1862 (some of these people may have been trying to avoid war service) there were 15,000 civilian invaders of Nez Perce land. General Benjamin Alvord of the District of Oregon came to inspect the volatile situation and decided that a military presence was needed, not so much to watch over the Indians as the whites. He wrote: "It can hardly be anticipated that even the virtues of this tribe and the establishment of the military post will prevent the natural consequence of such provocation, of whiskey and of contact with bad white men. . . . As the roads are now painfully infested by robbers and cutthroats, the presence of the military will materially aid the civil authority." Fort Lapwai was constructed twelve miles south of Lewiston, but it was already apparent that the original concept of the reservation had come undone.

In May 1863 the U.S. Superintendent of Indian Affairs, Calvin H. Hale, summoned the Nez Perce to a conference at Fort Lapwai, where he suggested reducing the size of the reservation. The Indians were offered compensation for lost land, but the Nez Perce were hard bargainers, and the chiefs from the Wallowa Valley in Oregon and other southern parts of the reservation flatly refused to give up their homes. The new boundaries that Hale had in mind encompassed land belonging to Lawyer and his followers. The southern bands, led by "Old" Joseph and other chiefs, had hard words with Hale and stalked out of the conference. Lawyer and 45 other chiefs signed the new treaty.

Years passed with the tribe divided into the Treaty Nez Perce, who lived within the new reservation, and the Non-Treaty Nez Perce, who stayed where they were, perhaps 4,000 members of the tribe in all. Despite the influx of settlers, the area was still not overpopulated, and the free Indians continued to live as they always had. Old Joseph died in 1871 and was succeeded by his son, also named Joseph, who took over leadership of that band. By this time the roughneck nature of the white presence had begun to transform into a settled community after women and families had arrived. "Wild" Indians were fast becoming an anachronism in the West and the pressure on Joseph and his compatriots to settle down had increased.

Among U.S. politicians and bureaucrats, the issue was hotly debated and the Non-Treaty Nez Perce had sympathizers. Agent John Monteith thought there was no reason in the world to prevent the Wallowa Valley, Joseph's domain, from remaining Indian territory. The place was remote and undesirable anyway, infiltrated by only 22 white homesteads. In his excellent study of the Nez Perce War, Mark H. Brown speculates that those few white settlers in the valley must have had nefarious motives in any event to have picked such a rustic place to live. To whites, the valley might have served the same purpose as Wyoming's Hole-in-the-Wall country, which was filled with desperados and people on the run. Monteith wrote to the Commissioner of Indian Affairs:

It is a great pity that the valley was ever opened for settlement. It is so high and cold that they can raise nothing but the hardiest of vegetables. It is a fine grass country and raising stock is all that can be done to any advantage. It is the only fishery the Nez Perces have and they go there from all directions. The valley is surrounded by high mountains and it is impossible to get a wagon in the valley until a road is built. If there is any way in which the Wallowa Valley could be kept for the Indians I would recommend that it be done.

While lobbying his government to let the Indians keep the valley, Monteith also urged the Non-Treaty Nez Perce to settle down. He did not need much imagination to foresee that, regardless of the fairness of the 1863 revision of the 1855 treaty, the days of hunter-gatherer tribes roaming freely in the United States were numbered. Another of his jobs was to prevent unscrupulous traders from selling alcohol to the Indians, since consumption of spirits seemed to have a pronouncedly negative effect on their behavior.

The government's point of view was that the 1863 treaty had been correctly signed by the tribe's designated leader, Lawyer, as well as by a majority of the tribe's council. A minority of Nez Perce, or at least those who had walked away from the negotiations before the decision was made, could not be allowed to defy the will of the majority. To the Non-Treaty Nez Perce, autonomous in their several bands, what Lawyer and his accommodationists signed had no relevance to them. They had never given up their land and no one could give it up on their behalf.

The Manifest Destiny side of the argument was expressed by Oregon Governor Leonard Grover, who declared:

The region of country in Eastern Oregon not now settled, and to which the Wallowa Valley is the key, is greater in area than the state of Massachusetts. If this section of our State, which is now occupied by enterprising white families, should be remanded to its aboriginal character, and the families should be removed to make roaming ground for nomadic savages, a very serious check will be given to the growth of our frontier settlements, and to the spirit of our frontier people in their efforts to redeem the wilderness and make it fruitful of civilized life.

Countering Grover's view, at this time in the Northwest there was an Indian prophet named Smohalla who founded a quasi-religious movement called the Dreamers. As opposed to communication between people and God taking place through prayer, Smohalla claimed that it took place through dreams, and the true deity was Mother Earth. The Dreamer philosophy held that whites, through agriculture, mining and dividing up land, desecrated the Earth and that if Indians returned to their ancient ways the whites would disappear. The Dreamer philosophy was attractive to Indians such as the Non-Treaty Nez Perce, and whites feared that it could cause a pan-tribal uprising throughout the Northwest.

In July 1874 a major figure in the upcoming conflict stepped onto the scene when General Oliver O. Howard took command of the Department of the Columbia. Howard had not been the luckiest general during the Civil War, beginning with losing an arm at Fair Oaks in the Peninsula (the Indians called him "Cut Arm"). After his recovery, he took part in the fruitless assault at Fredericksburg, and the following spring his XI Corps was surprised by Stonewall Jackson's decisive flank attack at Chancellorsville. (Rebel Colonel William Alexander commented that he hadn't seen such a wild panic since Bull Run.) Two months later, Howard's XI Corps was pushed out of the town of Gettysburg, losing 4,000 men as prisoners, though the Union fallback position on Cemetery Ridge turned out to be good defensive ground. Howard subsequently led a corps under Sherman in the Atlanta campaign, and after the war was named director of the Bureau of Refugees, Freedmen and Abandoned Lands. A staunch abolitionist who worked hard on behalf of ex-slaves in what was commonly called the Freedmen's Bureau, Howard was also an extremely devoted Christian. After his appointment to the Northwest, an army surgeon's wife named Emily FitzGerald mentioned in a letter to her mother, "I have heard the officers discuss him and he is not very popular among them. It is owing to his ferocious religion. He is one of those unfortunate Christians who continually give outsiders a chance to laugh and have something to make fun of. He says himself, 'You know, I am a fanatic on the subject.'"

Because the Wallowa Valley in Oregon was the major point of contention, Chief Joseph, who led one of the largest Nez Perce bands, became the central Indian figure in negotiations. He was tall and good-looking, and everyone was impressed by his dignified manner and intelligent mien. Even while speaking through interpreters he conveyed a certain eloquence and dexterity of thought. The whites came to view him as representing all the Non-Treaty Nez Perce.

At first Howard thought that it would be a "great mistake" to take the Wallowa Valley from the Nez Perce, since it was not worth fighting for and was evidently far more important to the Indians than to the whites. As the months passed, however, his attitude toward the Nez Perce hardened. In particular, the Dreamer religion seemed intolerable to Howard as an Indian rationale for not complying with United States demands. He held a series of meetings with Chief Joseph where the clash of cultures appeared insoluble. Mrs. FitzGerald wrote of one meeting: "This Joseph will admit no boundary to his lands but those he chooses to make himself. I wish somebody would kill him before he kills any of us. The Agent leaned over in the Council room yesterday and asked me if my hair felt on tight." Not to discriminate, in one of her gossipy updates to her mother she commented about the aftermath of another conference: "General Howard is promenading the porch quoting scriptures. Indeed, I think he is real good, but he is awfully queer about it."

In the autumn following Custer's Last Stand, the whites gave the Non-Treaty Nez Perce an ultimatum to report to the reservation by April 1, 1877. The deadline came and went, and, while General Howard called in troops from throughout the territory to make a show of force, in May another conference was called. This time the Indians chose as their spokesman a muscular, tough-looking chief named Toohoolhoolzote, who "had a heavy gutteral voice, and betrayed in every word a strong and settled hatred of all caucasians." After listening to a series of harangues from the chief, Howard lost his temper and had Toohoolhoolzote taken away and put in a guardhouse. He said, "My conduct was somewhat summary, but I knew it was hopeless to get the Indians to agree to anything so long as they could keep this old Dreamer in the lead." The intimidation worked. Joseph and the other chiefs picked out spots on the reservation where they would henceforth live, and Howard ordered them to be there in thirty days. He telegraphed Washington: "We have put all non-treaty Indians on reservation by using force and persuasion without bloodshed."

By the second week in June, Non-Treaty bands under Joseph, White Bird and Toohoolhoolzote had assembled at an old campground called Tepahlewam near the Salmon River. The camp was in an explosive mood and Howard, who had arrived at Fort Lapwai to oversee the deadline, received a worried update from the settlement of Mount Idaho, near the Indian camp:

> Yesterday [the Nez Perce] had a grand parade. About one hundred were mounted and well armed and went through the maneuvers of a fight—were thus engaged for some two hours. They say openly that they are going to fight the soldiers when they come up to put them on the reservation. . . . A good many were in town today and were trying to obtain powder and other ammunition. . . . They are evidently on the lookout for the sol-

diers. I believe it would be well for you to send up, as soon as you can, a sufficient force to handle them without gloves, should they be disposed to resist. Sharp orders and prompt action will bring them to understand that they must comply with the orders of the government.

On July 13, 1877, two days before the Non-Treaty Nez Perce were due on the reservation, three young men from White Bird's band rode to the home of a prospector named Richard Devine, an ex-Royal Navy sailor, and shot him while he lay sick in bed. Then they went to Henry Elfers' house and killed him and his two hired hands, Robert Bland and Harry Becktoge. After this they encountered and shot Samuel Benedict, who was out looking for stray cows. Benedict was wounded in both thighs, but escaped by playing dead.

The instigator of the murders was Wahlitits, whose father had been killed by a settler two years before, and who had been taunted the previous night for not avenging his father's death. Although some may wish to attribute the entire Nez Perce War to the rash act of an aggrieved son, Wahlitits simply provided the spark for a powder keg that was waiting to explode. The reason he was taunted so viciously for not having spilled white blood may have been a calculated effort by other young men of the Nez Perce tribe to prompt an act of violence. Once that door had been opened, the warriors could feel free to demonstrate their own fighting prowess, presenting their more cautious chiefs with a fait accompli.

When Wahlitits and his two companions returned to the camp, with captured horses and "fine rifles," to announce that they had been busy killing settlers, they were greeted as heroes. Other young men quickly began dressing and painting for battle. Now, just like dozens of Indian tribes before them, the proud Nez Perce would fight for what was theirs.

The next day an old settler named James Baker rode to his neighbor, Jack Manuel, to warn him of the uprising. The two men, with Manuel's wife, Jennet, six-year-old daughter and baby son, were trying to get to Baker's house when they were met by about twenty Indians. Baker was shot first and then Manuel was hit by an arrow in the neck and a bullet in the hip. He rolled down a hill and the Indians thought he had been killed. Little Maggie Manuel was shot by two arrows and also suffered a broken wrist as she fell from her horse. Jennet, already holding the baby, grabbed Maggie and ran back to the house where her father, George Popham, and a prospector named Brice had stayed. As the men watched, Jennet, who was a thin young woman with very long, blond hair, tripped and an Indian caught up to her and commited an "outrage." The Nez Perce told the men they would be unharmed if they gave up their guns, and, when Popham and Brice took the gamble, the Indians stayed true to their word.

That night the two men thought it would be best if they went out in the woods to hide, but the next morning Brice encountered the wounded six-year-old, Maggie. She said an Indian had come into the house during the night and stabbed her mother and killed her baby brother. That day, Indians burned the house. It has become a matter of controversy ever since whether Jennet Manuel's and her baby's remains were ever found.

A few miles away, Sam Benedict, who had successfully played dead earlier, got back to his home to warn his wife and kids. Some French prospectors came by and one stayed with the family. Benedict ordered his wife and children to flee out the back, but Indians were coming up so they returned to the house, just in time to see the Frenchman get shot dead in the doorway. Benedict tried to get away out the front but was killed for sure this time and his body fell into a creek. The Indians told Isabelle Benedict and her children to run.

At the Osborn homestead, two families holed up as a group of Nez Perce fired through a big window. The three men inside had just risen to fire back when an Indian volley got all three, killing two. When the Indians broke in, one of the women shoved a revolver to the wounded man, Harry Mason, but he couldn't lift it. He said to a Nez Perce, "Oh, shoot me!" and the warrior complied by firing a pistol into his head. The two new widows in the house later claimed that they had not been outraged, a contention that was politely accepted, but to the first settlers who encountered them after the Indians had left the claim seemed dubious.

That night, some twenty miles away, the Chamberlin and Norton families, along with Joe Moore and Lew Day, tried to get to Mount Idaho. Chamberlin drove the wagon carrying the women and children, while the other three men were on horseback. Nez Perce warriors found them and gave chase, finally shooting one of the wagon horses to bring the party to a halt. All the men were wounded but kept up a defensive fire while the attackers sniped from nearby cover. Jenny Norton got up at one point to plead with the Indians but was shot in the legs. At one point Chamberlin gathered his family and made a run for it, but shots and screams from that direction indicated they didn't make it.

Norton died and Day was immobilized by his wounds during the night, but Joe Moore kept firing. When he ran out of ammunition for his rifle, he found a shotgun and made powder charges that could at least make a flash and a noise. At first light, Jenny Norton laid still as horses approached and she heard someone cock a pistol above her head. Suddenly Moore shouted, "Don't shoot, for God's sake! It's us!" The newcomers were six settlers from Mount Idaho who had heard the gunfire, and who thought the dark-haired Mrs. Norton was an Indian. The men quickly hitched their horses to the wagon, just as a group of Nez Perce re-emerged to give chase. They would have been caught except another group of settlers rode out to join them and the Indians were forced off. Chamberlin was later found dead with his lifeless three-year-old in his arms. Underneath the two was the baby, still alive but with a knife wound in the neck and the tip of her tongue cut off. The pregnant Mrs. Chamberlin was found wandering hysterically after a number of outrages and with an arrow wound in her chest. One of the settlers said, "We had to run her down and surround her before we could make her believe we were friends."

Another party of settlers rode back to the Norton ranch, known as the Cottonwood House, to retrieve an abandoned wagon full of liquor, but they ran into a war party of some sixty Nez Perce. Settlers throughout the area congregated at Mount Idaho, which was prepared for defense. Altogether, including people who died shortly afterward of their wounds, 19 settlers were killed. One elderly Nez Perce warrior was lost when he fell from his horse and a band of chasing settlers came up. An ex-Confederate officer named George Shearer emptied two shotgun rounds into his chest and then broke the stock of his weapon over the man's head.

At the Indian camp at Tepahlewam, Chief Joseph arrived to hear the news that the Nez Perce were at war. He and his younger brother, Ollokot, had been on the other side of the Salmon River at the time of the outbreak, jerking some beef in preparation for the move to the reservation. When confronted with the situation, Joseph's advice was to speak to the whites and if necessary hand over the murderers. The leaders of the other large bands, however, White Bird and Toohoolhoolzote, thought that it was too late; too many warriors had been involved, and in any case the whites would be more likely to hang the chiefs. On the other hand, just as Howard's strong-arm tactics against Toohoolhoolzote at the May conference had temporarily intimidated the Indians, perhaps this flash of Nez Perce war prowess would have a similar effect on the whites. It was decided to wait a while to see what the whites would offer. In the meantime the encampment of some 700 men, women and children, would move several miles to the foot of White Bird Canyon, a place that was difficult to approach. Joseph's wife was about to give birth, and he and Ollokot stayed behind in the old camp. A few braves kept an eye on their tepees and, according to one account, shots were fired in their direction, just in case the brothers decided to go over to the whites.

It did not take long before lathered horses began riding into Fort Lapwai with news of the outbreak. Howard's correspondent from Mount Idaho, who had earlier warned of Indian maneuvers, now reported: "One thing is certain, we are in the midst of an Indian war." The fort had a company of the 21st Infantry and H and F Companies of the 1st Cavalry on hand. The senior commander was Captain David Perry of F Company. A tall, good-looking officer, he had not only compiled an excellent record during the Civil War but later in fights against the Bannocks and Modocs. His adjutant was Lieutenant Edward Theller, who had missed the Civil War by being stationed in California and was suspected of being a gambler.

The commander of H Company was Captain Joel Trimble, who had been wounded twice in the Civil War and was the officer to whom Captain Jack had surrendered four years earlier. Trimble's junior officer was Lieutenant William Parnell, a heavy-set Irishman who had fought in the Crimean War (claiming to have survived the Charge of the Light Brigade) and numerous battles in the War Between the States. Eleven Treaty Nez Perce volunteered to serve as scouts, and at first light on June 16, ninety-nine cavalrymen mounted up and headed south.

By hard riding they covered 60 miles during the day, and at night were met outside the settlement of Grangeville by a man named Ad Chapman, who had been riding around with armed civilians ever since the fighting began. The people of Grangeville wanted the cavalry to pause in their vicinity but Chapman warned Captain Perry that the Indians were nearby and might try to escape across the Salmon River. After only a couple of hours pause, the cavalry moved on, accompanied by Chapman and ten volunteers, until they were at the head of White Bird Canyon. There, as quietly as possible, the soldiers bedded down while their bone-tired horses got some rest.

In the Indian camp, it was still not clear whether they had passed the point of no return, and in any case about half of the 140 warriors were hopelessly inebriated from all the free liquor they had found. Nevertheless, some still had their wits about them and lookouts had been placed up the canyon approach route. With the first rays of dawn on June 17 a soldier lit his pipe, and cavalry survivors remembered that this unwise act was immediately followed by a coyote howl that echoed down the canyon.

In the early morning, Lieutenant Theller, with Ad Chapman, eight soldiers and a couple of Indian scouts, took the lead, followed by Perry's F Company in column of fours, and then Trimble's H Company, with the other scouts as flankers on the right and Chapman's volunteers on the left. The ground was broken by small, sharp ravines and scattered with brush and trees. At one point, the soldiers were astonished to encounter Isabelle Benedict and her two children, who emerged from hiding. They gave her a loaf of bread and told her they would get her on their way back.

When the lead party approached the foot of the canyon the soldiers were met by half a dozen Nez Perce. Indian accounts said they were carrying a white flag; a soldier remembered that their greeting was "What do you want?" Any ideas for a parley were destroyed, however, when a man in a big white hat—Ad Chapman—fired at the Indians and forced them to flee. Other Indians began firing back and Theller ordered his men to spread out.

When Perry came up he found Theller occupying a small bluff overlooking the village and already in a firefight with Indians to his front. He thought that a charge would place the cavalry on disadvantageous ground. F Company's bugler had been killed by one of the first Indian shots, but Perry was able to deploy Trimble's H Company on his right while his own men took the center. Chapman's volunteers manned a high knoll that protected the left. Every fourth man was a horse-holder in the rear. Unfortunately, Nez Perce warriors were already firing into the command from high ground on either flank. On the right, several Indians raced by, ducking like trick riders behind their horses to get behind the cavalry line.

Already suffering dead and wounded, Perry's line became unhinged when Chapman's volunteers on the left came under intense fire and ran. Indians scrambled onto the knoll and were able to enfilade the entire line, shooting six soldiers in a matter of minutes. Others got behind the cavalry on the left so that the soldiers were being fired at from every direction. Perry sought to maneuver his F Company closer to Trimble, but a few men were too hasty. Trimble's H Company, noticing the panic, began to abandon their line. Perry wrote: "From that time on there was no organized fighting, but the battle was confined to halting first one squad and then another, facing them about and holding the position until flanked out."

Everyone was retreating back up the canyon except for Sergeant Michael McCarthy and a small contingent, who were holding a good position up front. When they realized they were alone, they ran for their horses, but McCarthy's was shot and he had to run for it. Two of his men came back for him and he doubled up with a trooper until they found a riderless mount. His new horse was also shot, however, and McCarthy had no choice but to roll down a hill and hide.

One can imagine the terror of Isabelle Benedict, who now witnessed her cavalry rescuers fleeing back up the canyon oblivious to her pleas. Finally two troopers stopped and found a horse for her, pulling up her children on their own mounts. A few minutes later, however, Isabelle was thrown from her horse.

Lieutenant Theller had meanwhile led seven men into a small ravine, but they were all trapped there and killed. The bodies were found surrounded by spent cartridges, so they apparently did not go quickly. The Nez Perce warrior Yellow Bull, who was 21 at the time, described the annihilation of another group of five soldiers who had taken cover behind some rocks. He said the last soldier died when an Indian threw a stone at his head.

The Nez Perce warrior Wounded Head killed a soldier by using the only bullet he had, in an old pistol. He then took the soldier's carbine and ammunition, leaving his ancient handgun on the body by way of trade. Proceeding up the canyon, Wounded Head came upon Isabelle Benedict. He ordered her to get behind him on his horse and then took her back to the village. According to Isabelle, women in the village convinced him to set her free and a few days later she got back to Grangeville.

Captain Perry, who had also been unhorsed for a while, finally reached the top of the canyon to join Lieutenant Parnell, the Crimean War veteran, who seemed to have kept his head as well as anyone in the fight. To illustrate how, during a battle, minutes can seem like hours (or vice versa), Perry looked at his watch and declared that since it was seven o'clock, they might be able to hold that position until darkness. Parnell had to remind his superior that it was seven o'clock in the morning. The two officers and their twenty or so remaining men were almost out of ammunition, and there was no choice but to fight their way back to Grangeville. Once there they found Captain Trimble, Ad Chapman and other men, less thirty-five troopers who had been left behind in the canyon.

Of these, Sergeant McCarthy got back alive. Two Indian women had seen him go into his original hiding place and told warriors to search the area. McCarthy had stealthily crawled to a new spot, however, and though the warriors shot into the brush they eventually gave up the search. The other 1st Cavalry fugitives or wounded on the field were less fortunate. In a statistic that speaks for itself, the only three prisoners the Indians took in the battle were from the army's Nez Perce scouts, two of whom were sternly warned not to help the whites again, while the third joined the hostiles.

After the Battle of White Bird Canyon, which, after the Little Big Horn and the Fetterman Massacre, ranks foremost among the U.S. Cavalry's defeats in the West, both sides had to reconsider their strategies. General Howard had the advantage because he could simply call for more men from throughout the United States' various military departments. He estimated that he would need 500 troops in order to put down this latest Indian "rebellion."

The Nez Perce, on the other hand, faced a far more difficult road to success. Though the young men were euphoric over their easy demolition of two companies of cavalry (34 dead troopers at the cost of only three wounded), the older chiefs knew what they were up against. If they could defeat 100 soldiers in a battle, the next time it would be 300. If they defeated 300 soldiers, they would then have to face 600, and so on. Meanwhile, the tribe's women and children had to be protected. The primary question was: Where to go?

At Fort Lapwai, General Oliver Howard received reinforcements from the 1st Cavalry and 21st Infantry, plus two Gatling guns and a mountain howitzer, and after a week set out to engage the hostiles. Joined by the remnants of Perry's command and groups of settlers, the force consisted of over 300 men. Their first task was to bury the bodies in White Bird Canyon, which had been lying exposed to sun, rain and animals for nine days. In contrast to the aftermath of other Indian battles, notably against the Sioux, Perry's dead were not mutilated or scalped.

On the day after the battle, the Nez Perce were reinforced by a hunting party consisting of Rainbow, Five Wounds and other renowned warriors who had just returned from a trip to Montana. After debating what they should do next, the bands crossed the Salmon River into the hilly country beyond, where it would be difficult for the army to follow. Passing by a fortified settler outpost at Slate Creek, a few of the Indians approached under a white flag to pay a storekeeper their outstanding bills. When the army got to the river, dozens of mounted warriors charged down the opposite bank in what amounted to a colorful demonstration. It subsequently took Howard's force three days to cross the river.

On July 1 the whites committed their greatest blunder of the campaign by forcing the band of Looking Glass to join the war. Howard, acting on the advice of local citizens, sent two cavalry companies under Captain Stephen Whipple, with about twenty citizen volunteers, to bring the chief into custody. Looking Glass' land fell within the reservation boundary, but the chief was known to be sympathetic to the Non-Treaty Nez Perce bands. Some of his young men had already participated in the fighting and Howard was nervous about leaving the band astride his supply route to Fort Lapwai.

Whipple approached under a white flag, but Looking Glass refused to come out, instead sending an interpreter named Peopeo Tholekt. His instructions were to say, "Leave us alone. We are living here peacefully and want no trouble." A white settler jabbed the interpreter in the ribs with his rifle and demanded that Looking Glass appear. The whites were on the other side of a creek and the Indians, seeing the rough treatment given their representative, became wary. They set up a white flag near Looking Glass' tent and Lieutenant Sevier Reins accompanied Peopeo Tholekt across the stream to the village. Just then a settler fired his rifle and hit an Indian warrior in the thigh. Reins raced back across the creek while the Indians ran the other way, abandoning their possessions. When the cavalry charged in, they burnt several of the lodges and seized the band's herd of 700 horses. Looking Glass, who had sent a message to White Bird's band at the beginning of hostilities saying that they were fools, had now been forced to join the uprising. Some minor chiefs and warriors who had been on the fence followed Looking Glass into the hostile camp.

While Howard struggled through the mountains west of the Salmon River, the main body of Indians, in a move attributed to Rainbow and Five Wounds, recrossed the river and began heading back east across easy ground. The day after his bungling of the Looking Glass affair, Captain Whipple moved his two companies to the Cottonwood House where, unbeknownst to him, he was directly in the Indians' line of advance. On July 2 Whipple sent two civilian scouts in the direction of the Salmon River. They ran into the Nez Perce and one was killed, the other making it back. Whipple then sent out Lieutenant Reins with a dozen picked men as an advance guard while the rest of the cavalry followed. Reins' men were charged by a group of warriors under Rainbow, however, and six men were shot from their saddles. The remainder found shelter among rocks, but while one Indian pinned them down in front, the rest of the warriors attacked from the rear and the command was wiped out. The young brave Yellow Wolf said that the Indians marveled over one soldier who was shot once in the forehead and twice in the chest but still refused to die. An old warrior named Smoker shot him point blank twice more with no effect, prompting other warriors to make fun of Smoker's gun. Then the Indians began hitting the soldier over the head with clubs until he finally expired.

Whipple's two companies formed a skirmish line against a potential onslaught but the warriors had retired. The troops fell back to Cottonwood House. Captain Perry had meanwhile been dispatched by Howard to join Whipple with the two Gatling guns. On the morning of July 4, Whipple rode out to meet him and the combined companies of 113 men holed up around the ranch. Indians surrounded the place and both sides exchanged long-range fire.

On July 5 a party of 17 citizen volunteers under "Captain" D.B. Randall came up on a rise near Cottonwood and saw the entire Indian force stretched out for half a mile below them. Determined to break through to the cavalry, Randall ordered his men—thereafter known as the

"Brave 17"—to charge through the Indians. The Nez Perce opened their line to let the volunteers through, but then closed up and chased them, forcing the men to dismount and set up on a small hill for all-around defense. The cavalry under Perry and Whipple, a mile and a half away, could clearly hear the shooting but hesitated to come to the rescue. At one point two of the volunteers broke through the Indian cordon and rode up to the cavalry to ask for ammunition. Getting it, they rode back to their friends. This was too much for George Shearer, the ex-Confederate officer, who headed off by himself to join the surrounded group.

Finally Perry ordered his command to mount and they rode toward the fight. By the time they arrived, however, the Indians were leaving the field. The volunteers had suffered two dead and three wounded, one of whom would die later. Captain Randall had received a mortal bullet wound in the spine. Recriminations came fast over the cavalry's hesitation to back up the volunteers.

To the Nez Perce, who had successfully lost General Howard in the mountains on the other side of the Salmon River, the dash across the prairie with all their dependents and belongings was a gamble. They had no wish to get tied down in a protracted fight en route, and the warriors' job was simply to screen the helpless main column. By now filled with confidence over their ability to fight soldiers, the Nez Perce set up camp on the west side of the Clearwater River, at the southeastern tip of the reservation. There is little doubt that some young men of the Treaty bands joined the hostiles at this time. Raiding parties burned up to 30 homesteads and at one point warriors approached within 12 miles of Fort Lapwai. Annoyed at the slow movement of the soldiers, about 75 white volunteers rode up to the Indian camp, but were forced back onto a height and their horses were run off. After a few days of waiting for the cavalry, the volunteers retreated on foot, having named their position Mount Misery.

In the meantime, General Howard had found his troops unable to negotiate the Salmon River at the point the Indians had crossed, so he backtracked to his original ford near White Bird Canyon. He then crossed the smaller Clearwater and headed north. His command now consisted of 440 men with three guns, plus 50 packers and a number of scouts. On July 11 he caught sight of the Nez Perce camp on the other side of the Clearwater from a high bluff and opened fire on it with his artillery. The

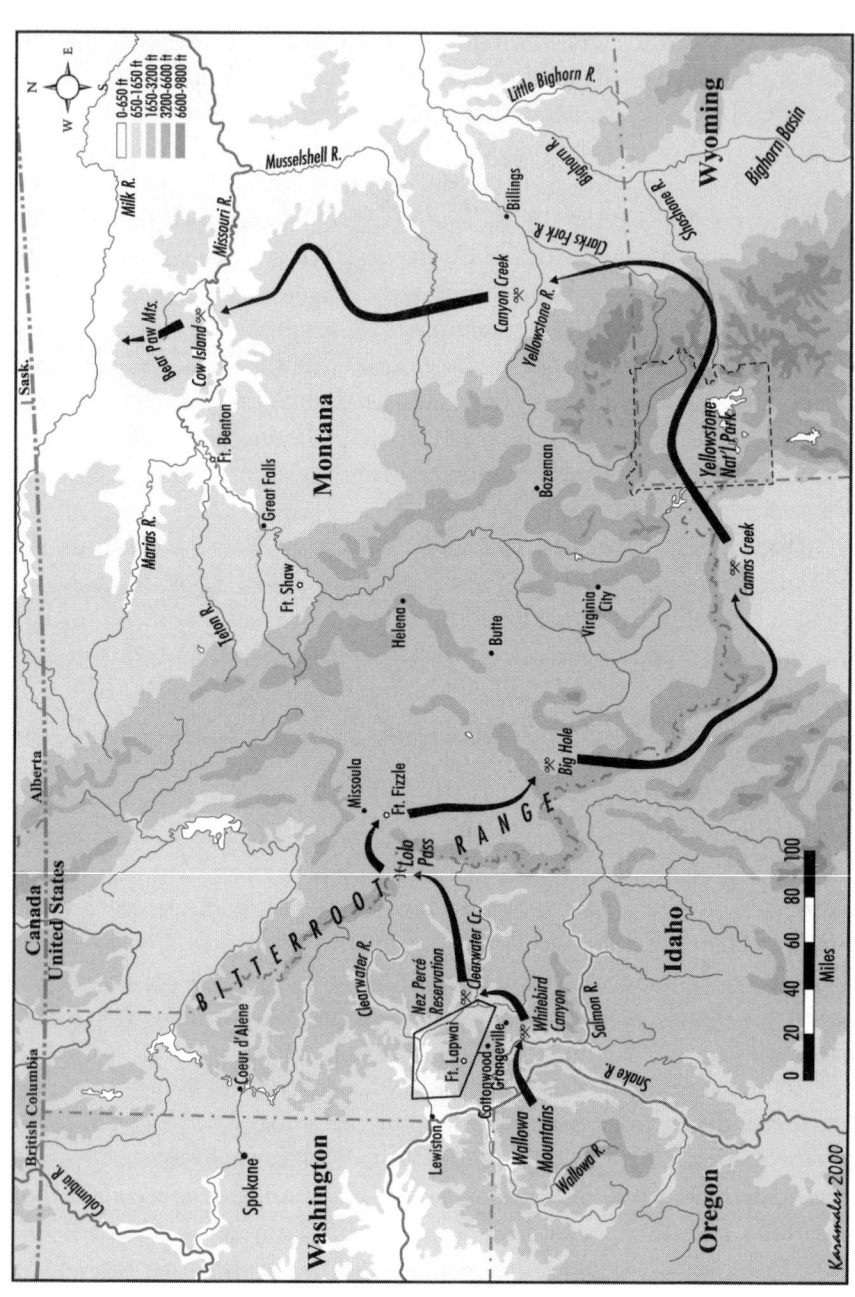

rounds did no damage but provided the Nez Perce with ample warning of the army's approach. They moved their pony herds into nearby gullies to keep them away from the shooting.

Howard came down from the bluff and began to approach the camp across flat ground; however, Toohoolhoolzote and 23 warriors ran up to a ridgeline and began firing at the soldiers, hitting several. More Indians crossed the river and took up firing positions to the army's left and right. Howard deployed his men in a wide semicircle. The soldiers tried to dig themselves into the ground with their bayonets. A newspaper reporter on the scene, Thomas Sutherland, wrote:

> Although we outnumbered the Indians we fought to a great disadvantage. The redskins were in a fortified canyon, shooting from the brow of a hill, through the grass, and from behind trees and rocks, while our men were obliged to approach them along an open and treeless prairie. At times a redskin would show his head, or jump up and down, throwing his arms about wildly, and then pitch himself like a dead man flat upon the grass, and these were the only chances our men had to fire.

In the morning a mule train came into camp and mounted Indians rushed it, killing two packers. A counterattack chased them off. In the middle of the afternoon, cavalrymen had to charge a ravine where warriors were so close they were picking off troopers with impunity. Three officers were badly wounded in the attack but the Indians were pushed back. During the confusion the artillery's howitzers and Gatling guns started firing into a company of soldiers by mistake, wounding a trooper in the thigh. At one point in the fight, a young Nez Perce rode his horse all the way across the soldiers' front, then turned and repeated the feat in the other direction. A trooper got him in the shoulder just before he was out of range.

In late afternoon one of the army's Nez Perce scouts suddenly jumped up and ran toward the Indian lines while everyone on both sides fired at him. He stripped off his army jacket, acquired a horse and then led a charge back against the soldiers. Ad Chapman, when he heard that soldiers on the firing line were running short of ammunition, mounted his horse and braved a hail of fire to bring up a cartridge box. At White Bird

Canyon, Chapman's failure to hold the cavalry's left was blamed for initiating the army debacle, and he evidently spent the rest of the Nez Perce War trying to redeem himself.

A big problem for the soldiers was that the only spring in the area was covered by Nez Perce sharpshooters, so the troops and their animals suffered badly from thirst. When darkness fell, the firing slackened off and some men were able to crawl over to get water. Fearing attack, the soldiers lay awake during the night, listening to the Indians across the river whooping, chanting and, at times, mournfully wailing.

The next day the long-range sniping began again, as warriors went back and forth from the camp to join the battle or to rest. At some times there were no more than 100 Indians manning the whole front, which covered two and a half miles. In the afternoon Howard ordered Captain Marcus Miller with a battalion of infantry to attack the Indian left. Before he could do so, a large pack train appeared, heading for the camp, and Miller was ordered to meet it and stand guard against Indian attacks. Just as the pack train reached the safety of the army's lines, Miller suddenly ordered a charge, catching the Indian left by surprise. Nez Perce warriors tried to flank Miller but were themselves outflanked by the rush of another company of soldiers. Miller began rolling up the Indian flank while the rest of the army cautiously advanced behind howitzer fire. Sutherland recalled watching the men

> stealthily crawling through the grass. Suddenly a voice called, "Cease firing with the howitzer." Then the stillness of the advance was broken by a single cheer, which, taken up, passed all down the lines, and the men sprang to their feet and rushed on. "To the river! To the river!" came the cry, and at the same moment a Gatling gun thundered past us, quickly followed by a howitzer. The gun by which I was standing rapidly limbered up and we all hurried to the bluffs.

The Nez Perce line collapsed and everyone ran to the village, which was just across the narrow Clearwater. With no time to pack up, the Indians had to leave behind most of their possessions. Sutherland said, "The hills on the opposite side of the Clearwater were swarming with flying Indians, stampeded ponies and frightened cattle." The infantry and artillery had trouble rushing across the river but fired at the village from the opposite side. Yellow Wolf, who had been on the Indian right and so was one of the last warriors to leave the field, arrived in the camp to find one of Chief Joseph's wives having trouble with her horse while her new-born baby lay on the ground nearby. Yellow Wolf helped calm the horse and then handed the woman her baby; together they rode through howitzer bursts to safety.

Captain Perry's cavalry were the first soldiers to cross the river, but there was no serious pursuit of the Indians. In the abandoned camp the troopers found a wide variety of booty, from feather headdresses to the contents of looted country stores. The soldiers had suffered 15 dead and 25 wounded. Thirteen Indians were thought to have been killed, though only 8 bodies were found. Some who were wounded might have died shortly after the fight.

Although the Battle of the Clearwater had the superficial appearance of an army triumph, Lieutenant Parnell, for one, termed it a "victory barren of results." The Nez Perce bands, now far less encumbered than before, were still intact and would need to be fought again. If the warriors had lost anything it was their overconfidence, and they would not again seek to take on hundreds of infantry and artillery in a set-piece battle. Howard's cavalry meanwhile, had still not inspired great respect from either the Nez Perce or the white infantry. When the Indians recrossed the river farther north and Captain Perry's horsemen approached, a rearguard of braves under Rainbow suddenly fired from behind cover, prompting the cavalry to dissolve in confusion. The soldiers brought up a Gatling gun to clear the far side, but were hesitant to force a crossing.

An interesting phenomenon of the Nez Perce War was that the whites—from individual soldiers to General Howard to the national press—thought that Joseph was the Nez Perce war leader. Just as the legend of Rommel, the "Desert Fox," was encouraged by the British in World War II, in part to cover the embarrassment of their defeats, the whites seemed to find comfort in the idea that their Nez Perce opponent was a military genius. Howard even termed Joseph the "Indian Napoleon." Six-year-old Maggie Manuel claimed for the rest of her life that it was Joseph she had seen stabbing her mother on the second night of the uprising.

Although Joseph had been the most impressive Indian leader in negotiations prior to the war, and was also head of the largest Non-Treaty band, he did not play a prominent role in the fighting. According to Indian survivors, Joseph was unhappy during the months of war and was usually assigned to take care of the women and children when the tribe was on the move. His younger brother Ollokot led the band's warriors in battle and soldiers who, as at the Clearwater, claimed to have seen Joseph dashing around giving orders, might have seen his brother. Of course, given the whites' misconception, Looking Glass, Rainbow, Five Wounds, Toohoolhoolzote or any other Indian warrior showing leadership in battle might also have been mistaken for Joseph.

According to a Nez Perce named Josiah Red Wolf, "Not only was Joseph hard to persuade to stay in the fight but he tried to drop out after the [Clearwater]." On July 15, Joseph sent a messenger to Howard to inquire about surrendering his band. Howard had been en route to Fort Lapwai at the time but backtracked six miles in order to meet the emissary. He said that he required unconditional surrender and that Joseph would be tried before a military court. During the meeting a warrior from the other side of the river fired at the men but missed. Joseph's messenger went over to the whites but Joseph and his people stayed with the hostiles.

That evening, the Nez Perce chiefs held a council to decide what to do next. Looking Glass emerged as the dominant chief in this meeting and won his argument to head through the Bitterroot Mountains to buffalo country in Montana. Nez Perce hunting parties had always enjoyed good relations with the Montana settlers and it was thought certain that the soldiers from Idaho wouldn't follow. Some of the chiefs hoped that if the bands could stay away for a year or more, the whites would forget about the whole thing. Another attraction of crossing the Rockies was that they might receive help from their old allies, the Flatheads and the Crows, and maybe even from their ancient enemies, the Sioux. If Montana turned out to be untenable they could always move north to the "Old Woman's Country" (Canada), where Sitting Bull resided.

To Joseph, who had not wanted war in the first place, the decision to become nomads in Montana was additional bad news. The question attributed to him, "What are we fighting for if not our country?" may not have been posed at the council, but it did summarize his feelings. By now, however, Looking Glass was in control, backed by the young warriors, and Joseph could probably not have defected at this stage if he had wanted to.

A band of 35 Indians under Red Heart did surrender to Howard at this time. They had recently arrived from Montana and had no wish to join the war. Howard had their hair cut off, marched them sixty miles on foot to Fort Lapwai, and from there to Fort Vancouver, where they were held in custody for a year. This treatment did little to convince the remaining hostiles to come in.

On the 16th, the great retreat began, as the Nez Perce started over the rough Lolo Pass through the Bitterroot Range to Montana. Howard's chief of staff, Major Mason, termed the pass a "narrow defile, densely wooded and almost impassable with undergrowth." An advance party of the army, consisting of five friendly Nez Perce scouts, civilian volunteers under Ad Chapman, and E Company of the 1st Cavalry followed. The Indian rearguard under Rainbow, however, was waiting under cover on both sides of the pass. The ambush could have devastated the column, but Rainbow lost his temper at the sight of the Treaty scouts and fired, hitting one in the shoulder. Another was wounded and a third was killed, gratuitously shot a number of times. The other two Nez Perce scouts were allowed to leave while the whites backed out of the pass.

General Oliver Howard was now satisfied that he had vanquished the hostile Nez Perce and brought peace to Idaho Territory. The Bitterroot Range comprised the border between the Military Departments of the Columbia and the Dakota, so the Indians would henceforth be Montana's problem. In the next few days, however, Howard received three surprises. After Mason had failed to pursue the bands on the Lolo Trail, 40 warriors slipped back to Idaho to raise havoc and steal hundreds of horses. These raids renewed fears of the possibility that Nez Perce success could prompt an uprising of Dreamers in other Northwest tribes. The second surprise was that after a visit to Fort Lapwai, Howard learned he was under severe criticism in the press for being slow and dilatory, and for not having defeated the bands. The third was that General William Tecumseh Sherman, head of the War Department, ordered Howard to pursue the hostiles regardless of departmental lines. When Howard inquired how his men, who had already been in the field for a month, were to be supplied outside his department, Sherman answered that they should sustain themselves like the

Indians. Howard was ordered to continue his pursuit of the Nez Perce wherever they went and until they were caught.

It is worth noting that while U.S. cavalrymen each had one mount, the Indians, like the steppe warriors of Asia in previous centuries, considered additional horses to be part of their equipment. This allowed them to outrun any cavalry unit, not to mention formations of infantry. The 800 Nez Perce crossing the Lolo Pass were accompanied by a pony herd of about 2,000 animals. And, as this large column passed gradually through the Bitterroots, on the Montana side everyone was in an uproar.

The nearest town on the eastern side of the pass was Missoula, where a fort had been authorized the month before. Captain Charles Rawn of the 7th Infantry, who had only 4 officers and 30 men, had received a message from Howard to prevent the Indians from coming out of the pass. The call went out for volunteers, and at a narrow spot in the valley that led from the Lolo Pass into Montana, Rawn constructed a long timber and sod breastwork.

On July 28, two days before Howard entered the pass, the Indians were already coming out on the Montana side, only to find Rawn's force waiting for them behind the new fort. Looking Glass and other chiefs approached the defenders and explained that the Nez Perce had no quarrel with the Montanans; they had left the fighting behind them in Idaho. Rawn had his orders, however, and refused to let the Indians through unless they gave up their guns and ammunition. Looking Glass had another, private, meeting with Rawn at which he stressed that the Indians had no quarrel on this side of the Bitterroots. Meanwhile, civilian volunteers had been walking off, in groups of up to a dozen, since they couldn't see the point of fighting the army's battles against Indians who didn't even want to fight. The next morning, Rawn and his remaining men could only look on helplessly as the entire Nez Perce column passed their fortification along the low hills to the north. Looking Glass had decided not to force his way through but just to go around Rawn's redoubt, which was henceforth known as Fort Fizzle.

Once the Indians were on flat country they headed south and, true to their word, didn't bother anybody. They arrived at the town of Stevensville and bought supplies from the local stores and traded with settlers. The Nez Perce had a lot of gold dust and money by now, so were able to pay top dollar. It was agreed not to sell whiskey to the Indians and

Looking Glass roamed the town to make sure everyone behaved. A number of Flatheads were also on hand to keep an eye on the Nez Perce. When one warrior acquired liquor and got drunk, Looking Glass dragged him back to the camp and put him under guard. The only violent incident was when some of Toohoolhoolzote's men, the worst roughnecks in the tribe, ransacked the home of a man named Underwood. Looking Glass made the warriors mark seven horses with Underwood's brand and leave them at the ranch by way of compensation. White settlers had thrown up more stockades in the area, and amused warriors approached to look at them and chat with the occupants. One of the stockades was called Fort Run and another was called Fort Skedaddle.

Although the Montana settlers have been exonerated by history for the Fort Fizzle affair and for their subsequent trading with the Nez Perce, one man who thought their behavior was disgraceful was Major Charles Gibbon of the 7th Infantry based at Fort Shaw. While the Indians leisurely strolled across southwestern Montana, Gibbon put together as large a force as he could—169 men—and began marching toward the hostiles.

After covering 100 miles at 12 miles a day, the Indians had crossed a ridge and come down to a stream-fed plain called the Big Hole. On this familiar ground, a sort of traditional way station between the buffalo country and the Rockies, the Nez Perce intended to rest for a few days. By this time, a few warriors were having bad dreams that included doom and sudden death. Five Wounds asked Looking Glass to send a scouting party back to see if they were being followed. Looking Glass, who knew Howard could not possibly be nearby, and who also feared what young warriors might do if they were out of his sight, refused. The Indians didn't know that Gibbon and his 7th Infantry had arrived from northern Montana, been buttressed by volunteers, and had picked up the Nez Perce trail.

On August 8, men of Gibbon's advance guard climbed a tree and saw the entire Indian camp below them on the Big Hole. Gibbon hastened his command forward and by dawn on the 9th it was spread out at the foot of the hills, several hundred yards from the Indian camp. As the soldiers quietly advanced, their orders were to wait for a gunshot to announce that the battle had begun. An Indian came out of his tepee in the early morning, mounted a pony and rode straight into the soldiers' skirmish line. Yellow Wolf said this was an older man with poor eyesight. He was shot by four troopers at close range, and the battle was on.

The soldiers ran toward the camp in the half-light, firing into the tepees. The Nez Perce were caught completely by surprise. Warriors, women, children, horses and soldiers mixed together in a confused melee, Gibbon's men shooting at anything that moved or, inside the tepees, at anyone trying to hide. The Indians didn't collapse, however. Instead of running in the opposite direction, many of them ran toward the wooded area of the streambed from where the soldiers had come. Warriors found their arms and took up position behind the nearest cover. Soldiers later said that the Indian women also did their share. According to a Nez Perce interpreter, Duncan McDonald:

> In a fight between an officer and a warrior, the warrior was shot down dead. The warrior's sister was standing beside him as he fell, and as he lay there his six-shooter lay by his side. The woman, seeing her brother dying, seized the six-shooter, leveled it at the officer, fired, shot him through the head and killed him.

This plus other accounts of women and young boys fighting did little to justify the brutality of the army's attack. McDonald said, "Many women and children were killed before getting out of their beds. In one lodge there were five children. One soldier went into it and killed every one of them."

Despite achieving surprise, Gibbon's attack soon began to founder. Looking Glass on one side and White Bird on the other were rallying their men, putting the troopers under a crossfire. One concentrated counterattack could have broken the soldiers' line. Gibbon, who had been shot in the calf, called for a retreat back to a timbered spot on the hillside. As the soldiers withdrew, a brave named Grizzly Bear Youth pursued, but a large citizen volunteer suddenly turned around. Both men swung their rifles at the other as clubs, the Indian suffering a cut on the head while the volunteer was knocked down. Grizzly Bear Youth jumped on him, but the volunteer got the better of the match and started choking the Indian. Just then, another warrior came up, stuck his gun into the volunteer's side and killed him. Since Grizzly Bear Youth had been holding on to the man at the time, he suffered a broken arm from the shot.

When the soldiers crossed back across the stream they saw a number of Indian women and children hiding in the water. Gibbon said that several women held up their babies to him as a plea for protection. A civilian volunteer, Tom Sherrill, said that he saw a young white girl in the stream, his statement coinciding with that of some settlers who reported seeing a fair-haired young woman with the Nez Perce. A man later claimed he found some long blond hair in a grave on the battlefield, giving rise to the notion that Jennet Manuel, whose body was never identified, had been taken prisoner by the Indians only to die at the Big Hole.

During the army retreat, Gibbon was able to quell some urges to panic among his men and get everyone back into the timber. There was then a lull in the fighting while the soldiers listened to the Indians as they returned to their camp.

> Few of us will soon forget the wail of mingled grief, rage and sorrow which came from the camp four or five hundred yards from us when the Indians returned to it and recognized their slaughtered warriors, women and children. Above this wail of horror we could hear the passionate appeal of the leaders urging their followers to fight, and the war whoops in answer which boded us no good.

The attacking party was now itself under attack; the soldiers hastily dug rifle pits and tried to set up defensible positions behind logs. Nez Perce sharpshooters set up in sniper nests around the command. One rifle pit of four soldiers was wiped out by an Indian shooting from a tree. Another warrior, firing from behind a pile of rocks, scored five hits before a trooper got him in the eye with a carom shot. In late afternoon Gibbon's command was terrified when a brush fire came roaring toward them. Gibbon yelled, "If the worst comes, my men, if this fire reaches us, we will charge through it, meet the redskins in the open ground, and send them to a hotter place than they have prepared for us!" At the last second, the wind-driven fire petered out just short of the soldiers' position.

That night the soldiers took their rations from a dead horse. The Nez Perce had meanwhile buried their dead and the tribe was leaving the battlefield. Ollokot led the warriors who stayed behind to pin the soldiers down. The Indians might have had more ambitious plans for the remnants of Gibbon's command but they had learned General Howard was on the way. On the evening of August 10, Ollokot's warriors fired two volleys and then left to join the retreating column.

When Howard came up to the position on August 11 he said it looked like a hospital. Of Gibbon's 191 men, 29 were dead and 40 wounded, two of whom would die later. The Indian losses were even more devastating: 83 dead, at least 30 of whom were fighting men. Among the slain Nez Perce was Rainbow, considered to be the tribe's mightiest warrior. Rainbow had always said that his "medicine" made him invulnerable in any battle that began in daylight, but for a battle that began in darkness he was in trouble. His best friend, Paphlot, was also killed, as well as Five Wounds, who had urged Looking Glass to send out scouts the day before. Wahlitits and Sarpis Ilpip, who had instigated the war, also died in the fight. (The third of the original trio was a boy of 17, who continued to fight.) Wahlitits died along with his pregnant wife near the creek. When soldiers approached, he fired at them while telling his wife to flee. He was killed by a shot to the face, however, and his wife came back and grabbed his rifle, shooting one of the oncoming troopers before being riddled with bullets herself. Her body was found stretched across her husband's.

Although the press in Montana was initially ecstatic about the battle—one headline referring to warrior, but not civilian, deaths as "Thirty More Good Indians"—cooler heads wondered whether it had been an army victory or defeat. Howard confined himself to speculating that if Gibbon had had 100 more men the war might have ended. One man's bravery was another man's "glory seeking," and though Gibbon's attack on the village was brave, it cost him over a third of his men as casualties. Nez Perce losses were primarily noncombatants, while the tribe's retreat had not been halted. Further souring the outcome, when Howard's Bannock scouts arrived they dug up the hastily dug Nez Perce graves and scalped and mutilated the bodies. White souvenir hunters also descended on the field, destroying any last vestige of dignity the battle might have represented for those who held the field.

As Gibbon's wounded were put into wagons and Howard waited for the rest of his command to catch up, the Indians proceeded south. The relaxed, carefree trek of the Nez Perce through southern Montana took on a more bloodthirsty aspect as the tribe meandered back across the border into southern Idaho. Everyone had lost friends or family members in the Big Hole fight. After the initial uprising, once their chiefs had taken command of the young men, the Nez Perce had passed up numerous chances to murder noncombatants. When the Indians had fought Howard's army in Idaho they had done so under more or less normal rules of warfare. Gibbon's indiscriminate attack on sleeping women and children in Montana came as a shock.

As the Nez Perce moved south from the Big Hole, white settlers congregated in towns and built up fortifications for defense. Some, however, refused to leave their homesteads. At the Montague-Winters ranch, four men were killed by a party of warriors. Another deserted ranch was ransacked and all the horses stolen. A settler commented, "They take no guns except breech-loaders, the muzzle-loaders they either destroy or leave." The next day a man named Alexander Cooper was accosted by some Nez Perce who claimed to be friendly. He was told to hand over his gun, which he did, but when his three friends refused to do likewise Cooper was killed.

On August 13, Looking Glass and White Bird held a conference with Tandoy, chief of the northern Shoshone. Just like the Flatheads, however, the Shoshone were unwilling to join the Nez Perce war against the whites. Two days later, Nez Perce warriors came upon a commercial wagon train that was accompanied by half a dozen whites and two Chinese. Unfortunately, along with flour and other goods, the wagon train contained a large quantity of liquor, which the Indians broke into. That night, the Chinese were ordered to scamper around the Indian camp on all fours. A man named Lyons took advantage of the debauchery to slip away through the woods. He was found a week later, starving and dehydrated. The other five whites were escorted out of camp and killed. One of them, Daniel Coombs, was found shot repeatedly and knifed, his hand still clinging to the heavy handle of a bullwhip that was covered with hair and blood. After the murders, the Nez Perce chiefs ordered the remaining liquor to be dumped on the ground, prompting a melee, including knife fights, to break out in the camp. A man named Ketalkpoosmin was shot in the back by a fellow warrior while he was helping to get rid of the liquor.

Howard, meanwhile, had been reinforced by volunteers from Helena, Butte and Virginia City, and was gradually catching up to the Nez Perce. On August 19, while Howard was camped in a place called the Camas Meadow, his Bannock scouts reported that the fugitives were only 15 miles away. Nez Perce scouts had likewise reported the enemy's proximity and the chiefs were influenced by a dream one of their warriors had had in which the Nez Perce would double back and steal the army's horses. In the dark early morning hours of the 20th, Ollokot, Looking Glass and Toohoolhoolzote led a party of about 30 braves up to Howard's camp. While most of the warriors covered the tents, a few braves crept among the animals, cutting loose their ropes. Suddenly a shot rang out and the whole camp erupted. Some Nez Perce opened fire while others rode among the herd, chasing the animals away from the camp. A number of civilian volunteers panicked and ran among the tents of the soldiers. No one could see what to fire at, but it was soon apparent the Indians had gone, along with several hundred animals. When daylight broke, Ollokot and his men were annoyed to find that instead of capturing the army's horses they had captured the mules of the pack train. And now three companies of cavalry were on their heels.

Lieutenant Norwood's L Company of the 2nd Cavalry was in the center, and when the troopers ran into a line of Nez Perce deployed along a lava ridge, they soon found themselves alone. Their flanking companies ran in the face of an Indian counterattack and Norwood's men dug in on a hill. For four hours they held their ground against Nez Perce sharpshooters until Howard came to their rescue with the main body of infantry and volunteers. The cavalry had lost one man killed and eight wounded, two of them mortally. In the entire engagement the Indians had suffered only a few minor wounds. When word of Camas Meadow got out to the national press, Howard found himself under additional vicious attacks, while the legend of Joseph, the perceived war leader of the Nez Perce, continued to grow.

The real leader of the Nez Perce at this time was no longer Looking Glass, whose reputation had diminished after the surprise at the Big Hole, the elderly White Bird, the radical Dreamer Toohoolhoolzote, or Joseph, who continued to take care of the women and children, but a man who had joined the tribe after it crossed the Bitterroots: Lean Elk, or, as the whites called him, Poker Joe. A half-breed who spoke English, Poker Joe was a Nez Perce who had resided on the Montana side of the Rockies, but who joined the exodus with his six lodges when Looking Glass led the rest of the Non-Treaty Nez Perce over from Idaho. Poker Joe was not only a skilled fighter but had an intimate familiarity with the buffalo country and keen knowledge of the whites. Though not undisputed leader of the bands, Poker Joe became the supervisor of the retreat and was most responsible for keeping everyone together inside hostile, and confusing, territory.

In 1872, Yellowstone National Park had been created at the northwestern corner of Wyoming. Though still bereft of roads and facilities, the park was already being traversed by tourists, its beautiful, rustic scenery meant to preserve a slice of the primeval American West. On August 23, 1877, the Nez Perce headed into the park, providing more realism than many of the tourists wished for.

A vacationing party of seven men and two women woke up on the morning of the 24th to find three Nez Perce warriors on horseback calmly looking at them. One of the Indians, Yellow Wolf, said he had been inclined to kill them all but then a man came up and shook his hand, which softened his feelings. The whites were nevertheless taken back to the Nez Perce camp and interviewed by Poker Joe. After being divested of their guns and supplies and being forced to trade their horses for worn-out ponies, they were allowed to leave. Suddenly, outside camp, they were surrounded by a party of braves.

Young Umtillilpcown (Swan Necklace), the third member of the original trio that had started the war, rode up to George Cowan and shot him in the thigh. Cowan fell from his horse and rolled down a hill, his wife Emma and her younger sister following to stand by him. Another man, Al Oldham, was shot in the side of the face, while a third, Emma's brother Frank Carpenter, hurriedly made the sign of the cross as an Indian took aim. The warrior held his fire. Down the hill, Emma Cowan was hysterically trying to cover her husband as Indians approached, but she was pushed away as a warrior drew a pistol and shot him in the forehead. Cowan slumped and everyone left him for dead, Emma's last glimpse being of an Indian bashing him in the head with a rock. At that point, Poker Joe came riding up and the shooting ceased. Emma and her brother and little sister were taken back to the Indian camp.

Emma spent the night around Joseph's campfire and later wrote, "The Chief sat by the fire, somber and silent, foreseeing in his gloomy meditations possibly the unhappy ending of his campaign. The 'noble red man' we read of was more nearly impersonated in this Indian than in any I have ever met. Grave and dignified, he looked a chief." The problem was that the Nez Perce leaders continued to have difficulty controlling their young warriors, who, especially since the Big Hole, were out for white blood.

George Cowan, meanwhile, woke up late in the afternoon. Feeling the bloody back of his head where he had been hit with a rock, he first assumed that the bullet in his forehead had gone all the way through his skull. As it turned out, the pistol shot had been undercharged, or perhaps the powder was damp, so the bullet had lodged in his forehead. He tried to move but then saw an Indian watching him from 25 feet away. The warrior calmly got to one knee, took aim and shot him in the right hip. Fortunately his latest antagonist did not follow up. Over the next four days, Cowan, with a badly wounded head, and bullets in his left thigh and right hip, crawled to safety, hiding from both Nez Perce warriors and the army's Bannock scouts. Though he was unaware the Bannocks were working for the army, it's probably just as well he avoided them. Within a year the Bannocks would be embroiled in their own war with the whites and, unlike the Nez Perce, they placed great pride in taking scalps.

Emma and George Cowan were reunited a month later and both lived to a ripe old age. The Nez Perce had meanwhile picked up an elderly prospector named Shively, who spent ten days with the tribe, his assignment to help Poker Joe navigate the bewildering terrain of the park. On August 25, the Indians picked up a soldier who had been released from duty and was on his way home. This man pointed them to a nearby party of ten tourists from Helena, whom the braves found the next day.

Two of the vacationers, Weikert and Wilkie, rode out to explore the area of their camp but they ran into Indians and Weikert was shot in the shoulder. Shortly after, shots were fired at the main camp, but the men thought it was Weikert and Wilkie fooling around. When they realized they were under Indian attack everyone scattered. Two men named Stewart and Kenck ran off in one direction, Kenck in the lead and Stewart already creased in the leg. Stewart recalled:

> I ran about 50 feet farther when I received the wound in my hip that dropped me. In a very few minutes two Indians ran past me after Kenck; soon after, I heard two shots fired and Kenck exclaim, "I'm murdered." One of the Indians then came back to me, keeping his gun pointed in my face. . . I instinctively threw up my hands and begged for my life. Setting his gun down on its

butt, he asked me if I had any money. I said, "I have a little." He
. . . rolled me over, put his hand in my righthand pocket, where
he found $263 and a silver watch. By this time the other Indian
had come down and they opened my roll of money and had a
great laugh over it, seeming very much elated at getting so much.
They then examined my wound which, at that time, was bleeding profusely, told me they would not kill me, and walked off,
leaving me lying in full view of our camp, which I saw the Indians
plunder.

The other Helena tourists got away; however, a music teacher named
Richard Dietrich refused to leave the area until he was sure of young
Joseph Roberts' safety. Dietrich had been responsible for convincing
Roberts' parents to let him come along on the trip and had vouched for
his safety. Roberts had escaped in the first few minutes, while Dietrich
found his way to a nearby house to wait for his charge. Unfortunately,
Nez Perce came by the house and the music teacher's body was later found
lying in the doorway, a bullet in the heart.

It appears that for some time in Yellowstone Park, the Nez Perce got
lost. Their captive prospector, Shively, said this occurred when Poker Joe
decided to rely on the advice of a local Shoshone chief, whereupon
Shively kept mum while the tribe started going in a circle. Meanwhile, the
indefatigable Oliver Howard was slowly catching up while other U.S.
Army units converged on the park's exits. From the south a force under
Colonel C.C. Gilbert was approaching, and from the southeast Wesley
Merritt's 5th Cavalry was in place. In the northeast, meanwhile, the
famous 7th Cavalry had arrived.

Custer's command had been rebuilt since the disaster of the previous
summer and was now commanded in the field by Colonel Samuel
Sturgis, who had lost a son at the Little Big Horn. Helped by Crow
scouts, Sturgis placed himself next to the Heart Mountain astride the
Clarks Fork River, a spot that overlooked the most logical route for the
Nez Perce to emerge from the rugged Yellowstone hills onto the plains.
Sturgis' position overlooked miles of territory on either side and particu-
larly the canyon that extended down from the park. What happened next
was either bad judgment on the part of Sturgis or the most brilliant Nez
Perce maneuver of the war.

The park had already been getting crowded, white and Indian scouts from four separate formations buzzing around the Nez Perce column of 700 men, women and children. Every time Howard tried to send a courier to Sturgis to warn him of the Indians' approach, however, the man was killed by the Nez Perce. Sturgis, in turn, tried to get two men to Howard but they were caught and killed 16 miles from his camp. The eastern part of Yellowstone also contained a number of prospectors. While survivors of the tourist parties went on to immortalize their experiences, no records have been kept of individual panners who fell into the path of Toohoolhoolzote's or Ollokot's roaming "flankers."

Sturgis and the 7th Cavalry, spoiling for a fight, were afraid that the Nez Perce would get around them. They had arrived at their station a few days early and, as time passed without sighting the hostiles, their tension grew. Finally, Crow scouts reported that the Indians had been sighted to the south near a river called the Stinking Water (later renamed the Shoshone River). The large dust clouds indicated that it was the entire Nez Perce column. Without wasting a moment, Sturgis mounted up his 400 men and rode toward the Stinking Water. The next morning the Nez Perce came pouring out of Yellowstone at exactly the spot Sturgis had vacated. It was a small party of Nez Perce braves with extra horses that had made the feint toward the Stinking Water, trying to stir up as much dust as they could. They then backtracked and rejoined the tribe. White newspapers had already been admiring the brilliant trap laid for the hostiles; however, the Nez Perce, against all odds, had made it onto the open plains.

One can't help wonder whether the 7th Cavalry's Crow scouts had proven completely reliable in causing Sturgis to abandon the one position that could have trapped the Nez Perce once and for all. The Crows had served the army well against their traditional enemies, the Sioux, and some credit them for holding off Crazy Horse at the Battle of the Rosebud, saving General George Crook from an even worse defeat than he suffered prior to the Custer massacre. The Crows, however, had always been allies of the Nez Perce. Once out of Yellowstone, Looking Glass went to meet with their chiefs but, just as with the Flatheads and Shoshone, could enlist no support. The best he could get was a promise from one chief that the Crows wouldn't actually fight the Nez Perce but only seek to steal some horses. Nevertheless, to the whites, Indian loyalty was a tenuous factor in the campaign. Howard's Bannocks had already mutinied and tried to steal some army horses. The Nez Perce themselves had served as army scouts during the Rogue River uprising of the Columbia River tribes in the 1850s. Even though the Crows worked for the army at this stage of the Nez Perce War, the "miraculous" escape of the fugitives from Yellowstone Park, insofar as it depended upon Crow scouts, raises an eyebrow.

On the open plains, the young Nez Perce warriors found their legs and rode out on either side of the column, burning homesteads and killing isolated whites. At one point they found a stagecoach and several braves hopped aboard. According to one account, "an Indian mounted each horse and with one on the box, off they went tearing and whooping like lunatics." The stagecoach's lone passenger had been a female "entertainer" named Fanny Clark, who was successfully able to hide before the Indians arrived.

When Sturgis realized the Nez Perce were behind him, he immediately pointed the 7th Cavalry north and set off in pursuit. After days of forced marches he caught up to the unwieldy column, spotting the hostiles entering a wide valley along a stream called Canyon Creek. The 7th Cavalry charged against the Indian rearguard led by Looking Glass. Sturgis, whose determination to win a victory may have exceeded his tactical expertise, ordered the regiment to dismount once they encountered resistance—a decision that frustrated many of his troopers. Captain Benteen of Custer fame, commanding H Company, launched two assaults on the Nez Perce rearguard, without success. On the Indian side, Yellow Wolf was mystified that the soldiers were so cautious. "We did not line up like soldiers," he said. "We went by ones, just here and there entering the canyon." He claimed that once the Nez Perce had gotten through, "Only one warrior, Teeto Hoonod, was there doing the fighting. His horse hidden, he was behind the rocks holding a line of dismounted soldiers back. He was shooting regularly, not too fast."

When darkness fell, Sturgis considered it prudent to halt his pursuit. He had suffered three dead and 11 wounded, while the Nez Perce may have sustained a few wounds. The next day General Howard, having emerged from Yellowstone Park, put himself at the head of 50 cavalrymen and arrived on the scene. He recorded:

It was the most horrible of places—sagebrush and dirt, and only alkaline water, and very little of that! Dead horses were strewn about and other relics of the battlefield! A few wounded men and the dead were there. To all this admixture of disagreeable things was added a cold, raw wind, that, unobstructed, swept over the country. Surely if anything was needed to make us hate war such after-battle scenes come well in play.

Sturgis continued his pursuit of the Nez Perce, but after a few days reported that he was unable to continue. "I find it impossible for my command to gain upon them, and their direction is taking me further and further from supplies. I have . . . reluctantly determined to abandon a hopeless pursuit before my horses are completely destroyed or placed beyond recuperation."

The chase would once more be left to General Howard, who had been on the Nez Perce trail since the first week of the war. By now the tribe was in central Montana, approaching the Missouri River, heading due north for Canada. Though Howard, together with the freshest troops of Sturgis' command, doggedly persisted in his task, there was no hope that he could overtake the Indians. There was one other army unit, however, that might possibly have a chance to cut off the retreat. Messages were dispatched to Colonel Nelson Miles at his encampment on the Tongue River, southeast of the fleeing bands.

Miles, who had been a major general at age 27 in the Civil War, was an ambitious officer in the Custer mold, and had picked up where the latter had left off in the war against the Sioux. On the first day of 1877, Miles had found Crazy Horse on the Powder River and sent him reeling back in a series of skirmishes. He received Howard's message about the Nez Perce on September 17, and, even though he was 170 miles away from an intercept and the hostiles were an easy march from Canada, by the next day his command was on the move. He had three troops each of the 2nd and 7th Cavalry, six troops of the 5th Infantry (mounted on captured Indian horses), a Hotchkiss gun and a Napoleon cannon. He also had 30 Cheyenne scouts, along with some Sioux, bringing his total force to around 400 men.

The Nez Perce, meanwhile, had outdistanced Sturgis and Howard only to be harrassed by Crows trying to steal their horses. In fact, everyone was having problems with the Crows, who were turning out in increasing numbers. After Canyon Creek, army scout A.J. Fisher wrote: "We all went back to where the soldiers were camped and there learned that the Crows had stolen my pack animals, clothing, bedding, etc., as well as a number of pack and saddle animals from others. The Crows took no part in the fight but staid in the rear and stole everything they could get their hands on."

On September 23 the Nez Perce crossed the Missouri River at Cow Island, a traditional ford and also the last stop for steamboats during the low-water periods. When the Nez Perce arrived, there were 50 tons of supplies on the bank meant to be shipped on to settlements and outposts in western Montana by wagon train. The supplies were guarded by Sergeant William Moelchart and a dozen soldiers plus four civilians. Two English-speaking Nez Perce, one of whom must have been Poker Joe, approached Moelchart and asked for food, only to be refused. A while later warriors returned, holding up gold dust and money in order to purchase the goods. Surprisingly, in view of army pay scales, the stubborn Moelchart turned them down again, though he did put a side of bacon into a bag with some hardtack and gave it to the warriors. Unfortunately, this was not enough for hundreds of men, women and children.

That evening the Indians attacked. Moelchart later claimed he held off seven charges, but Yellow Wolf said the warriors fired only enough to keep the soldiers pinned down. Meanwhile, the rest of the tribe approached the giant pile of food under the cover of a ravine and took what they wanted, setting the rest on fire as they left. The soldiers lost a man killed, and two of the civilians were wounded.

The next day, warriors found a wagon train full of goods upstream. The Indians professed friendship and a teamster recalled, "One young buck who staid around the wagons asked us repeatedly, in good English, when we were going to have dinner." The Nez Perce went away after dark but the next morning were back. A small mounted force was approaching from Fort Benton in the west, and while warriors held them off at long range, others ransacked and burned the wagons. One of the teamsters and one cavalry volunteer were killed.

The key factor that caused the Nez Perce to be milling around at this time instead of making haste for the Canadian border was that Looking Glass had supplanted Poker Joe as the "trail boss" of the retreat. Now that a guide was no longer needed, Looking Glass objected to the strenuous marches. The old people couldn't keep up and the children were becoming alarmingly frail. There was no need to run the tribe ragged so that just as they reached Sitting Bull they would appear like a mob of beggars. Looking Glass, whose reputation in battle had been won against the Sioux, may also have been concerned as to exactly how the tribe's traditional enemies should be approached. On September 29 the Nez Perce set up camp along a tributary of the Milk River just north of the Bear Paw Mountains. Scouts were continually searching to the south to make sure Howard was not nearby. Looking Glass could not have known that Howard had been intentionally slowing his marches, hoping that the Nez Perce would follow suit.

Nelson Miles, who had hoped to catch the Nez Perce before they crossed the Missouri, had the good fortune to encounter the season's last steamboat, the *Fontenelle,* during his march along the river's southern bank. He used the boat to transfer the 2nd Cavalry to the northern side and was about to continue on when a small boat containing the wounded from Cow Island came downstream. The men informed him that the Nez Perce had already crossed the river. The *Fontenelle* was by now a mile downstream but Miles ordered his Napoleon cannon to fire a shot beyond its bow. The steamboat got the message and headed back to transfer Miles' entire command to the north side of the river. By the evening of the 29th, the army's Cheyenne scouts had found the Nez Perce trail.

At two in the morning on September 30 the cavalry mounted up and advanced on the Bear Paw camp. By eight o'clock they were seven miles away, approaching from due south. The camp was in a semi-basin, lined with jagged ridges from the north slanting down across the east side and bending to the south. Across a winding creek, the west side was open prairie, where lay the tribe's pony herd. The camp itself was cut with numerous small depressions and ravines. Some Nez Perce were already packing their ponies for the day's march when two braves who had seen the army's scouts came tearing into camp. Looking Glass tried to keep everyone calm, saying it was a false alarm, there was plenty of time.

When the cavalry was within two miles of the camp, they broke into a trot and then a gallop. Miles recalled, "This gallop forward, preceding the charge, was one of the most brilliant and inspiring sights I ever witnessed on any field." To the Nez Perce the beating of 1,600 hoofs on the cold, hard ground sounded like a buffalo stampede. But it soon became clear that the Indians' worst nightmare had arrived. A scout rode to high ground, waved a blanket, fired his rifle and yelled, "Enemies right on us!"

Miles ordered the three troops of his 2nd Cavalry and his Cheyenne scouts to flank the Nez Perce right in order to seize the pony herd, while the 7th Cavalry was ordered to attack head-on. Yellow Wolf said, "Hundreds of soldiers were charging in two wide, circling wings." While some warriors hurried to get women and children on horses and to safety, others rushed to the ridges to take up firing positions in defense of the camp.

West of the creek, the Cheyenne and 2nd Cavalry rode into the Nez Perce pony herd and stampeded it. There was a wild melee as defenders fired back into the swarm of attackers while women and children tried to reach their mounts. L.V. McWhorter, who interviewed Nez Perce survivors, recorded two recollections centered on a conspicuous Cheyenne. The Nez Perce Shot in Head told him, "A strange Indian chief wearing a great-tailed war bonnet, was pursuing a woman on a cream-colored horse. . . . I heard her begging for her life. They passed over the hill out of sight and I know not what there happened." Ealahweeman, who was a young boy at the time, related to McWhorter:

My father told me to run out from there; to skip for my life. . . . Everybody caught whatever horse they could. . . . I ran my horse about half mile when I thought of my little brother, that I had left him. I find him, and getting him up behind me, we go. The Cheyennes and soldiers are shooting at us as we pass.

One woman is ahead of us. I saw her shot and fall from her horse; the work of a bonneted Cheyenne on a spotted horse.

Our warriors are passing shots with the cavalrymen. They are close together, mixing up. Soldiers continue sending shots at us, but they cannot stop us. My little brother holding right to me, has one braid of hair shot off close to his ear. Two soldiers pursue us but are driven back before they catch us.

The gallant Ollokot, who led the warriors of Joseph's band, was killed during this stage. Joseph was with the herd at the start of the battle along with his 12-year-old daughter. He gave her a rope and told her to catch a horse and flee, while he ran back to the rest of his family. "I dashed unarmed through the line of soldiers. It seemed to me that there were guns on every side, before and behind me. My clothes were cut to pieces and my horse was wounded, but I was not hurt. As I reached the door of my lodge, my wife handed me my rifle, saying, 'Here's your gun. Fight!'"

About 200 Nez Perce on horseback streamed north to safety. G Troop of the 2nd Cavalry pursued but 50 Nez Perce warriors turned to counterattack, forcing the troopers back. At the northern edge of the camp Toohoolhoolzote led a group of warriors onto an outcrop of rocks from where he thought they could fire down at the soldiers. Unfortunately he was assailed by the 2nd Cavalry from all sides and shot dead along with five of his men, three others wounded. Poker Joe was killed sometime during the morning because in the confusion a Nez Perce mistook him for a Cheyenne.

From the south, the 7th Cavalry had gotten within 200 yards of the camp when their charge was shattered by a wall of fire from warriors along the ridgeline. So many horses and riders went down that all three companies halted; they then veered right to try the camp from the east. Captain Owen Hale's K Troop was in the lead and had to dismount to advance up the ridge. They had just reached the top when they were surrounded by warriors counterattacking through ravines. After close-quarters fighting, K Troop pulled back, leaving its dead and wounded near the top of the ridge.

Riding to support Hale, Captain Ed Godfrey of D Troop saw a Nez Perce taking aim and then found himself tumbling to the ground in a somersault as his horse dropped dead. The warrior was about to shoot again when trumpeter Thomas Herwood dashed up, only to take a shot in the chest. Godfrey found the horse of a dead sergeant, but moments later was back on the ground, shot in the side. "I thought it singular," he said, "if I was wounded, that I didn't bleed. In order to investigate I loosened my belt and the instant I did I felt the warm blood running down my body. So I 'cinched' up again as quick as I could."

Owen Hale was kneeling behind his troop's firing line reloading his pistol when a bullet tore through his neck, killing him instantly. His body was found next to his adjutant's, Lieutenant Biddle. The third 7th Cavalry captain, Myles Moylan, was wounded in the thigh. All three first sergeants were dead and total casualties in the three companies were fifty percent. A few minutes later, Lieutenant Edwin Eckerson reported to Nelson Miles, "I am the only damned man of the Seventh Cavalry who wears shoulder straps alive!"

Miles had meanwhile ordered the 5th Infantry to attack from the south, where the 7th Cavalry had originally tried. The 2nd was still off chasing fugitives and rounding up the Indian pony herd. Miles thought he had one last chance to force a quick decision, and ordered Lieutenant Henry Romeyn to organize a charge by combined companies of the 5th Infantry and the now-leaderless 7th Cavalry. The men lined up and re-loaded their guns, waiting for the signal. Romeyn stood up and waved his hat, but was immediately shot in the lung. The rest of the men cheered and raced forward a few steps, only to meet a withering fire and fall back down. A small party of soldiers, however, got over the ridge and into the Nez Perce camp. Chief Joseph recalled, "Ten or twelve soldiers charged into our camp and got possession of two lodges, killing three Nez Perces and losing three of their men, who fell inside our lines." The remainder retreated to safety.

There were no more attempts to rush the camp and both sides fired at each other from behind rocks until darkness. Losses had been about even, with over 20 dead and 40 wounded on a side. The Nez Perce had lost their horses, but the warriors, except in a few instances, had success-fully kept the attackers away from their women and children. Over 200 people had gotten away and no more than 80 braves still held the lines. Miles ordered soldiers to circle the camp to the north to prevent addi-tional Indians from escaping.

During the cold gray dawn of October 1, Nelson Miles might have reflected that there was a good reason why this tribe had been able to fight its way through everything the U.S. Army had been able to throw against it. For once, the old Custer tactic of a surprise attack on a sleepy village hadn't worked. Miles' dilemma was that even while he pinned down the remaining Nez Perce, Sitting Bull was a hard day's march away with as many as 1,000 battle-hardened men. Nez Perce messengers had by now reached the Sioux. Miles dispatched scouts to warn against an onslaught of warriors from the north who could easily engulf his weakened com-mand. He may have picked up the mantle of Custer, but he had no wish to share his predecessor's fate. He also sent a courier to the south, to Howard or Sturgis, for help.

On the second day of the Bear Paw battle, everyone was excited to see two columns of horsemen coming over the hills. On closer view, however, these turned out to be buffalo, strangely roaming in parallel columns. That morning an Indian was seen riding from the north and Looking Glass, the war leader who most anticipated help from Sitting Bull, rose up from his firing pit on the east side of the camp to get a better view of the rider. He was shot in the head by an army marksmen, however, and killed. The rider turned out to be a Nez Perce warrior returning from the north to rejoin his fellows, or perhaps to rescue a relative.

With the demise of Looking Glass following the deaths of Ollokot, Toohoolhoolzote and Poker Joe, Joseph and the 70-year-old White Bird were now the only chiefs still alive. Ironically, the whites' notion that Joseph had been the tribe's war leader all along would soon appear to be vindicated. That afternoon Joseph accepted an invitation from Miles to cross the lines and discuss surrender. Miles demanded unconditional surrender, while Joseph insisted that the tribe needed to retain at least half its weapons in order to hunt for food. In another account, Joseph was agreeable to surrender but mentioned that he did not control the entire tribe. In a disgraceful move that Miles later tried to cover up, he held Joseph as a prisoner in his camp. Some say that Joseph was manacled hand and foot and placed among the army's mules; it's more likely that he was wrapped tightly in a blanket and put under guard. Unfortunately, one of Miles' officers, Lieutenant Lowell Jerome, had ventured too near the Nez Perce lines that afternoon and had likewise been taken captive. On the third day Miles was forced to exchange Joseph for Jerome, the two men shaking hands as they met each other in no-man's-land.

By the third day of the siege the soldiers had been reached by a wagon train and were well supplied with blankets, tents and food. The Nez Perce, most of them noncombatants and wounded, were hungry, and further beset by freezing rain and snow and a cruel wind that blasted in from the northwest. The soldiers had elevated their Napoleon cannon so that it was able to fire like a mortar into the Indian camp. On that day a shot landed by a dugout filled with women and children. A 12-year-old girl and her grandmother were buried alive.

On the evening of the fourth day, General Oliver Howard arrived, having ridden through a snowstorm all day with 17 men. He brought his interpreter, Ad Chapman, and two elderly Nez Perce named Old George and Captain John, who had daughters in the camp. Colonel Miles had chased Crazy Horse throughout the previous winter, only to see General Crook get credit for the Sioux champion's surrender. He wasn't overly pleased when Howard, a superior officer, arrived on the scene, but Howard magnanimously assured Miles that he would not assume command. At that point, Miles became more congenial.

It was Howard's arrival, however, together with lack of word from the Sioux, that comprised the last straw for the remaining Nez Perce. It is not clear whether White Bird and his people left camp on the last day of the siege or during the preceding night, but they were able to get through the army lines to Canada. Shortly after the battle, Canadian authorities counted 140 Nez Perce males and 93 females in residence. Tragically, a number of other Nez Perce were murdered by Assiniboines and Gros Ventre Indians while attempting to escape north. A white scout found five Nez Perce bodies on a trail and then a corresponding five scalps in an Assiniboine tepee. The number of murdered Nez Perce refugees has been estimated as high as 34.

On October 5, 1877, the fifth day of the Bear Paw siege, Chief Joseph surrendered, impressively reflecting the courage and pathos of the Nez Perce warriors who had fought to keep the tribe's hopes alive. The last sentence of his capitulation statement, conveyed by Captain John, interpreted into English by Ad Chapman and jotted down by Howard's adjutant, Lt. Charles Wood, was: "Where the sun now stands I will fight no more forever." The soldiers watched with respect as 418 freezing, starving Nez Perce, including wounded warriors, old people and women and children, emerged from their jagged holes on the windswept plains of northern Montana into army custody. After a trek lasting 90 days and covering over 1,500 miles through the land of Manifest Destiny, they had been caught just short of the Canadian border.

The great irony of the Nez Perce War is that despite the suffering of the tribe it engendered a higher regard for the Western Indians—even if they were "Dreamers"—than had existed among the white population prior to the war. William Tecumseh Sherman's unvarnished comments on the affair reveal how attitudes had begun to change:

Thus has terminated one of the most extraordinary Indian wars of which there is any record. The Indians throughout displayed a courage and skill that elicited universal praise. They abstained from scalping; let captive women go free; did not commit indiscriminate murder of peaceful families, which is usual, and fought with almost scientific skill, using advance and rear guards, skirmish lines, and field fortifications. Nevertheless, they would not settle down on lands set apart for them, ample for their maintenance; and, when commanded by proper authority, they began resistance by murdering persons in no manner connected with their alleged grievances.

Midway through the campaign, Sherman had ordered Howard to care for his prisoners in his own department; thus, during surrender negotiations Howard and Miles informed Joseph that his people would be returned to their home territory. Lt. Wood wrote, "I am very sure that no matter what the exact words were, everyone there, including General Howard, understood and fully expected the final disposition to be the return of these prisoners to the Department of the Columbia—that is to say, to the Lapwai reservation in Idaho." Joseph was betrayed—and Howard and Miles dishonored—when the Nez Perce were instead sent to Indian Territory (Oklahoma), where many died of malaria and malnutrition. In 1885, after public protests led by the Presbyterian Church, the Nez Perce were returned to Idaho, though Joseph and a few others, for their own protection, were resettled in Washington State.

The citizens of Idaho, who had borne the brunt of the initial, undisciplined uprising, remained the only whites in America unimpressed by the tribe's subsequent achievements. To people elsewhere, the Nez Perce struggle had touched on chords integral to American principles: freedom to live as one chooses and the willingness to fight against great odds. In terms of the reconciliation of starkly disparate cultures, the great retreat achieved far more than the Nez Perce could have realized at the time.

Chapter 4

THE ALLIES AT DUNKIRK

A pilot remembered that on the beautiful, clear morning of May 10, 1940, small children in the Dutch countryside waved cheerfully at his plane as it sped over their fields. They had probably never seen such a big, sleek aircraft. It may be assumed that the little arms soon got tired of waving, however, because the stream of airplanes with black crosses emblazoned on their wings were suddenly soaring over by the hundred—fighters, bombers and transports. Germany was once again on the move.

Winston Churchill, among others, opined that World War II in Europe, together with the Great War, comprised one grand struggle against the Germans, interrupted for two decades by a peace that was really a period of rest and recuperation. Others, subscribing to the "great man" theory of history, believe that the demonic figure of Adolf Hitler was primarily responsible for the second bloodbath. In either case, by the time the rematch had been declared, there were more military wild cards in the deck. As the twentieth century approached its midpoint, motor vehicles, on the ground and in the sky, had drastically altered the battlefield. Massive firepower could now be combined with fast movement, and the time allotted commanders to react to developments had correspondingly decreased.

Aside from sieges of cities, which had become increasingly rare after the development of the cannon, the history of warfare had always featured mobile operations. It came as a surprise that World War I in the West quickly devolved into a static slugging match, trench warfare essentially comprising two parallel siege lines. The skill of government bureaucrats at raising armies of millions of soldiers, combined with the advanced firepower technology with which they were supplied, had matured before transportation technology allowed the lethal masses to move. Each side could only take cover, and otherwise hope that the other would crack under the weight of heavy but spacially limited offensives. As the conflict developed, Russia fell out of the war and the Germans had a brief opportunity to swing superior force against the West, but then America came to the rescue of the Allies and the Germans could no longer keep pace. The nascent German colossus was subsequently dismembered and its armed forces put under heel—a situation that did not last.

The Great War had swept away monarchical empires that had endured for centuries, causing the political landscape to also become more complex. The remaining empires, turned parliamentary democracies, France and Britain, now had to ponder whether to declare war against Germany over countries like Czechoslovakia, which had never even existed before. The enduring problem was that the German people, in Nazi ideology the *Volk*, existed to a large degree outside the political boundaries of the German state. Beginning with Bismarck, Germany had embarked on a mission to create a "Grossdeutschland," encompassing all its ethnic members. The job of the other European powers was to prevent this consolidation, which, if successful, would result in a state that could dominate all of Europe. As late as 1941 few people realized that the Nazi political party, in fact, had larger, more sinister ambitions.

When Germany chose to force the territorial issue anew, the long-suffering nation of Poland found itself once more the frontline. Russia, the natural ally of the West in maintaining a balance of power against Germany, had turned Bolshevist, frightening Europe's ruling elites and preventing other nations from making secure alliances with the Kremlin. The only country that was willing to join hands with the Soviet Union was Nazi Germany, and the two totalitarian giants concluded a non-aggression pact on August 20, 1939. Once the signatures were dry, Poland's fate was sealed. Germany provided the military expertise; Russia occupied its allotted slice of Polish territory. In the aftermath, the Soviets also took the opportunity to grab the Baltic states and part of Romania. They also tried to grab some land from Finland but, unfortunately for the Red Army, the Finns fought back.

World War II thus began on September 1, 1939, the Polish invasion triggering declarations of war against Germany by Britain and France.

When Poland fell quickly, the German Army moved west to face its previous antagonists. The Great War was on again.

A positive factor for the Allies was that, as opposed to the Kaiser's army, Hitler's forces had been scrambled together in a hectic buildup lasting only five years. Hitler himself was an ambitious ideologue, but he was trying to overturn the verdict of the previous war with a military that had only recently consisted of 100,000 men, with no tanks, planes or capital ships. Meanwhile, the victors of World War I had maintained their infrastructure, equipment and sufficient troop levels. France had the largest army in Europe; Britain held unquestioned dominion over the European seas. While the Allies had no inclination to go on the offensive against Germany—the Polish campaign had ended before some troops even received their mobilization orders—there was no lack of confidence among the French and British that they could hold the Hitlerite resurgence once the Boche again came west.

For years the French had anticipated a new German threat by spending enormous sums to construct the Maginot Line, a system of fortifications along France's eastern border with Germany. The line extended from Switzerland in the south to Luxembourg and Belgium in the north. It was practically impregnable to frontal assaults and the only fear was that the enemy would try to flank it through the neutral states. In the south, Switzerland, a porcupine of a country where marksmanship was the national sport and the terrain was mountainous, discouraged the Germans from trying to get beneath the line. Where Luxembourg and Belgium paralleled the French border, the dense, hilly Ardennes Forest served as a veritable extension of the French defenses. Northern Belgium was the logical place for the Germans to flank the Maginot Line, and was the route they had taken to invade France in World War I. In 1914, the Schlieffen Plan had called for a gigantic right hook through Belgium to sweep along the Channel coast and fall on Paris. The strategy had nearly worked until the French Army counterpunched at the last minute, helped by Parisian taxi drivers, and had just barely forestalled the Germans at the First Battle of the Marne.

At the beginning of World War II, all eyes were on Belgium. Hitler was presented with a plan by his General Staff that called once more for a sweep to the north that would simultaneously seize the Channel coast and threaten Paris. The strategy was not original, but perhaps this time superior German technique could make it work. The dictator urged for a prompt offensive but his generals kept putting him off, postponing one date after another due to bad weather. Hitler was in a rush partly because the sudden German arms buildup had provided him with an army that, only temporarily, was more modern than the Allies'. His other incentive stemmed from mistrust of Joseph Stalin. The Soviets had decided to adopt a defensive policy that called for an immediate counteroffensive into the territory of an invader, and were building up their forces, especially armor, accordingly. Hitler's worry was that such a defensive posture was indistinguishable from an offensive one and, with the disappearance of Poland, Germany now bordered the Soviet Union. Hitler may indeed have had better-than-average insight into the thinking of his fellow megalomaniac, and he knew that pretexts for war could be trumped up quickly. As long as the German Army was forced to remain in the west, Stalin could stare at a nearly defenseless frontier that presented opportunities for political blackmail, if not military success. The Germans needed to force a decision in the west; France had to be dealt with as quickly as possible.

The French, meanwhile, had no doubt that Belgium was the key to the coming battle. It would have been undiplomatic for the French to have extended the Maginot Line along their border with neutral Belgium; this would have been tantamount to recommending that their neighbor become a German province. On the other hand, the French had no wish to fight the Germans again on their own soil. If the enemy attacked, the French and their British allies would cross the border to confront the invaders on Belgian territory. The Belgians were somewhat cooperative in discussing these plans (the French were secretly discussing a similar contingency with the Swiss in the south), but were trusting more to their army and the principle of neutrality than to the might of the Allies. From the French point of view, it would be unfortunate if the Germans once again tried to flank their defenses through Belgium; on the other hand, once Hitler violated Belgian neutrality, he would activate a 700,000-man Belgian army that would fight alongside the Allies. If he violated the Netherlands, another 200,000 men would come in against the Wehrmacht. In any case, the French were not going to sit still and absorb the blow as they had in 1914. If the Germans wished to repeat the Schlieffen Plan, conducting a northern right hook, this time the Allies would rise to meet them. Not only the Belgians and Dutch would fight back, but the French 1st, 7th and 9th Armies, plus the British Expeditionary Force (BEF) of over 200,000 men, would advance across the border to meet the Germans head-on.

It can be comforting, or perhaps not, to know that global consequences affecting millions of lives can be influenced by the actions of ordinary individuals. On January 10, 1940, a staff officer named Helmut Reinberger, who had disobeyed orders by hitching a ride on a small plane, accidentally came down inside Belgium with the German General Staff's plans in his briefcase. He hurriedly tried to burn the papers but his efforts only attracted the curiosity of Belgian soldiers. Brought to a headquarters he again tried desperately to destroy the papers by throwing them in a stove, but it was unclear whether he had burnt them all. In Berlin it had to be assumed the Allies had become cognizant of the plan. The affair provided an opening for another obscure individual, a German staff officer named Erich von Manstein, to submit an entirely different plan to conquer France.

As chief of staff of Gerd von Rundstedt's Army Group A, Manstein had considered that existing plans were too limited in scope. Even if the offensive through Belgium succeeded and the Channel coast were seized, what could come next except another long war of attrition? Instead, a more dynamic thrust might completely destroy the Allied position. Manstein thought that the offensive into Belgium should only be a feint; the true thrust should take place through the Ardennes Forest, the one stretch of land where the Allies least expected an attack. If an irresistible mass of armor were to sneak through the forest and cross the Meuse River, the Allied front would be fatally severed at its weakest point. Manstein showed his plan to Germany's leading armor expert, General Heinz Guderian, who affirmed that panzer divisions could negotiate the ground. In the early months of 1940, however, the plan went nowhere as the General Staff jealously protected its own prerogatives. The persistent Manstein earned the enmity of his superiors and was reassigned to take command of a reserve infantry corps.

Guderian and Manstein, however, found an ally when the latter received a formal interview with the head of state before taking on his new command. The former NCO Adolf Hitler was still far from the stage where he could dictate strategy to the German General Staff; however, upon hearing Manstein's plan he became excited and initiated a series of wargames to test the theory. Guderian wrote that Chief of the General Staff Franz Halder called the plan "senseless"; the commander of XVIth Army, General Busch, stated at a High Command conference, "I don't think you'll cross the river in the first place." General Rundstedt came out against it. Guderian wrote of "the hard task ahead, in whose successful outcome nobody at that time actually believed, with the exception of Hitler, Manstein and myself. The struggle to get our ideas accepted had proved exhausting in the extreme."

The risk of the Manstein plan is clear when one remembers that neither the Allies nor the Germans knew for certain at the beginning of the war what role armor would play in operations. The Allies distributed their tanks throughout their formations, thereby strengthening all their divisions; in addition, by placing independent tank battalions at disparate points they could be more flexible in countering enemy thrusts anywhere along the front. The French had belatedly begun to create three armored divisions, with a fourth on the drawing board, but these were more collections of tanks than integrated combined-arms formations. In May, the British were still in the process of creating their 1st Armoured Division. In contrast, the Germans had not only chosen to concentrate all their tanks into divisions, but consolidated those divisions into armored corps, and, in the upcoming battle, gathered most of their corps into a single, huge panzer group. Under Manstein's plan they could ensure irresistible strength at one point, but at that point only. If they did not assess correctly where their great armored phalanx should be aimed, the offensive would fail ignominiously.

The idea of launching the German armor through the Ardennes Forest, in particular, was fraught with peril. What if Allied reconnaisance saw the columns on their three-day approach march through the trees? Wave after wave of bombing attacks could decimate the panzers before they even reached the front. Simultaneously, of course, the French could have prepared a strong greeting party at the Meuse. The panzers could all be trapped in one massed group, stymied and destroyed. French reinforcements could then pour into the sector, placing the attackers at an ever-increasing disadvantage. Hitler, lacking the political support for another long war of attrition in the west—aside from having to look over his shoulder at Stalin—would have had no choice but to request another armistice, if he weren't overthrown in an army coup.

Nevertheless the Manstein plan was adopted, and given the code-name Operation Sichelschnitt (Sickle Stroke). Fedor Bock's Army Group B would attack through the Low Countries to make the Allies believe his was the main thrust. He had one panzer division for the Netherlands and two for northern Belgium. In the south, von Leeb's Army Group C would demonstrate against the Maginot Line with infantry and artillery and feint toward Switzerland to hold French attention. The real attack, von Rundstedt's Army Group A, of 1,800 tanks with follow-up motorized corps and infantry, would try to sneak through the Ardennes, in between the Maginot fortifications and the main Allied strength.

While at the beginning of World War II no one was quite sure how to employ armor for best strategic results (the Germans turned out to be correct, prompting the Russians, among others, to hastily reorganize their tank units), the same question applied to airpower. Planes were far faster and stronger than during the Great War and all kinds of theories existed about their future impact on battles. The Italian Giulio Douhet had prophesied that future wars would be decided by bombers alone. His theory left open the question of fighters and antiaircraft fire, and prematurely dismissed the impact of ground and naval efforts. Britain and the United States, two countries unsusceptible to land invasion, nevertheless proceeded to establish large strategic bomber fleets. For tactical ground support, the British possessed the Fairey Battle and the Blenheim. The French built both bombers and fighters, but their air force failed to become powerful due to a lack of industrial infrastructure. They imported American Douglass and Glen-Martin light bombers to buttress their own weak force. The single-minded Germans disdained the long-term strategy of strategic bombing and quickly built the largest air force in Europe, one that was devoted exclusively to aiding their army.

All nations attempted to develop a high-performance fighter, and the British not only had a good one in the Hurricane but an even more impressive work in the Spitfire. France's best fighter, the Dewotaine 520, only went into production in May 1940, slightly too late; they meantime depended on the Morane 445 and the American Curtis P-36, which were both a notch below. The Germans had large quantities of the Me-109, whose performance was bested only in some respects by the Spitfire.

The epitome of the German air philosophy was represented by the Ju-87, commonly called the Stuka. The purpose of this slow, short-range aircraft was to pulverize enemy ground positions by dive bombing—a far-ranging form of artillery that could soften the path of advancing infantry and tanks. Though an ungainly-looking aircraft in profile, its bent wings and fixed undercarriage made it resemble a vicious bird of prey in a dive. The Stukas and their bombs were fixed with sirens that made an unnerving high-pitched noise when descending. They typically carried two 500-pound bombs each that could be dropped with an exceptional degree of accuracy. Fast fighters and heavy bombers represented the trend in airpower, but the homely Stuka ended up the symbol of blitzkrieg (lightning war). Its only mission was to assist the troops on the ground.

In overall strength, the Allies had no reason to be intimidated. The French Army by itself was as large as the German; with the addition of the British Expeditionary Force and the Belgians and Dutch, the Allies were superior. They also had 3,500 tanks to the Germans' 2,800. The French Char B was the best tank on either side, and the French Somua and British Matilda were comparable to the most powerful German tank, the Mark IV. At this stage of the war, both sides employed hundreds of tanks that mounted only machine guns. The German Luftwaffe held a slight advantage over the combined Allied air forces in number of planes.

The real advantage held by the Germans as the Battle of France approached was that they were the aggressors. The German General Staff would decide where the crucial actions would take place, and they could mass their forces accordingly. As events would prove, the Germans had also come up with better ideas for the employment of the new weapons of armor and aircraft (and antiaircraft). For this, the campaign in Poland had proven invaluable for identifying weaknesses in technique. As one example, Poland had taught the Germans that light armored divisions were a useless compromise. Three light divisions were converted into the 6th, 7th and 8th panzer divisions (using captured Czech tanks). Poland had also provided valuable lessons for the new generation of Germans about march timetables, infantry–armor and air–armor coordination, as well as simple battle experience.

In retrospect we know that Poland was obliterated by the first classic example of blitzkrieg; however, in 1940 the Allies were yet unconvinced that they would be facing a new form of warfare. After all, a Polish army that fell in three weeks could not be confused with the united might of France and Britain. Once the Germans came west they would be facing the victors of Versailles. The Maginot Line would funnel any German attack through Belgium; if the Germans could be stopped in their initial rush, the war would once again become a matter of resources, and the Allies would once again prevail.

The "Phony War" (in German, the "Sitzkrieg"), when the opposing armies in the west sat looking at each other across the German border, lasted for eight months. During that time there was enough excitement. The German pocket battleship *Graf Spee* was trapped and scuttled in an Argentinian harbor. The British battle cruiser *Hood* was blown up by the battleship *Bismarck*, but then the *Bismarck* was hunted down and destroyed. A U-boat managed to sneak into the British naval base at Scapa Flow and sink the battleship *Royal Oak*. On April 20, 1940, the Germans occupied Denmark and simultaneously invaded Norway by sea and air. Finland finally gave in to the Soviet Union. Off the Norwegian coast the German Navy was decimated by British ships, but German troops air- and sealifted into the country held the advantage on the ground.

Meanwhile, millions of people on both sides, including soldiers on the front line, hoped that another European cataclysm on the scale of the Great War could be avoided. What was the point, and why should another generation be thrust into a charnelhouse of blood and destruction? Still, the Germans had their reasons and the Western democracies were determined to resist aggression. The only missing element, particularly in France, was a certain enthusiasm that had existed at the beginning of the Great War. The task of the French in the inter-war years had been to ensure that theirs was the strongest army in Europe in order to forestall German designs. The British had backed them up and had once again shipped an army to the continent to help. Now the Hun was once more at their throats and the prevailing attitude was, "Here we go again."

On May 10, 1940, the Germans launched the Battle of France. It began, obliquely enough, with massive air fleets aimed at the Netherlands and Belgium. The Dutch, who had stayed neutral during WWI, were the initial target. They had thought that their system of rivers, canals and dikes would stall an invader, but they hadn't considered the advent of air transport. German paratroopers seized the bridge over the Rhine at Arnhem, over the Meuse at Maastricht, and airfields near The Hague. Belgium, too, came under attack by novel means. The Belgians had constructed a vast, state-of-the-art fort called Eben Emael that guarded the strategically important confluence of the Meuse River and the Albert Canal. Early on the morning of May 10, wooden gliders landed on top of the fort and 80 German commandos poured out. The fort was intended as a formidable obstacle against frontal attack but had no defense against men running around above throwing demolition charges through the weapons slits. By midday the 1,200-man garrison of the fort had surrendered.

As news of the German tricks and surprises reached the Allied High Command, there was almost a sense of relief that the ambiguity of the Phony War had vanished. The Low Countries' ill-advised attempt to adhere to "neutrality" had only made them the Germans' first prey. From north to south, the Allies' three best armies, the French 7th, the BEF, and the French 1st, promptly entered Belgian territory to confront the invaders, the French 7th assigned to cross through along the coast to help the Dutch. In front of the Ardennes, the French 9th Army, comprised of reservists, also crossed the Belgian frontier to take up positions on the Meuse, though little was expected to happen in that sector.

The initial task of the German ground troops was to seize bridges or, failing that, force river crossings. In the Netherlands all was confusion as the Dutch fought paratroopers in their midst and assaulting troops to their front. Some bridges were blown but others taken. The Dutch had no answer to the 9th Panzer Division except to hold it up at water lines. The Belgians were subsequently forced to defend not only their border with Germany but their border with the Netherlands. The 3rd and 4th Panzer Divisions, under Erich Hoepner, were relentlessly chewing their way through the Belgian front. When troops dug into a strong defensive position, they would notice a small Fiesler-Storch spotter plane circling above. The craft would leave, and then minutes later squadrons of Stukas would come screaming down, pulverizing the line. While the melee in Holland approached its foregone conclusion, the Belgians were forced back faster than they anticipated. Their cavalry divisions in the Ardennes were ordered to pull north if they encountered difficulty, leaving that wooded sector to the French. The Belgians were more concerned with protecting their cities in the north.

The British Expeditionary Force marched forward to arrive at the Dyle River on May 12, a sector that protected Brussels and Antwerp, only to find the Belgians had failed to fortify the line. Nevertheless the British, led by Lord General John Gort, dug in and prepared to receive the Hun. Farther north the French 7th Army was being cut up by air attacks, and encountered Germans in the Netherlands who stalled its advance. On the BEF's right, the elite French 1st Army had charged into Belgium and was already grappling with the Wehrmacht. On May 13 the first major tank battle of the campaign was fought at the Gembloux Gap, when General Prioux's Cavalry Corps met the German 3rd and 4th Panzer Divisions. The French lost 117 tanks and withdrew slightly, but there was no breakthrough. By this time it was already apparent that Allied air support was lacking. The Germans were hitting hard at crucial spots with waves of planes; Allied craft, when seen at all, were spreading bombs across the landscape with little effect.

To the Germans, the first three days of the offensive had gone as well as they could have expected. Von Bock's army group in the Low Countries, backed by the Luftwaffe, had made good progress, though the BEF had yet to be tested and the French armor comprised a serious obstacle. It's an unusual fact that the German offensive into the Netherlands and Belgium succeeded as well as it did, when it was not even the main effort. The real German problem was a gigantic traffic jam in the Ardennes, where they had bunched seven panzer divisions—by far the greatest concentration of tanks in history—which were trying to sneak up undetected.

While battles raged in the north, skirmishes were reported to the south as small Belgian and French units encountered the great panzer armada in the forest. The three German columns stretched back 50 miles east of the Rhine while a screen of Me-109s flew overhead to chase off French or British air reconnaisance. Reports came in of strong German forces in the Ardennes but the Allied high command remained oblivious to what was about to hit them.

On the fourth day of the offensive, the leading elements of von Rundstedt's Army Group A reached the Meuse. All the bridges had been blown and the French 9th Army lay on the other side. Briefly, the fortunes of the panzer group depended on a few grenadiers in rubber boats. On the right, the 7th Panzer Division, under an unproven general named Erwin Rommel, found an unguarded weir near the small village of Houx and was able to get some men over the river. At his main crossing site at Dinant, Rommel's boats were badly shot up and he ordered his Mark IVs to drive up and down the river bank with their turrets at a 90-degree angle to plaster the far side. Soon he had a company across and his bridgehead began to take shape.

In the center, General Hans Reinhardt's corps, consisting of the 6th and 8th Panzer Divisions, became stalled on the river at Montherme. This sector had the toughest terrain and, instead of "B-class" reservists, was held by the 102nd Fortress Division comprised of colonial troops. Indochinese machine-gunners swept the river with fire and tenaciously held on against the panzers' best efforts to blast them out of their bunkers. The Germans earned a foretaste of what was to come when, after the war, the French, and then Americans, tried to subdue the Indochinese on their home ground.

On the German left, Heinz Guderian commanded the primary cross-ing opposite Sedan with the 1st, 2nd and 10th Panzer Divisions. Wave after wave of Stukas, Dorniers and Heinkels built a wall of explosions across the river until it seemed that no French could have survived the bombardment. Nevertheless, Guderian's boats were destroyed or turned back when they tried to cross the river. A few French bunkers held out and controlled the river with machine guns and accurately spotted artillery fire. Finally, a dozen men got across and set about attacking the French bunkers from the rear. A number of Knight's Crosses were earned, many of them posthumous. Amid the fury of blasts and smoke and man-gled bodies near the river, Guderian met Colonel Herman Balck of the spearhead 1st Rifle Regiment of 1st Panzer Division. Balck, paraphrasing a strict instruction Guderian had issued weeks earlier during training on the Moselle, told the general, "Joy riding in canoes on the Meuse is for-bidden!" Slowly another bridgehead began to form.

That night, while the Germans erected pontoon bridges, a nervous tremor shook the French High Command. If the reports from the Meuse could be believed, the entire Allied strategy had been fatally misguided. Hopefully the reported breaks in the Meuse front were only cracks that could soon be filled. Orders went out for reserves, including two of France's three armored divisions, to converge on the crossing sites near the Ardennes. At daylight on May 14, Allied bombers came soaring in against the German bridges while 250 fighters flew overhead to fend off Messerschmitts. The Germans had prepared for this reaction and had sur-rounded their bridges with a cordon of AA guns. The French pilots found that they could avoid the 20mm and 37mm flak by dropping their bombs from high altitude. The British came in low in order to hit the small targets. Some squadrons of Fairey Battles were completely wiped out attempting to destroy the bridges, their pilots earning admiration from German gunners for their bravery. The British lost 40 of 71 bombers that day. Including French and fighter planes, total Allied losses were closer to 100 aircraft. Nevertheless, the attempt to sever the bridges failed.

On the ground, the constant waves of air attacks were discomfiting to the invaders and Guderian, for one, had a close call. He had made his headquarters in a chateau decorated with hunting trophies.

Suddenly there was a series of explosions in rapid succession; another air attack. As though that were not enough, an engineer supply column, carrying fuses, explosives, mines and hand grenades, caught fire and there was one detonation after the other. A boar's head attached to the wall immediately above my desk broke loose and missed me by a hair's breadth; the other trophies came tumbling down and the fine window in front of which I was seated was smashed to smithereens and splinters of glass whistled about my ears. It had in fact become very unpleasant where we were and we decided to move elsewhere.

Throughout May 14, Reinhardt's panzer corps in the center continued to be held up by colonial troops. The Germans got infantry over the river but each attack was followed by a vicious counterattack that reclaimed the ground. On the right, however, Rommel had passed 30 tanks of the 7th Panzer to the west side and was already pushing cross-country. Impetuously leading from the front, Rommel nearly got killed when the Mark III in which he was riding was hit twice and disabled by French anti-tank and artillery fire. Rommel was bleeding from the cheek, and when his armored command vehicle came up it took a hit in the engine. Stukas swooped in to deal with his antagonists. In the meantime Rommel was almost captured by onrushing colonial infantry. He had run into the 4th North African Division, dispatched to plug the gap at Dinant.

In front of Guderian, the French launched two battalions of tanks, each backed by an infantry regiment against the growing German bridgehead. French troops fleeing from the Meuse panicked at the sight of the vehicles, thinking they were German. The attack was uncoordinated, and after the French 7th Battalion lost 11 of 15 tanks engaged in a clash with 1st Panzer, the units went on the defensive. Meanwhile, the French 3rd Armored Division, backed by the 3rd Motorized Division, was advancing on Guderian's bridgehead from the south. The 1st Armored was ordered to hit Rommel's penetration on the German right.

Farther north, after its advance along the North Sea, the French 7th Army had been rebuffed near Breda in Holland and its elements were being pinned against the coast, cut off by attacks along its line of advance. To its right the BEF held the Dyle River line, phlegmatically waiting to be seriously tested. The French 1st Army had taken heavy losses but was holding its own in thrusts and counterthrusts against 3rd and 4th Panzer. The Allies outnumbered the Germans in the north and there were no signs of collapse there among the Allies' elite forces.

Opposite the Meuse bridgeheads it was an entirely different story. As the German armor streamed across the bridges, whole French regiments were calling it quits, throwing away their arms. On Guderian's front it was the French artillery that gave up first, shaken by Stuka attacks, leaving their front units in the lurch. Near Sedan was a village called Stonne on a height that dominated Guderian's crossing points from the south. The French 3rd Armored Division arrived at this vital position and then, disastrously, spread its tanks in small groups—each of one heavy and two lights—across a wide front instead of mounting a concentrated attack. Nevertheless, the French badly bloodied the German Grossdeutschland Regiment, which barely held on to the village.

At this point, during the afternoon of May 14, Guderian made the crucial decision of the campaign. With strong French armor arriving from the south in the anticipated counterattack, Guderian asked the commander of 1st Panzer whether he thought his division should halt its advance in order to protect the bridgehead, or continue into France where it was obvious the enemy was showing signs of collapse. 1st Panzer's chief of staff, Major Wenck, responded with Guderian's motto, "*Klotzen, nicht kleckern.*" ("Slam them, don't tap them.") That decided it. 10th Panzer, which had just come over the Meuse, was assigned to reinforce Grossdeutschland and hold off the French 3rd Armored. 1st and 2nd Panzer would begin the dash to the Channel. The next day would decide the fate of France.

Winston Churchill had become prime minister of Great Britain on May 10, the very day of the German offensive, and had so far not been overly worried, due to continued reports from the BEF. He was shocked, however, on the morning of May 15 to receive a phone call from French Premier Paul Reynaud that began, "We have been defeated." Churchill, not a morning person, responded, "Surely it can't have happened so soon?" Reynaud explained how the front had unexpectedly caved in at Sedan. The prime minister, showing no more insight into blitzkrieg warfare than his French counterpart, tried to instill confidence. "All experience shows that the offensive will come to an end after a while. I remember the 21st of March, 1918. After five or six days they have to halt for supplies . . ." After the prime minister's exhortation, Reynaud could only repeat himself: "We are defeated; we have lost the battle."

On that May 15th, French General Huntziger, convinced that the German intention was to roll up the Maginot Line from behind, pulled back his 2nd Army from the Meuse to form a defensive shoulder. To his left, General Corap ordered his 9th Army, which was rapidly disintegrating anyway, to abandon its positions on the river. The center corps of panzers, Reinhardt's, had finally overwhelmed the brave 102nd Fortress Division and was now streaming across the Meuse. The colonial fortress troops didn't have motor transport, so instead of being able to join the retreat were killed or captured. The isolated German crossing sites, which at one time consisted of thin pontoons protected by desperate toeholds of infantry, had linked, and ballooned into a huge bulge. Seven panzer divisions were on the French side of the Meuse, with follow-up motorized and infantry divisions continuing to pour out of the Ardennes. There was nothing to prevent them from continuing into France.

Of the three French armored divisions, the 3rd, which had voluntarily broken up into small groups, was steadily being fought down around Stonne. The 1st Armored, after a clumsy approach march that caused its tanks to use up their fuel, had arrived on the other side of the enemy penetration, where Rommel's 7th Panzer was racing west. In view of his fuel problem the French commander safeguarded five of his six artillery batteries by moving them out of harm's way. Rommel, accompanying his leading regiment, saw the French tanks to his right and slugged it out with them briefly. Then he resumed his advance, leaving them for his following regiment to deal with. He probably didn't recognize the famous

French 1st Armored. An hour later, the 31st Panzer Regiment, attached to Rommel from the 5th Panzer, came up the road and the 1st Armored was still there, hurriedly attempting to refuel. On seeing that many of the French tanks were heavies, the German commander attacked them with his Mark IVs. As usual the French B tanks were almost impossible to knock out unless they were hit in the tread, but it helped that most of them were already immobile.

As for the French 2nd Armored Division, this vital formation became ineptly scattered to the wind before it even engaged the enemy. Initially intended to advance east to support 1st Army in northern Belgium, it had then been pointed toward the Meuse breakthroughs to the south. By the time trains and transport had been arranged, the enemy spearheads were to the west. With the rail network a shambles from Stuka attacks, the roads clogged with refugees, and contradictory orders being issued every few hours, the division lost its cohesion. Its isolated battalions continued to show up, sometimes in the form of abandoned equipment, but the units never had a chance to fight in unison.

Before the war the French had not correctly assessed how to employ armor for best impact on the battlefield. In tanks they had both quality and quantity, but they mistakenly viewed them as powerful defensive weapons, thus negating their primary assets. Part of the problem might have been lack of training since the French had only begun creating armored divisions in January 1940; on the other hand it never occurred to most French military leaders that the value of an armored division was its ability to strike a massive blow at a point of its own choosing. Used defensively, the tank was little more than motorized artillery. As the war progressed and the Germans found themselves attacked by superior armored formations, they considered anti-tank guns to be the proper defensive answer to armored assault. Their own tanks were husbanded for aggressive counterattacks so that the value of mobility would not be squandered. In May 1940 the French did have one officer who had foreseen the future of armored warfare, and he was ordered to hastily put together enough strength so that he could call his command the 4th Armored Division. The commander of the new division was a mere colonel named Charles de Gaulle.

On the German left, meanwhile, the 1st Panzer Division had run into the tough French 14th Division under General de Lattre de Tassigny, but its partner, 2nd Panzer, had leapfrogged ahead. In the afternoon, 2nd Panzer found itself running neck and neck with Reinhardt's 6th and 8th Panzer Divisions, which had made up for their delay at the river by advancing 40 miles that day. Prisoners were coming in by the thousands. The scenes on the roads consisted of, to the right, long columns of frightened civilian refugees trundling west; to the left, long columns of defeated, disarmed French soldiers trudging east; on the roads themselves, German armored columns charging toward the Channel. The farther the panzers traveled into France, the weaker the opposition became. It was a complete breakthrough: the panzer corridor was growing by the hour.

The momentous 15th of May saw other developments. In the Netherlands the fighting had ended and Dutch troops were handing over their arms. The final straw had come on the previous day when the center of Rotterdam was obliterated by Heinkel-111 bombers in a controversial attack. The Germans had tried to call off the attack, about half of the bombers receiving word, but their original intent of forcing a military decision by killing civilians resonated darkly, especially among the leaders of Britain's Bomber Command. It's ironic that the Germans, who had constructed a tactical air force and not a strategic one, would be the first nation to force a strategic decision through air attack. The effect of Rotterdam was greatly magnified at the time because first reports held that 30,000 civilians had been killed, further panicking refugees and creating fear of the Luftwaffe. The number of fatalities was later confirmed to be 814.

In London that day, a conference took place that decided the role the RAF would henceforth take in the Battle of France. So far Britain's Fighter Command had only grudgingly contributed ten squadrons of Hurricane fighters to the battle, holding back its Spitfires. The Hurricanes, though providing German fighters their toughest competition in the air, had been decimated during the first few days; between their losses and those of Battles and Blenheims only 202 of 474 British planes on the continent remained operational. The French air force was unable to hold its own against the Luftwaffe, and the clamor—from both the French and the BEF—for more British fighters had become deafening. At the meeting of the War Cabinet that day, however, Air Marshal Hugh Dowding of Fighter Command made a dramatic case that unless the flow of Hurricanes dispatched to oblivion in France was cut off,

Britain itself would soon be defenseless. As historian John Terraine commented, "Fighters alone could confer that air superiority which a 1940 battle required, but Dowding clung to his fighters as a miser clings to gold; of his 19 Spitfire squadrons he never let go." The War Cabinet decided that four more squadrons of Hurricanes could be committed to the battle.

The other important decision concerning Britain's airpower was that Bomber Command could now fulfill its long-cherished dream of attacking the Ruhr. The island nation viewed its bomber force as a weapon of strategic defense; as with the principle of deterrence which held sway during the postwar nuclear age, the theory was that any enemy who dared to attack would soon find its infrastructure and vital industries destroyed. The War Cabinet also assumed that by unleashing the mighty sword of Bomber Command against enemy rear areas, the Germans would be forced to transfer significant quantities of fighters and AA guns from the front for home defense.

Thus, while Fighter Command was trying to avoid committing any planes to the continent, Bomber Command was overflying the battle to hit targets hundreds of miles beyond. In France and Belgium, where the fighting took place, the Luftwaffe continued to wreak havoc on Allied positions, reserves, marching columns, communication centers and airfields. The British Advanced Air Striking Force, flying from airfields in France, was quickly being whittled down. The French air force had simply not been ready for the war. To French troops, one of the demoralizing aspects of the battle was that they would flee the carnage wrought by panzers at the front, only to find greater carnage, inflicted by level bombers and Stukas, in their rear.

Bomber Command went into action that night with 96 planes aimed at the Ruhr. Afterward, 24 crews claimed to have identified their targets, though this is not to say any were damaged. Encouraged by the loss of only one plane on the raid, Bomber Command launched similar missions against Germany every night for the rest of the month. Luftwaffe strength, including flak guns, remained committed to the battle.

On the morning of May 16, Rommel mounted a set-piece attack against a network of French fortifications. After breaking through he took a battalion of tanks and raced a further 50 miles, destroying a stray battalion of French tanks en route and capturing the 1st Armored Division's five artillery batteries that had earlier been pulled back, never having had a chance to fire. By the following daybreak he was in Avesnes, throwing the French rear into chaos and putting himself at great risk. The panzers were nearly out of gas and ammunition, and Rommel was isolated deep inside enemy territory. Historian Guy Chapman wrote, "The rest of the division had not followed up as ordered, and between them and their general stood numbers of unguarded French soldiers who had surrendered. Other French units, unaware of the situation, passed up and down behind him during the morning." Rommel formed his lead tanks and their tired crews into a hedgehog and then, in his command vehicle, accompanied by a Mark III, went back to find the rest of his division.

To the northeast in Belgium, meantime, the British Expeditionary Force was efficiently discouraging German attacks with well-aimed artillery fire. Its commander, Lord Gort, had won the Victoria Cross in the Great War and struck many as the epitome of a British officer: big, brave, unflappable and thoroughly dedicated to his men. Still, he was beginning to get nervous about the French. Something was obviously going wrong to the south, but General Gaston Bilotte, who had been assigned to coordinate the efforts of the French, Belgians and BEF, was not keeping him in the loop. One thing Gort knew was that Britain's Advanced Air Striking Force was nowhere to be seen on the BEF front— its aircraft were being launched against the Meuse sector and precious few were returning. Early on May 16, Gort sent a staff officer to Bilotte's headquarters and only then discovered that the French had decided to pull back the northern armies. (The Belgians found out even later.) For an army to abandon its lines in the face of an aggressive enemy meant an extremely dangerous series of maneuvers. Instead of entrenched troops and artillery emplacements, the Germans would be looking at thin rear-guards protecting long columns of guns and men with their backs turned. Confusion was inevitable, as were increased casualties. In the BEF's case it also meant a degree of humiliation. On their way to the front they had been cheered by French and Belgian civilians as saviors against the German threat. Now they had to traipse back through those same towns in retreat, failures through no fault of their own.

Generals Alan Brooke of I Corps and Michael Barker of II Corps leapfrogged their divisions, while divisional commanders alternated their brigades, one passing through the defensive screen of another. The retreat began at nightfall on the 16th in order to disengage without harrassment from the Luftwaffe. The next day, two British tank battalions, the 15th and 19th Hussars on the BEF's left, were trapped and crushed by German pursuit when neighboring Belgian units precipitously gave way. In the BEF's center, Bernard Montgomery's 3rd Division and Harold Alexander's 1st Division fell back in seamless tandem. Alan Brooke wrote, "It was intensely interesting watching [Alexander] and Monty during those trying days, both of them completely imperturable and efficiency itself."

Winston Churchill had meanwhile arrived in Paris to meet with the French leadership and receive a report from General Maurice Gamelin, commander-in-chief of the French Army. Disconcertingly, while the meeting took place officials were burning wheelbarrows-full of state archives in the courtyard outside. According to Churchill, "Utter dejection was written on every face. In front of Gamelin on a student's easel was a map, about two yards square, with a black ink line purporting to show the Allied front. In this line there was drawn a small but sinister bulge at Sedan." After hearing Gamelin's report of the German breakthrough, there was a long silence and then Churchill asked, "Where is the strategic reserve?" Gamelin replied, "*Aucune.*" ("There is none.")

In his memoirs Churchill states that he was dumbfounded at this response and goes on to describe how "one can always have, one *must* always have," a strategic reserve. Regarding "*Aucune,*" he continued:

> I admit this was one of the greatest surprises I have had in my life. Why had I not known more about it, even though I had been so busy at the Admiralty? Why had the British Government, and the War Office above all, not known more about it? . . . We had a right to know. We ought to have insisted. Both armies were fighting in the line together. I went back again to the window and the curling wreaths of smoke from the bonfires of the State documents of the French Republic.

The ingenuous aspect of Churchill's exchange with Gamelin is that, in the Prime Minister's telling, Operation Sichelschnitt could easily have been defeated if the French had adhered to basic military principles. The British are seen to be shocked that France, despite millions of men and thousands of tanks along and near the front, did not have an additional elite army with hundreds of tanks in deep reserve to counter unexpected developments. In contrast to Churchill's "surprise" at the strategic situation, the German von Manstein, who had far less knowledge of Allied troop deployments than the British War Cabinet, had correctly assumed months earlier that the thrust through the Ardennes would shatter the Allied front. Later in the meeting Churchill had another exchange with Gamelin that was more to the point, and more honest:

> Presently I asked General Gamelin when and where he proposed to attack the flanks of the Bulge. His reply was: "Inferiority of numbers, inferiority of equipment, inferiority of method"—and then a hopeless shrug of his shoulders. There was no argument; there was no need of argument. And where were we British anyway, having regard to our tiny contribution—ten divisions after eight months of war, and not even one modern tank division in action?

The French, in fact, had been throwing reserve divisions against the panzers since the Meuse crossings began on May 13. The reinforcements had become absorbed into the maelstrom of refugees, fleeing troops and enemy motorized units cutting in all directions. In terms of flexibility and combat doctrine, it had been proved that French armored divisions were simply not as good as German ones. The French air force was outclassed while the British air force refused to commit its main strength to the battle. The French had placed adequate reserves behind the front in Great War terms, but they had been unprepared for the speed of blitzkrieg warfare. By the time a reserve division reached its start line, panzers were already behind it and the Luftwaffe was inflicting serious casualties, unchallenged in the sky. The Germans, of course, were also reinforcing the crucial sector, and had the advantage of knowing from the beginning where their additional motorized and infantry divisions should be aimed.

It might have comforted the Allied leaders if they had known that a large portion of the German High Command was also infused with a Great War mentality, and that the panzers' dash deep into France was making them uneasy. If Churchill could scarcely believe that the French were not poised to deliver a powerful counterblow, his sentiment was shared by Hitler, von Rundstedt and von Kleist, commander of the panzer group. On everyone's mind was the First Battle of the Marne, where the French had snatched victory at the very last moment from between the closing jaws of defeat. History is full of examples of attacking units advancing too far too quickly, making themselves increasingly vulnerable to counterattack.

At an early planning conference for Sichelschnitt, Guderian had explained to the German army group commanders how he planned to be across the Meuse by the fifth day of the offensive, after which Hitler asked, "And then what are you going to do?" According to Guderian, "He was the first person who had thought to ask me this vital question." It was Guderian who made the decision to head west on May 14, after consulting with one of his divisional commanders, not OKH (German Army High Command). On May 16 Guderian found his corps rushing into the French town of Montcornet at the same time as 6th Panzer Division of Reinhardt's corps. "Since the Panzer Group had laid down no boundary between the two corps, we soon agreed on one among ourselves and ordered the advance to go on until the last drop of petrol was used up." To a significant degree the velocity of the German attack was being determined by the panzer leaders themselves, while the High Command could only worry that they weren't getting carried away.

On May 17, Hitler issued the first of his "panzer halt" orders. The reckless advance of the armor had to cease until infantry could arrive to hold the flanks. When sternly informed of the order by von Kleist, Guderian resigned his command, only to be immediately reinstated by von Rundstedt. After much discussion Guderian received permission to conduct a "reconnaisance in force," but his corps headquarters had to stay where it was. This was enough of a fig leaf for Guderian to order his 1st, 2nd and 10th Panzer Divisions onward. He wrote, "Corps headquarters remained at its old location in Soize; a wire was laid from there to my advanced headquarters, so that I need not communicate with my staff by wireless and my orders could therefore not be monitored by the wireless intercept units of the OKH and the OKW."

It was fortunate that Guderian kept his privacy, because on that very day he was hit from the south by the one French officer whose philosophy of tank warfare resembled that of the Germans: Colonel Charles de Gaulle of the so-called 4th Armored Division. The 4th Armored had been hastily assembled during the past week from reserves and had never trained as a unit; nevertheless, 200 tanks churning unexpectedly into the flank of 1st Panzer Division comprised a serious challenge.

The French moved at dawn in two columns from their staging area at Laon toward Montcornet, 20 miles away. De Gaulle had a heavy battalion of Char Bs, two battalions of Renault R-35s and a company of D-2s, with infantry support by a battalion of (motorized) Chasseurs and a unit of Dismounted Dragoons. At first the French tanks overran everything in their path, littering the approach roads with burned-out German motorcyles and armored cars, chasing the infantry before them. They found an enemy supply column and shot it up end to end, while bearing down on the headquarters of 1st Panzer. In Montcornet, which was crammed with soft-skinned ammunition trucks, German staff officers frantically tried to set up artillery and flak guns to hold off the swarm. A few panzers were in the vicinity and others made it out of repair shops to do battle.

The French overran the village of Lislet and left it a blazing ruin. Outside Montcornet, German artillery hammered the attackers from across the Serre River, but at one point twentyfive French tanks effectively severed 1st Panzer's advance elements from their supply. As darkness approached, de Gaulle had driven a wedge into the "panzer corridor," but his tanks were running out of gas, his infantry support was inadequate and he had no artillery at all. His biggest problem was that instead of prompting the enemy to collapse and flee, as the panzers had done in prior days, German soldiers were holding fast at selected points, counterattacking his flanks and peppering his armor with small arms, AA guns and artillery. A French tank officer wrote: "The enemy platoon commanders have a terribly enterprising air about them." Further, more panzers were coming on the scene and Stukas were arriving overhead. There was no choice but to withdraw, having at least given the Germans a taste of their own medicine.

Guderian took his time reporting de Gaulle's attack so as not to increase the nervous tension already apparent in higher headquarters. His superior, von Kleist, only found out about it the next day and considered it a local problem. On the 19th de Gaulle, reinforced by a battalion of Somua tanks and some artillery, struck out again from Laon with the intention of cutting the German bulge at the neck. This time, however, his approaches were mined and covered by a thick screen of enemy anti-tank guns. Not merely the 1st Panzer, but the 10th Panzer and 29th Motorized Divisions were waiting for him, while Stukas dove at his three attacking columns the moment they left their start-line. The 4th Armored tenaciously grappled with the enemy through the night and next day before finally pulling its disparate units back across the Aisne River. De Gaulle's attacks were meant to have been matched by similar thrusts from the north by the 1st and 2nd Armored Divisions, but these units had already been broken up. Guy Chapman rendered a verdict on the efforts from the south that were to remain the best the French could offer against the panzer corridor:

> There had been no miracle. No Frenchman had ridden the whirlwind. Colonel de Gaulle had appeared, reappeared and most disconcertingly disappeared at intervals, thus giving a false air of unity to what in fact was little more coherent than a succession of uncoordinated armed scuffles with flank guards.
>
> These indecisive skirmishes were inflated by Parisian editors laudably anxious to put heart into the civilians: stories which at least did not end in disaster could do something to stiffen opinion. In Colonel de Gaulle, the champion of the armored fighting vehicle, they now had a figure they could build up as a commander who knew how to fight . . .

A sense of relief swept across Paris when it became apparent the panzer hordes were not inclined to descend on the capital; however, this was matched by the concern that gripped London when the Germans' true intent was revealed. On May 20, Guderian's panzers reached Abbeville at the mouth of the Somme River, which flowed into the English Channel. It was as if a sword had been driven clear through the body of the Allied front; one twist and the head would fall. As the panzer divisions arrived at the sea, all that remained was for them to turn right and sweep up the coast—Boulogne–Calais–Gravelines–Dunkirk–Ostend—and the Allies' elite armies would be surrounded and destroyed.

While the specter of another Miracle of the Marne played havoc with the nerves of the German High Command, the same precedent continued to influence Allied thinking. For the outcome of the Great War to be reversed in a campaign lasting mere days was unthinkable; surely the fortunes of war were due to even out. Everything had to be tried to find a weak spot in the enemy. Among the virtually trapped northern armies, the Belgians were hugging the coast in the one corner of their territory yet unconquered. French 7th Army had been disbanded after its painful foray into Holland, its surviving units spread among other formations. French 1st Army had done the hardest fighting to date and suffered the greatest losses. The most powerful formation remaining—the only one that could still knife through the strangling cordon of panzers—was the BEF. Lord Gort, as always, would do his duty, his dilemma being whether that meant saving the British Army or trying to save France.

On the 19th Gort warned London that the BEF might have to be evacuated. No such contingency appealed to his superiors. The next day, Chief of the Imperial General Staff Edmund Ironside flew across the Channel to get a counteroffensive going. Gort calmly informed him that almost the entire BEF was already at grips with the enemy; nevertheless, his 5th and 50th divisions could put something together. Ironside then proceeded to French headquarters to meet Bilotte, commander of the army group, and Blanchard of French 1st Army. "All in a state of complete depression," Ironside observed of the French generals, "no plan or thought of plan." At one point the six-foot-four British officer shook Bilotte by the buttons of his tunic. The French agreed to commit two 1st Army divisions, which would attack south on the 21st along with the British on either side of the city of Arras.

Ironside returned to England shaking his head, jotting in his diary, "God help the BEF." His view was that even though it had not yet suffered heavy combat casualties, the French and Belgian failures on the BEF's flanks meant that its only hope was to break southwest through the panzer corridor to join the main front. The Advanced Air Striking Force was incapable of supporting the attack because it was in the process of evacuating its airfields in France. The French air force was asked to participate but was unsure of the operation's location and timing. Around midnight on the 20th, Lord Gort was hardly surprised to receive word that the two French divisions on his left wouldn't be ready for another two days. On a positive note, General Prioux of the French Cavalry Corps offered to support the BEF's right with some 60 light tanks. The courageous Prioux was having great difficulty reconcentrating his armor after he had been forced to disperse it among infantry divisions.

Although Ironside and Gort shared a dismal appreciation of how the French were conducting the battle, in contrast to his superior Gort had no faith in the BEF's moving south. He was already thinking in terms of forming a bridgehead on the coast closer to home. Nevertheless, he would follow his orders and attack south on the 21st, whether or not French 1st Army was ready to join in. If Gort had seen a breakthrough to the south as the BEF's last hope, one can be sure the entire 5th and 50th Divisions as well as other units would have been found to launch a full-blooded thrust. Instead, the British employed the 4th and 7th Royal Tank Battalions—74 tanks in all, including 16 heavy Matildas—each with a supporting infantry battalion and a battery of artillery and anti-tank guns. The whole was known as "Frankforce," after Harold Franklyn, commander of the 5th Division. On the spot the attack would be led by Giffard LeQ Martel of the 50th Division, one of Britain's leading tank experts. They didn't know who or what they would encounter, but their limited objective was to clear the roads west and south of Arras.

The previous day the 7th Panzer Division had arrived south of Arras, and on the 21st its lead tank regiment proceeded west. Rommel, for a change, was not with his spearhead but was driving around prodding the rest of his division to close up. The SS motorized division Totenkopf (Death's Head) had arrived from OKH reserve to take position on his left. The 5th Panzer was behind and to the right, due east of the city.

In the afternoon Frankforce crossed its start line in two columns, Martel with the lead tanks in an open car. The right-hand column immediately ran into Rommel's motorized infantry and blasted its way through. Scores of Germans surrendered while others were cut down or overrun by the tanks, their anti-tank fire having bounced harmlessly off the thick-skinned Matildas. The Totenkopf Division had its initial taste of combat in the campaign and its first line broke before the onslaught. Having taken a series of small villages, the British arrived at the hamlet of Wailly, where Rommel saw the defense was in chaos, vehicles jammed together in the streets and not enough guns shooting back. Rommel, who always did seem as comfortable commanding a battalion as an army group, succeeded in erecting a curtain of fire. "I brought every available gun into action at top speed against the tanks," he said. "Every gun, both anti-tank and anti-air-craft, was ordered to open rapid fire immediately and I personally gave each gun its target." The right-hand column was soon stopped.

The British left-hand column had a shorter route to circle beneath Arras, and inflicted even more damage. They found a German column on the road and left it a fiery wreck as enemy troops fled across the open ground. Isolated German panzers and anti-tank batteries that tried to duel with the 4th Royal Tanks were left smoldering in the wake of the advance. Rommel saw the men of one howitzer battery abandon their guns to join a stampede of infantry. Many other Germans, however, stuck to their weapons, pouring a methodical fire into the British. The Matildas finally met their match in German 88mm flak guns, hastily brought up, that fired over open sights.

By nightfall the British were forced to retire back the way they had come. In the dark a few anti-tank gunners were killed when their battery accidentally engaged Prioux's tanks on the right. Otherwise the French were able to fend off counterattacking panzers as Rommel's lead regiment doubled back to help in the battle. By the end of the day, the Germans had lost over 400 men, 9 medium and a number of light tanks, plus dozens of trucks and guns. Word spread through the German Army that the Totenkopf Division had panicked under fire, though there was little evidence the men behaved differently than any other infantry surprised by tanks. At this stage in the war, SS units were seen as an unwelcome aberration in the Wehrmacht, and Totenkopf was all the more odious because its commander and many of its personnel were drawn from the Nazis' concentration camp network. Any unkind rumor about their performance would be readily accepted, even as the numbers of prisoners lost— 173 from 7th Panzer and only 2 from Totenkopf—argued that the SS had not given up easily in the fighting.

The British lost 14 Matildas and 32 light tanks. It can be said that the attackers could not hold on to their penetration because they hadn't brought enough infantry along; however, Gort had no intention of extending his front farther south in the first place. He had been ordered to attack and he had done so. Now he could return to pondering how to save his army. On the British left, the belated French attack took place the next day but consisted of only one reinforced regiment, which fought well but failed to dent the panzer corridor.

In the German High Command the action at Arras rekindled anxiety, partly because Rommel thought he had been attacked by five divisions. Not only Hermann Hoth's corps of 5th and 7th Panzer stayed in the area to counterattack, but Reinhardt's corps of 6th and 8th Panzer was ordered to hook north toward Bethune above Arras. Guderian's corps was even stripped of 10th Panzer to be held in reserve against a renewed British effort. Since Guderian, in his drive up the coast, had originally earmarked 10th Panzer to advance upon the port of Dunkirk, the latter decision had far-reaching consequences.

At this juncture, the French, British and German high commands were united in their point of view: the only chance of retrieving the Allied position in France was to sever the panzer corridor, cutting off the German spearheads to reunite the Allied front. To Lord Gort and his subordinate commanders in the BEF, however, evidence had mounted that the British had inadvertently tied themselves to doomed allies. With depressing regularity, both French and Belgian formations melted away in the face of German attacks, untrained or unwilling to face the onslaught. Not being susceptible to moral or physical collapse, the British indeed possessed the only troops who might have cut through the German lines, but to commanders on the spot there was no point in doing so. The French were well en route to the greatest military catastrophe in their history. The BEF needed to disengage while it still could.

French commander-in-chief Gamelin had meanwhile been replaced by Maxime Weygand, who, at the age of 73, exuded confidence and fighting spirit. Churchill felt that under his leadership the legendary strength of the French Army would at last reappear. While Frankforce was slugging it out with 7th Panzer and its supporting troops, Weygand was in the northern pocket conferring with General Bilotte and King Leopold, commander of the Belgians. Elaborating on the recurrent theme, Weygand devised a plan for eight divisions of the northern armies to attack toward a powerful French group coming up from the south. They would meet in the center of the panzer corridor, which was barely 20 miles wide at key points. Just a short advance from both sides would reunite the Allied front. Thereafter it would be Guderian and the German spearheads isolated and pinned against the coast—the tables would have turned. For reasons that have remained vague, Lord Gort was unable to attend the meeting; but such was the poor state of Allied communications that he could plausibly say he didn't know when or where it was taking place.

At the English Channel, Guderian spent May 21 sightseeing in

Abbeville while waiting for further orders. Paris was ripe for the picking to the south, as was the Channel coast to the north. In the evening his next objective was confirmed: roll up the coast and seize the ports. 2nd Panzer was promptly dispatched to take Boulogne while 1st Panzer attacked Calais. When 10th Panzer returned to Guderian's command after the scare at Arras, it was devoted to Calais while 1st Panzer and the Grossdeutschland Regiment, backed by the SS regiment Leibstandarte Adolf Hitler, were sent on to Dunkirk. On the right of Guderian's corps, not only Reinhardt and Hoth's panzer corps were pressing against the rear flank of the BEF but Hoepner's corps, comprised of the 3rd and 4th Panzer Divisions, had arrived to join them from Bock's Army Group B. Nine of Germany's ten panzer divisions were in the same area, attacking the coast or the rear flank of the Allied pocket. In the east Bock's infantry, with Reichenau's Sixth Army its cutting edge, steadily fought its way forward against the Allied front.

The BEF, which prior to Arras had suffered only 500 combat casualties, was now fighting on all sides. The Royal Navy landed more infantry and tanks to help the French hold Boulogne and Calais. In the east German Sixth Army pressed the British—who were struggling to maintain contact on their left with fleeing Belgians and on their right with retreating French—and gained a bridgehead over the Scheldt at Oudenarde. Above Arras, Rommel, like a hunter who had been surprised by a wildcat, was now stalking his prey, attempting to circle behind the British 5th and 50th Divisions while 5th Panzer attacked the city head-on. On the west side of the Allied pocket, Gort was attempting to reinforce the ad hoc units of drivers and supply troops he had scratched together to hold off Guderian, Reinhardt and Hoepner. At the port of Boulogne the fighting resembled a Napoleonic battle. According to Guderian:

> The attack on the town itself assumed a curious form, since for some time neither our tanks nor our guns managed to penetrate the old town walls. By the use of a ladder from the kitchen of a nearby house, and with the powerful assistance of an 88mm flak gun, a breach was at last made in the wall near the cathedral and an entry forced into the town itself.

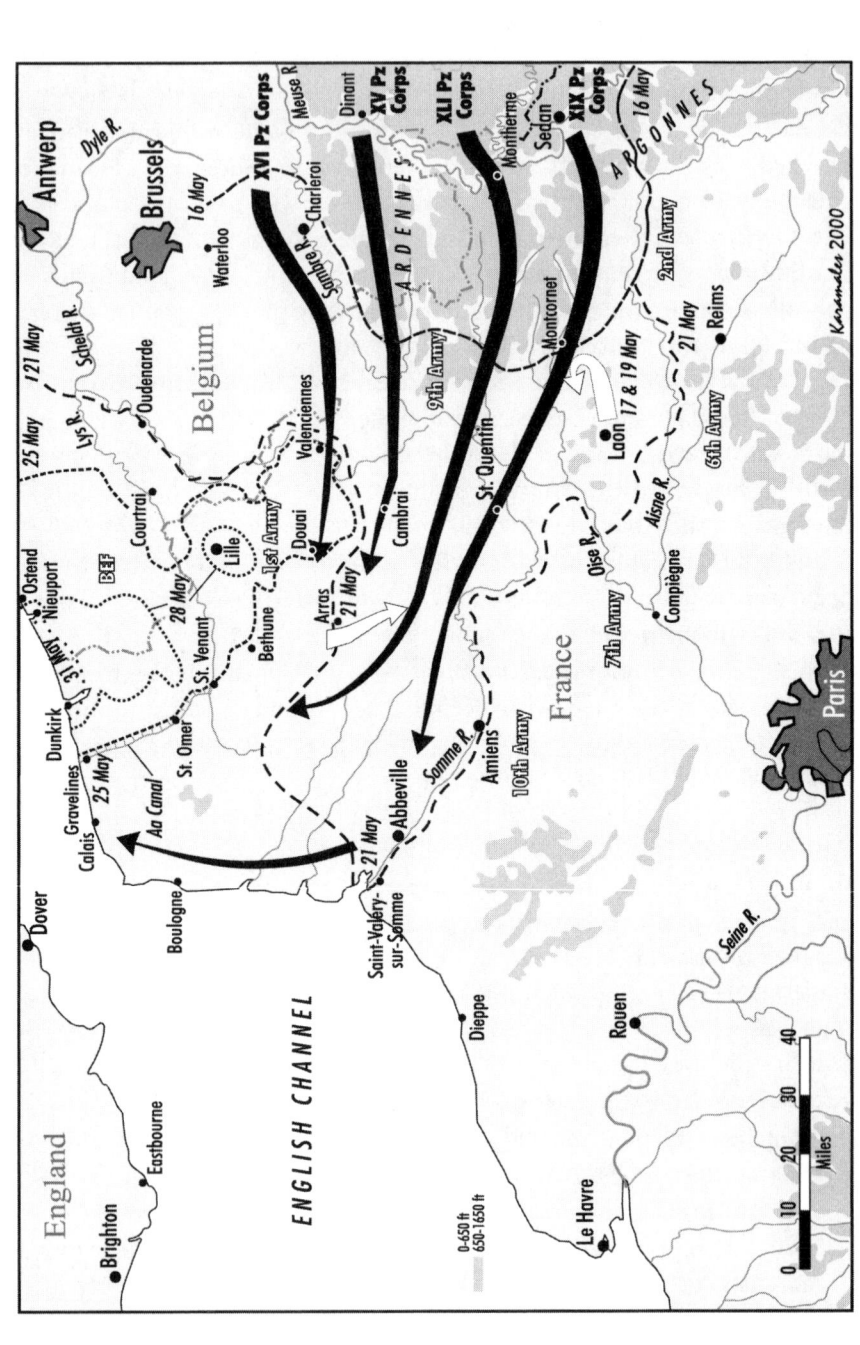

German troops clambered through the breach and scaled the ladder to get atop the walls. At the harbor, British destroyers arrived to take off the garrison, though a motor torpedo boat was sunk and other vessels damaged by German tank fire. Churchill rued the hasty evacuation of British troops from Boulogne and subsequently ordered the garrison at Calais to hold fast. Even if the BEF couldn't make its way to Calais, Brigadier L.N. Nicholson and his Guards could tie down 10th Panzer Division, keeping it away from Dunkirk. Guderian told the commander of 10th Panzer to take his time and avoid casualties, since he would soon be reinforced by the heavy artillery freed up at Boulogne.

General Weygand, meanwhile, had sold his grandiose plan for a concentric attack on the panzer corridor to the French and British leadership; however, it was a fantasy. On the night of the 23rd Lord Gort ordered his men to retreat north from Arras, abandoning the supposed start line of the attack to the south. An alarmed Premier Reynaud immediately wired Churchill for an explanation. Churchill replied, "We have every reason to believe that Gort is still persevering in southward move." To the French and Belgians, however, the suspicion that the British were intent on bugging out, leaving the continental armies to their own devices, formed a dark cloud over Allied relations. A confusing factor was that there was still no evidence of Weygand's massive push north into the German cordon. The French south of the Somme seemed defensive-minded in the face of numerous German bridgeheads, more determined to protect Paris than to attack north to reach the separated part of the front.

As for the northern army group, at the Aa Canal, which ended at Gravelines just 15 miles from Dunkirk, the British were unable to hold the Germans. The 1st Panzer, 6th Panzer and SS Verfugungsdivision (later renamed "Das Reich") all forced bridgeheads across the waterway. The last natural barrier on the side of the Allied pocket that faced the panzer group was penetrated. After the confusion of their retreat, the BEF had no firm line of resistance in its rear. The men were fighting courageously but had no means of stopping the panzer breakthroughs. One more leap and the Germans would hold the coastline, trapping the BEF in the interior. General Alan Brooke, commanding its II Corps, wrote in his diary on May 24, "Only a miracle can save the BEF now."

On May 24, as if in answer to Brooke's prayers, Hitler issued his second "panzer halt" order. The Germans were henceforth strictly forbidden to cross the Aa Canal and instead, according to the dispatch, were to "make use of the period of rest for general recuperation." The panzer commanders on the spot, the smell of final blood in their nostrils, were nonplussed. Chief of the General Staff Halder and commander of the Army, Walter Brauchitsch, were beside themselves with disgust at Hitler's interference in the battle just an inch short of its climax. At the time of the order, 1st Panzer Division was closer to Dunkirk than the bulk of the BEF, while French 1st Army was even farther to the south of the 40-mile-long salient, concentrated near the city of Lille. It's possible that if the stop order had come one day later the entire Allied pocket would have been surrounded, Britain's only standing army marched into German POW camps.

The "panzer halt" order has, in the decades since, become the "Who killed Kennedy" of World War II literature. So many clues and theories abound, from simplistic to exotic, that the definitive answer becomes more elusive the more one looks at the question. The most prosaic reason put forth for the order was that Hitler had been advised the area around Dunkirk was bad tank country. Crisscrossed by canals and dotted with marshes, it would indeed have made sense, time permitting, for the pocket to be reduced by newly arriving infantry divisions and artillery rather than by Germany's valuable panzers. The greater part of the French nation still stood intact south of the Somme, and the panzer divisions—their vehicles already reduced by up to 50 percent—would soon be needed elsewhere.

On the exotic side, Hitler has been quoted at this time extolling the virtues of the British Empire, comparing it to the Catholic Church for providing a stabilizing force in the world. If Britain, a racially-compatible nation in Nazi terms, were conquered, its overseas possessions would only fall to the Americans and Japanese; instead, Hitler desired peace with Britain and considered that by letting its army go he would be making it easier for London to honorably accept an armistice. Another theory in the geopolitical vein, also stemming from the Nazi leader's musings, was that Hitler desired the final battle of annihilation to take place in France, not Belgium. The Allied pocket straddled the border between the two countries; however, if the panzers, having gotten into the Allied rear, continued pushing east the Allies would be forced entirely onto Belgian territory. Hitler was self-conscious over violating Belgian neutrality, this theory goes, viewed Belgium as a potential model state in the New Order, and therefore desired that the Allies be obliterated on the geographic territory of France.

The reason for the order explained to commanders at the time was that the pocket could be left to the Luftwaffe, without ground troops having to suffer unnecessary casualties. This gets more to the point because Hermann Goering, according to all accounts, interjected himself into the issue of how to finalize the victory over the Allied elite. Airpower was an unknown quantity at the beginning of the campaign, yet the Luftwaffe had quickly gained a fearsome reputation for obliterating its objectives. Given a stationary target, again with time permitting, Germany's bombers alone could force the Allied holdouts to their knees.

Although after the war all fingers were pointed at Hitler alone, the genesis of the "panzer halt" occurred on May 23 with a call from Gunther von Kluge, commander of 4th Army, to von Rundstedt, in which he said, "The troops would welcome a chance to close up tomorrow." Rundstedt, an elderly general who had not yet appreciated firsthand blitzkrieg at its cutting edge, was more than willing to agree. The freelancing panzers had been hard enough on the nerves of Army Group headquarters when they were breaking through open country; now that a set-piece battle was to take place in which the Germans held a preponderance of force, there was no longer a need for the panzer leaders to advance recklessly on their own initiative. On May 24 Hitler visited von Rundstedt and placed his considerable weight behind the decision. Instead of a vague directive from Army Group of the kind Guderian had sidestepped on May 17, the panzer halt became a "Führer order" that no one could disobey.

Hitler's motives may have extended beyond merely rooting the German Army on toward the most stunning victory of the century. The Army was his only remaining rival for prestige in Germany, and its breathtaking success against France and Britain carried mixed implications. What Goering was whispering into Hitler's ear was that the Luftwaffe, a younger and more Nazified service, should for political reasons apply the coup de grace. At the same time, the panzers had already done their job and would be needed intact for the second stage of the battle against France south of the Somme. To Hitler, the geopolitical factors of eventual peace with Britain and sparing Belgian sensitivity might even have been part of his thinking.

Of course, since the "panzer halt" turned out to be one of the worst German decisions of the war, prompting an endless series of dismal consequences for the Third Reich, in retrospect no combination of reasons suffice to justify it. The big mistake was that it simply hadn't occurred to Kluge, Rundstedt, Goering or Hitler—or, in fact, to many Allied leaders—that the trapped armies could be evacuated by sea. Such an operation, the flip side of an amphibious invasion on the scale of the future Normandy landings, was inconceivable on such short notice. By May 24, 90 triumphant German divisions lined the narrow pocket and the Luftwaffe ruled the skies above. In the view of the landlubbing Germans, on the verge of a battle of annihilation, how could such an evacuation possibly be executed?

Since May 20, in a cave dug deep into the white cliffs below Dover Castle, British Admiral Bertrand Ramsey had been working day and night, calling in warships of the Royal Navy while rounding up channel ferries and seagoing barges. The BBC had also issued a request for private owners of boats 30 to 100 feet long to register their craft with the Admiralty. The idea that the very survival of the British Army might depend on Ramsey's efforts seemed almost surreal in his bleak, concrete-lined cave. The office from which the Admiral attempted to patch together a rescue fleet had once housed an electrical generator for Dover Castle—in British usage, a "dynamo."

On the continent, while the panzers waited, British 5th and 50th Divisions were still earmarked for an attack south in accordance with General Weygand's orders. As part of the plan, the Belgians and French 1st Army had shortened and then extended their fronts in order to free three more British divisions, the 2nd, 44th and 48th. On May 25, however, Reichenau's Sixth Army broke through the thinly held Belgian right near the town of Courtrai, at the junction between their army and the BEF. The Germans pouring through the gap threatened to link with the panzers, should they move again, and cut the Allied pocket in half. That day Lord Gort, on his own initiative, made the most crucial British decision of the campaign. Pressure from Weygand and the French, as well as from his own superiors, would henceforth be ignored: his job was to save the BEF. Gort abandoned all pretense of attacking south and ordered his 5th and 50th divisions north to plug the gap at Courtrai. Simultaneously, his 2nd, 44th and 48th divisions were dispatched west to fortify the "canal line" against the German armor. The rest of the BEF was ordered to head for the Allies' only remaining port: Dunkirk.

Not just following the campaign but during the last week of it, John Gort slipped into obscurity. During the final days of the BEF's struggle in France he lost touch with his units and after the battle was not reappointed to field command. It was as though his defiance of his own and French political leadership by uttering the simple word "retreat" caused him to employ his last reserve of courage. Fortunately, the BEF possessed other individuals, including future field marshals Brooke, Alexander and Montgomery, who could implement Gort's concept in the face of German pressure. The BEF had received its mandate from its own commander: "escape." By May 26 the War Office in London had come around to Gort's thinking and ordered Operation Dynamo to commence. The ships were on their way.

On the afternoon of May 26 the German High Command, seeing the chaotic Allied pocket suddenly transforming into a hardened shell, lifted the "panzer halt" order. Attacks began immediately, though now the opposition was tougher. The lead regiment of the SS Verfugungsdivision, which had seized the town of St. Venant astride one of the BEF's withdrawal routes, was knocked back by a fresh British brigade. Attacking panzers were held up by British strongpoints or assailed by counterattacks. On the coast Calais finally fell, but it had been a bloody, time-consuming struggle. Guderian said, "We took 20,000 prisoners, including 3–4,000 British, the remainder being French, Belgian and Dutch, of whom the majority had not wanted to go on fighting and whom the English had therefore locked up in cellars."

On May 27 the entire canal line erupted with full-blooded attacks by German panzer, motorized and infantry divisions. At Gravelines on the coast a French division was able to hold up Guderian. Inland, bypassed British units held on to strongpoints in the German rear, fighting desperately to protect the general retreat. The BEF's 2nd Division lost two-thirds of its strength, but nowhere allowed the panzers their customary free rein. The Germans' only clear success was won by Rommel's 7th Panzer, which cut through the Allied front near Bethune and got above Lille, where half of the French 1st Army was grouped, fighting to the east and south.

That day everyone's favorite division, Totenkopf, committed the worst atrocity of the campaign when one of its battalions gunned down 100 British prisoners near a town called Le Paradis. A detachment of the 2nd Royal Norfolks had been surrounded by the SS at a farmhouse yet refused to give up, inflicting over 60 casualties on their attackers. Finally, with the majority of men wounded and the rest out of ammunition, the British surrendered, only to be lined up against a barn wall and shot. SS moved among the wounded to finish them off, but two Brits survived under the pile of bodies. Their reports to German Army medics caused an uproar, and then a larger one later at the Nuremberg Trials, after which the Totenkopf battalion commander, Fritz Knochlein, was hanged.

The Luftwaffe, meanwhile, attempted to fulfill its promise to destroy the BEF from the air, and on the 27th, Dunkirk was plastered by waves of bombers. Over 1,000 people were killed in the town, and the harbor facilities were wrecked. Oil tanks near the shore were blown up, and for the next week a gigantic funnel of black smoke rose thousands of feet into the sky, an ominous sight to troops marching to the port from inland but a handy signal marker for ships. Admiral Ramsey's destroyers, ferries and barges were now streaming toward France to rescue the BEF. Less than 8,000 troops were brought off that day, not nearly enough, but Royal Navy commanders decided to use the miles of beaches stretching east from Dunkirk for further evacuations.

On the morning of the 28th the Belgian Army, which had been holding the east flank of the Allied pocket by the sea, surrendered to the Germans. King Leopold was harshly criticized by Churchill for this move, but then the Belgians, having already lost their country, would have been rare saints to continue bleeding in order to allow the British time to flee. Fortunately, Bernard Montgomery, who had executed a complicated night march to take position on the BEF's northern flank, anticipated the dilemma and received permission from Brooke to extend his men further to cover the new gap in the lines. While the Germans assimilated two hundred thousand new prisoners, Montgomery, along with assorted French units, was able to prevent exploitation of the Belgian collapse. Brian Horrocks, commanding the 3rd Division's machine-gun battalions, later wrote that the battle around Dunkirk was more difficult than the fighting he later encountered at Normandy.

On the 28th, 6,000 men were taken off the beaches and nearly 12,000 more from Dunkirk harbor, but the seaborne evacuation was still proceeding too slowly. Gravelines fell to 1st Panzer Division and, at the other end of the pocket, Ostend was lost and Nieuport was under attack as a result of the Belgian capitulation. Worse, General Prioux, the new commander of French 1st Army, was refusing to devote himself to retreat. The next morning Rommel's panzers joined hands with Reichenau's Sixth Army to slice off five French divisions around Lille. The French continued fighting, but the effect was to halve the Allied pocket. Toward the coast, a huge flow of men rode or hiked through an expanse of abandoned equipment and supplies. The BEF had begun the campaign as the only fully mechanized combatant, but now British bulldozers churned down the roads, knocking abandoned lorries and armored cars into ditches on either side. One of the Germans in pursuit, Kurt Meyer of the SS Leibstandarte, wrote, "The roads to the north are completely blocked. Endless columns of English trucks, tanks and guns leave them useless for any other traffic. The amount of equipment left behind is enormous." As the front contracted, Germans probed through woods and fields for each new crack in the lines. Both sides sprung sudden ambushes, and murderous firefights flared on the perimeter.

At the harbor the British discovered that even though docks and facilities had been destroyed, ships could still pull alongside the moles to take on troops. The BEF was leaving all its equipment behind, and it was only necessary for men to scramble aboard. Royal Navy Commander John Clouston ruled Dunkirk's eastern mole with an iron hand, forming the evacuees into disciplined groups and assigning them ships. (Clouston, who was killed four days later, shepherded about 200,000 men to safety.) The primary vessels were destroyers, coastal ferries, cargo steamers and 40 Dutch "skoots"—large, motorized barges—that had made it to England from the the Netherlands. Each of these vessels could take on 600–1,500 men at a time. Pulling men off the beaches was the more difficult problem since deepwater ships couldn't get close enough. They'd stand off-shore while their lifeboats went back and forth to get men.

Hermann Goering's belief that the Luftwaffe by itself could prevent an evacuation was being challenged on several counts. First, bombs dropped on sand tended to have little effect; more important, the fine weather that had held up through the month of May had by now turned intermittently bad. German planes were frequently grounded while British ships continued to ply their trade. Goering's biggest problem, however, was that German air superiority was not as great as he had thought. For the first time, British Bomber Command was making tactical strikes on advancing German troops; coastal defense squadrons were crossing the Channel; and, with the most impact of all, British Fighter Command had finally entered the battle with all its resources.

The Germans had done an impressive job advancing their airfields throughout the campaign, but these were still no closer to Dunkirk than the RAF's airfields in England. Air Marshal Dowding, who had always said, "The continued existence of the nation, and all its services, depends upon the Royal Navy and Fighter Command," was now proving his point. Once stingy at releasing squadrons of Hurricanes for the Battle of France, Dowding was throwing in every available squadron of both Hurricanes and Spitfires for the rescue of the BEF. The British reserve squadrons were not only fresher than their German counterparts, who had been flying nonstop missions since May 10, but in quantity and quality the Luftwaffe was meeting its match. The fearsome Stukas were found to be alarmingly vulnerable to fighter attack. German Dornier and Heinkel level bombers were repeatedly dispersed by waves of British fighters tearing through their formations. German Me-109s were not exactly overwhelmed after meeting the true strength of Fighter Command, but they had nevertheless encountered a serious challenge. Dogfights swirled over the Allied bridgehead whenever the weather allowed. Regarding Dowding's release of his Spitfires, John Terraine commented:

> It would be agreeable to record a stunning immediate success for these superb aircraft, the pride of the Command; that, however, would be an exaggeration. In this high-speed, split-second combat, experience was probably the most important single factor. And those who had it fought at an advantage over those who lacked it. A fair number of Hurricane pilots now had some very useful experience, and in due course steps were taken to diffuse this throughout Fighter Command, but on May 27, when Dynamo began, there was not yet time for this to happen.

On the ground, as the bridgehead shrank and the density of defenders correspondingly increased, the panzer divisions were no longer measuring their success in dozens of miles but in terms of orchards, farmhouses and streams seized against ferocious opposition. The Allied troops were slugging it out with an intensity rarely seen in the heady days of breakthrough. As Telford Taylor wrote, "The panzer generals who had waxed wroth at the stop-order now turned decidedly cool to the further use of their tanks." German heavy artillery had already been brought within range of Dunkirk; the offensive south of the Somme was to begin in less than a week. On the 29th, the panzers were pulled from the battle. Once the panzer spearheads were withdrawn, the German offensive—now comprised of newly arrived infantry divisions under separate army commands—lost much of its savage impetus. In contrast, the British evacuation, which had begun amidst the confusion of rear echelon troops and stragglers, became more disciplined as the BEF's front-line fighting men fell back on the bridgehead, and naval personnel arrived to coordinate movement on shore.

On the gray morning of May 29, the Royal Navy officers at Dunkirk harbor were beginning to feel their first sense of optimism. Thousands of troops were coming off each hour; the predictions of British leaders that only 30–50,000 men of the BEF could be pulled to safety had already been exceeded. That afternoon, however, for the first time in two days, the skies cleared and engines revved to life on a score of German airfields. The east mole was lined with a dozen ships when the bombers arrived overhead. The first bomb landed right on the mole, killing a number of soldiers. The destroyer *Jaguar*, packed with troops, pushed off but was chased by bombers and disabled. The destroyer *Grenade* took a direct hit and became a blazing hulk. She drifted off, threatening to block the harbor channel, but a trawler pulled her away, leaving her to burn for several hours before finally vanishing in an explosion. Back at the mole, a large steamer and some trawlers were sunk at their berths while another steamer, *Crested Eagle*, got underway only to be bombed and strafed end to end. Any feelings of optimism were put back on the shelf.

On the beaches there was no protection from the Stukas and fighters that came racing in, blasting holes in the long lines of men with bombs and machine-gun fire. There was no avoiding, either, the sight of limp bodies floating in on the tide from wrecks offshore, only to come to rest in ghastly piles at the waterline. A giant cargo vessel called the *Clan McAllister* attracted the Luftwaffe's attention and was hit repeatedly. It sank in such shallow water, however, that the abandoned wreck continued to attract German planes in the days to come.

On May 30 the weather thankfully closed in again and the perimeter continued to hold, but at any moment German ground troops could break through, bringing the evacuation to a halt. Admiral Ramsey realized that the Royal Navy, despite the fact that all warships in home waters had been assigned to the evacuation, needed to call upon its auxiliaries. That afternoon, British troops, queued in endless curling rows on the beaches, were astonished to witness the approach of an odd fleet. The small ships had arrived. Yachts, fishing boats, dredgers, tugs, coal-haulers and every sort of civilian craft able to traverse the Channel were coming to rescue the BEF. Many of them were sailed by their owners and original crews but others were manned by Navy seamen. Some of the craft loaded up and sailed troops back to England, but the majority shuttled between the beaches and larger ships waiting offshore. On that day 30,000 troops were taken off the beaches while another 24,000 were pulled from the harbor. It was on that day, too, that the Germans finally realized that their prey, once considered finished, was escaping en masse. German landpower had run up against British seapower, and, as enemies from Philip II to Napoleon had learned over the centuries, Britannia ruled the waves.

Alan Brooke, whose masterful handling of II Corps was primarily responsible for keeping the BEF front intact during the last week of May, was ordered back to England, tearfully relinquishing his command to Montgomery. In a meeting at Gort's headquarters, I Corps' Michael Barker was ordered to hold the western flank while II Corps evacuated, and, if worse came to worse, he was authorized to surrender his men to the Germans. According to Montgomery,

> The effect on Barker was catastrophic; he was incoherent at first and then relapsed into silence.
>
> I knew Gort very well, personally.
>
> After we had broken up I got him alone and told him that we could not yet say it was impossible to get I Corps away; but that it would *never* get away if Barker was in control, and that the only sound course was to get Barker out of it as soon as possible and give I Corps to Alexander.

Harold Alexander took command of I Corps and skillfully held off the Germans in the west while steering a steady flow of troops to the harbor and beaches. On May 31, 68,000 men were evacuated. As the BEF methodically poured across the Channel, the retreat increasingly owed its continuing life to the fierce resistance of French 1st Army, stubbornly holding its ground on the perimeter. At a meeting in Paris, Churchill was embarrassed to admit that so far 90 percent of the evacuees were British; henceforth, French and British would be taken off in equal numbers. Also, as Churchill recorded, "His Majesty's Government had felt it necessary in the dire circumstances to order Lord Gort to take off fighting men and leave the wounded behind." (Many wounded were brought off nevertheless.) On June 1 another 64,000 men were rescued, making over a quarter of a million in less than a week.

That day the skies were clear again and the Luftwaffe exacted a cruel toll from the rescue fleet. The RAF had decided that, instead of constant small patrols, it would send over fighters in larger numbers but only at intervals. The Germans were able to hit Dunkirk in between those intervals and 31 ships were sunk, troops on shore similarly bloodied. Of 41 British destroyers committed to the evacuation, only 9 remained afloat unscathed. Almost defenseless against German planes, British soldiers became resentful at the absence of the RAF, but most of Fighter Command's efforts took place out of sight of the beaches. After Dunkirk there was no trace of contempt for the RAF in accounts from Luftwaffe pilots.

On June 2 the last 4,000 men of the BEF's rearguard were brought off, though by now evacuations had been suspended during daylight hours. At night the great improvised rescue fleet suffered confusion and collisions, and on the way back to England encountered German motor torpedo boats that only came out after dark. Mines also took a terrible toll of vessels, and off Dunkirk there were so many wrecked and half-sunken ships that navigation became a nightmare. But Operation Dynamo was winding down. 70,000 French were brought off in the last few days until, on the morning of June 4, German troops entered the burning town. The remaining rearguard, 40,000 men of French 1st Army, laid down its arms.

While the Germans surveyed the material wreckage of a once-great army, a London newspaper expressed the British mood with its headline "Bloody Marvellous!" In nine days, 338,000 troops, two-thirds of them British, had been rescued from certain destruction. Aside from huge relief at the escape of British sons, brothers and husbands, the participation of the small ships provided an almost joyful boost to British morale. It was as though the entire seafaring nation had risen as an indefatigable whole to defy the Nazi war machine. It fell to Churchill to remind his jubilant countrymen: "We must be very careful not to assign to this deliverance the attributes of a victory. Wars are not won by evacuations." Still, the BEF's successful retreat, climaxing with the island nation extending its strong arm across the waves to bring back its men, signalled the prospect of future victory for those who refused to recognize defeat. As historian Walter Lord wrote, "History is full of occasions when armies have rushed to the aid of an embattled people; here was a case where the people rushed to the aid of an embattled army."

And the cost had been high. Of 861 vessels engaged in the evacuation, 243 were lost. Due to the fact that many vessels were filled to the brim with passengers when they were hit, the total of sunken ships represented an even more gruesome sum of lives. The BEF suffered 68,111 casualties overall. The RAF lost 177 planes during the rescue stage, of which 42 and 57 were, respectively, Spitfires and Hurricanes of Fighter Command. The Luftwaffe lost 132 planes attempting to stop the evacuation. The most serious loss to the British was all the BEF's tanks, trucks, guns and supplies, which meant that the army would be temporarily defenseless should the Germans attempt to cross the Channel. France, in addition, had lost its best army, the 1st, with all its equipment; however, the 120,000 French pulled off at Dunkirk were hurriedly put back aboard Royal Navy vessels and returned to their homeland to face the second phase of the German offensive, about to be launched from the line of the Somme.

On June 5, the day after Dunkirk fell, the German drive to conquer the remainder of France commenced. The British continued to reinforce their ally with amphibious landings of troops, but the panzers had once again found their element. The French fought back with renewed vigor and gallantry, but they no longer had the strength to resist the German onslaught. Mini-Dunkirks soon had to take place all along the French west coast as the British evacuated their 1st Armoured, 52nd Lowland, 1st Canadian Division and other units. The 51st Highland Division was trapped and captured by Rommel before it could be taken off. In a tragedy kept secret for years, the ocean liner *Lancastria* was destroyed off St. Nazaire with the loss of 3,000 men. Nevertheless the British had clearly shown that, despite their retreat, they were still in the war, determined to fight any way they could. Four summers later, many of the same British soldiers evacuated in 1940 would return to the continent, at Normandy, to liberate France from Nazi occupation.

On June 18, Adolf Hitler danced a jig as he waited to confront French representatives with their offer of surrender. He insisted that the capitulation take place in the same railroad car (preserved by the French) in which the Germans had conceded defeat in the Great War. One can forgive the Germans if, despite the escape of the BEF, they considered their campaign against the Western powers a success. France had surrendered and Britain had been rendered irrelevant. In the late summer, Goering was given a chance to subdue the British Isles with his Luftwaffe, but the task proved beyond his means. Hitler soon returned to his primary concern, the East, where, after the Battle of France, Joseph Stalin was forced to cling harder than ever to his non-aggression pact with Germany. Or, as Churchill commented:

> While to uninformed Continentals and the outer world our fate seemed forlorn, or at best in the balance, the relations between Nazi Germany and Soviet Russia assumed the first position in world affairs. . . . The Soviet leaders had been shocked at the fall of France, and the end of the Second Front for which they were so soon to clamor. They had not expected so sudden a collapse, and had counted confidently on a phase of mutual exhaustion on the Western Front. Now there was no Western Front!

The speed of the German victory, combined with the near-miraculous seaborne evacuation of the Allied northern pocket, meant that Western populations, as they had hoped, were spared a prolonged bloodbath on the model of 1914–18. However, by the end of 1940 Hitler's appetite for aggression had still not been sated. The most terrible phase of the greatest war in history was still to come.

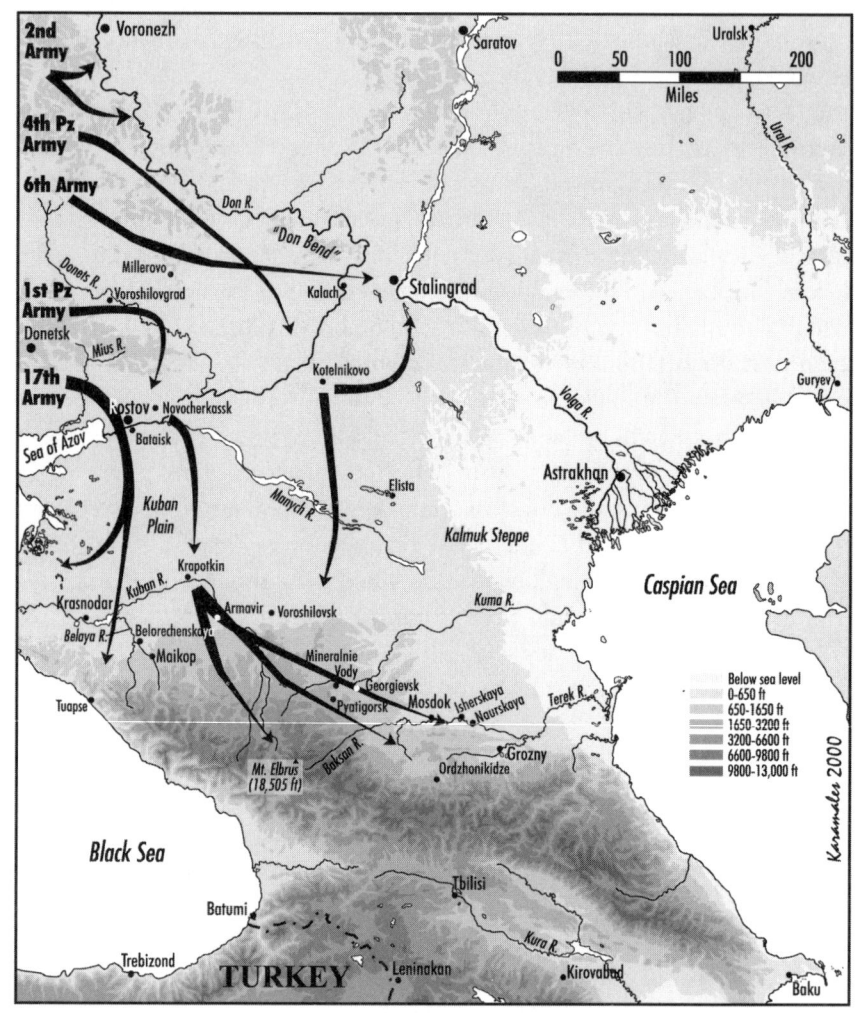

2nd Army • Voronezh

4th Pz Army

6th Army

Saratov

Uralsk

0   50   100   200
Miles

Ural R.

Don R.

"Don Bend"

1st Pz Army
Donetsk

Donets R.   Millerovo
Voroshilovgrad   Kalach   Stalingrad

Mius R.

17th Army

Kostov • Novocherkassk

Kotelnikovo

Guryev

Sea of Azov   Bataisk

Volga R.

Kuban Plain

Manych R.

Elista

Astrakhan

Kalmuk Steppe

Krapotkin

Kuma R.

Caspian Sea

Krasnodar   Kuban R.   Armavir • Voroshilovsk

Belaya R.   Belorechenskaya

Maikop   Mineralnie Vody

Tuapse   Georgievsk
Pyatigorsk   Mosdok   Isherskaya
Naurskaya   Terek R.

Mt. Elbrus
(18,505 ft)   Baksan R.   Grozny
Ordzhonikidze

Below sea level
0-650 ft
650-1650 ft
1650-3200 ft
3200-6600 ft
6600-9800 ft
9800-13,000 ft

Black Sea

Tbilisi

Batumi

Trebizond   Leninakan

TURKEY   Kura R.   Kirovabad

Baku

Karandes 2000

Chapter 5

FIRST PANZER ARMY
IN THE CAUCASUS

Once the Battle of France was won, Adolf Hitler controlled the "Old World," with the single exception of the island nation that had barely been able to pull its men away on fishing boats from beneath the muzzles of his panzers. Together with Italy—now a junior partner, along with Hungary, Romania, Bulgaria and Finland—and with the acquiesence of Spain and Portugal, Hitler held sway over a greater European empire than Napoleon had achieved, or Charlemagne or Caesar. During the following spring, Germany added Yugoslavia and Greece to its domain. Sweden was left alone as long as it continued to provide copper and iron ore; Switzerland could be dealt with at any time, though no one in OKW (German Armed Forces High Command) was anxious for that eventual war in the Alps. Nazism had succeeded not as a pan-European philosophy, but as an engine that propelled Germany to unprecedented control of Europe through conquest.

Hitler, however, realized that conquest of Europe alone was a geopolitical anachronism. In the industrial age, power depended upon access to natural resources, and Europe, despite being a focal point of military ambition throughout recorded history, was a poor prize to win in the mid-twentieth century. Britain could still summon more raw materials through its commonwealth and colonies than Germany could on the mainland. America had unlimited resources of its own, as did the Soviet Union. Even Japan had thinly defended sources of nature's riches within its grasp. If Germany rested on its laurels as the dominant power over such as Denmark, Holland and France, without establishing a pipeline to the natural wealth of the rest of the world in oil, nickel, chromium, manganese, rubber and other essential materials, its empire would be drastically short-lived.

Part of the euphoria in Berlin over the quick defeat of France and Britain in 1940 was due to the fact that Germany had taken a huge gamble on its army's ability to win short wars. Given the increased mechanization of battlefields since 1914, the country was not able to stockpile industrial output, metals and fuel for extended conflicts. Blitzkrieg tactics had validated the risk, but in the long term Hitler's state would need raw materials, in addition to military skill, in order to survive. Now that the "war to end all wars" was being reprised, every empire outside Europe could draw on the resources of the Americas, Asia, India or Africa to accelerate its industrial program. The only empire that had no means of employing such wealth was the one with the most powerful army—the Third Reich.

On June 22, 1941, Germany invaded its new neighbor to the east, the Soviet Union, which possessed all the oil and natural resources the Third Reich would need in perpetuity. Since evidence had mounted that Joseph Stalin was employing those resources for a vast arms buildup of his own, Germany was also seeking to erase a growing military threat, as well as a social and economic philosophy—Communism—that the Nazis had successfully obliterated in Germany, but which had become institutionalized in the huge Soviet state.

From a purely military perspective, Hitler had the same problem with Russia as Napoleon had following the Treaty of Tilsit. Both dictators were at war with Britain after subduing the rest of Europe, but the British were not a serious offensive threat. Russia on the other hand—huge and nearly unreachable on the fringe of the continent—possessed the strength to intervene in Europe whenever it wished. The dominant power in central Europe could scarcely feel comfortable while the Eastern behemoth looked on, harboring its ambitions while clandestinely conspiring with "perfidious Albion." The Germans, nevertheless, were well aware of Napoleon's experience and had no wish to repeat it. The difference between Napoleon's and Hitler's invasion of Russia is that Germany was not seeking a diplomatic solution or a lasting intimidation of a foreign ruler; the Germans needed to seize and control Russian resources. The natural wealth of Ukraine and the Caucasus would ensure the Third Reich's enduring power and at the same time strip the Soviet Union of its military potential.

Though Germany had better reasons to go into Russia than had France in 1812, the job still remained not to share the fate of Napoleon. Fortunately for the Germans they didn't need to rely on Joachim Murat and horse cavalry as their spearhead; they trusted instead to Guderian, Hoepner, Hoth and Kleist, commanding mechanized panzer groups that could cover a hundred miles a day. They also had air fleets under Kesselring and von Richthofen that could wreak more havoc in one hour than Napoleon's artillery train could achieve in weeks. In the age of mechanized warfare the Russian state was more vulnerable to attack from the west than it had been since Peter the Great had destroyed the Swedes over 200 years before. And Peter had enjoyed more institutional support than Stalin, whose Bolshevik regime held power over the vast country through a revolutionary bureaucracy that was barely twenty years old.

The premise of the invasion, which was the biggest military operation of all time, was that the Soviet Union would collapse quickly, or at least be forced to shrink behind the Urals, as the result of one gigantic campaign. Hitler had said, "Kick in the door and the whole rotten structure will fall down," and his view was seconded by most Western observers. When, incredibly, the 144 German divisions employed at the start of Operation Barbarossa achieved surprise, obliterating the Soviet border in hours, the fate of Stalin's regime seemed sealed. Within weeks of the assault, German advances were being measured in hundreds of miles, and Soviet casualties were being measured in millions.

In the decade prior to the invasion, Stalin had executed or imprisoned the bulk of the Russian officer corps, imposed draconian economic policies on the countryside, eliminated the merchant middle class, attempted to stamp out the practice of religion, and then held together the disparate nationalities of the Soviet Union through state-run terror directed from Moscow. It was difficult to imagine the Bolshevist state resisting the most powerful armed forces in Europe from such a foundation; a government that so ruthlessly controlled every aspect of its citizens' lives would never have been tolerated by a Western population in the first place. What Hitler failed to consider, however, was that Soviet Communism, which would indeed collapse of its own accord before the end of the century, did have one redeeming value: it was the perfect system with which to wage total war.

It is debatable whether the German Army could have performed any better than it did in the summer and early fall of 1941. For every tactical failure or delay, there was an unexpected, breathtaking success. In terms of ground taken and casualties inflicted within that time, the Wehrmacht set a standard that, due to the advent of nuclear weapons, will probably never be challenged. Nevertheless, the invasion failed, and by the end of the year Leningrad was still unconquered, though cut-off, and Moscow remained defiant, though burnt out German reconnaissance vehicles lay in its suburbs within binocular sight of the Kremlin. Soviet counterattacks were by then taking place all along the 1,000-mile front.

Among the surprises the Germans encountered once the invasion was underway was the primitive state of the Russian road network, which inhibited tactics depending on quick, precise movement. To panzer commanders who had trained and campaigned on European roads, the soggy dirt tracks of Russia, barely traced on their maps, created frustrating delays. The Russian weather, which could by itself defeat mobile operations as well as inflict casualties, affected the invaders as much as the Red Army. Although "General Winter" is commonly believed to have been Stalin's best ally in 1941, German generals invariably gave more credit to "General Mud." Even before the crippling cold arrived, supplies could not be delivered to the spearheads of the attacking armies. Once the ground froze the Germans could move again, but the terrible freezes that followed made many of their weapons useless. The Soviets, in contrast, already knew how to clothe their troops, what lubricants to use for their weapons, and how to design vehicles that could negotiate bad roads and snow.

In addition, the Soviets had far more tanks, guns and planes than the Germans had expected, as well as seemingly limitless reserves of manpower, mobilized by Stalin from the civilian population wherever he chose. If Stalin desired 100,000 Muscovites to dig anti-tank ditches, his order would be obeyed. If he wished a barely trained infantry division to attack a German panzer regiment, the men would do so. The Red Army was directly connected to the will of Stalin not only through the military chain of command but through the system of commissars—Communist political agents—who stood at the elbow of unit commanders to ensure faithful service to the state. Since the commissars' power was derived from the Soviet government rather than from the vicissitudes of battlefield fortune, they were more powerful than the army's officers. The Soviet leadership, thanks to its elimination of any competition for power within the state, also had the luxury of not having to care about how many casualties the Red Army suffered.

If, as many believe, the Germans' inability to topple the Soviet Union was the fault of their own bad planning, rather than due to false premises about Soviet vulnerability, the problem can be ascribed to indecision in their high command once the invasion was underway. Astonishing as it may seem, after the Wehrmacht had crushed the huge Soviet armies on or just behind the border by midsummer 1941, the German leadership was unsure what to do next.

Of the three army groups launched into Russia—North, South and Center—Army Group Center was the strongest, with two panzer groups flanking its infantry formations, as opposed to one panzer group for the others. In accordance with the opinion of the German General Staff and most of the Army's generals, Moscow was to be the primary target of the invasion. However, once the Germans had broken in, Hitler increasingly disparaged the Russian capital as a main objective. Napoleon, after all, had marched straight to Moscow and conquered that city, only to find himself holding a half-destroyed metropolis of no strategic value while the true strength of Russia swarmed around him. Hitler had no wish to repeat such an elementary error.

After two weeks of debate in August, Hitler was able to gain the support of his most illustrious armored commander, Heinz Guderian, to split off the panzer groups of Army Group Center to implement a "broad front" strategy. Guderian's 2nd Panzer Group would peel to the south to help achieve the greatest victory in the history of land warfare, centered on Kiev. Heinrich Hoth's 4th Panzer Group would maneuver north to help neutralize Leningrad. Afterwards, they would rejoin on the central front to mount a drive on the Soviet capital.

The military decision not to head straight for Moscow in the summer of 1941 may rank as the most momentous of the 20th century; on the other hand it may have been irrelevant. If Army Groups South and North had not received timely assistance from the panzer groups of Army Group Center in late summer, their offensives would have stalled; Kiev and the bulk of Ukraine would have remained in Soviet hands, and the efforts of Army Group North would have represented a tenuous stab deep into forbidding territory. A German-occupied Moscow, with long supply lines overlooked by intact Russian armies on its flanks, might have been haunted throughout the winter by the ghost of Napoleon scolding the new invaders for not having learned the first lesson about taking on the Russian state.

In retrospect, the greatest German error was to propel the advance forward on a broad front and then reassemble to take Moscow once the summer had ended. This is when they encountered debilitating weather as well as Soviet reserves raised quickly from the population or hurriedly transshipped from the east. The depleted Wehrmacht persisted in attacking well beyond the point of prudence in keeping with the theory, persistently held in Berlin, that the Soviet Union was on the very brink of political, moral or material collapse.

In the winter of 1941–42 the Red Army hammered the invaders with vicious counterattacks all along the front. Its first major victory came at Rostov in the far south, where First Panzer Army was forced to withdraw. In the north, 100,000 Wehrmacht troops were surrounded at Demyansk and could only be supplied by air. In front of Moscow, German commanders urgently begged Berlin for permission to fall back. The troops were freezing, outnumbered, outgunned and being flanked by Siberian troops cutting them off from sources of supply. Hitler was hardpressed to argue against a single request to withdraw, given that German soldiers were suffering and dying on a frontline perilously near disintegration. Nevertheless, he refused to follow in the footsteps of Napoleon and sanction a retreat. He overrode his General Staff and ordered all troops to "stand and fight" where they were.

Hitler's "stand and fight" order to his soldiers deep inside European Russia is generally credited with saving the German Army, particularly on the central front, as opposed to letting it perish in a desperate withdrawal as occurred in 1812. His refusal to admit defeat and insistence on holding ground, in this instance, were validated by the inability of the Soviet High Command to coordinate and supply long-range operations. In practice, moreover, the "stand and fight" order inspired a massive flurry of movement on the part of German units to barricade themselves in towns and villages, to ensure warmth and defensible positions. German frontline troops, who would have abandoned their guns and heavy equipment in a full-scale retreat, kept their weapons and headed to the nearest shelter to wait out the winter. The Soviet counteroffensive lapped around

George Washington assembled his largest army of the Revolution to defend New York in 1776. The stunning British victory at Brooklyn, however, prompted a headlong series of retreats and a perilous decline in American strength.

"Wars are not won by evacuations," as Winston Churchill observed, but in practice a retreat can require as much fortitude as a victory. Above, Washington successfully orchestrates the American withdrawal to Manhattan, under cover of darkness and fog.

Former British officer Charles Lee in a contemporary caricature with one of his ever-present hounds. "I must have some object to embrace," he said. "When I can be convinced that men are as worthy objects as dogs, I shall transfer my benevolence."

Despite his eccentricity, Lee was considered by many in 1776 to be America's most dynamic military commander.

Banastre Tarlton, shown here in a portrait commissioned by his mother, went on to become the most notorious cavalry leader of the Revolution. He achieved his first exploit at age 21, surprising General Charles Lee in a daring raid behind American lines.

When Napoleon invaded Russia he sought a climactic battle with the enemy. At Borodino the Russian army finally chose to fight, and after suffering heavy losses ceded Moscow to the French. Above, Napoleon is depicted at the battle by a Russian painter, Vasily Vereshchagin.

The Russians set Moscow aflame after it had been occupied by the French. "A demon inspires these people," Napoleon exclaimed. "They are Scythians!"

Matei Platov, leader of the Don Cossacks. After the French began to retreat, Cossack involvement in the campaign increased dramatically until they soon resembled a plague to the weakened French columns.

Joachim Murat, the flamboyant leader of Napoleon's cavalry. An observer of Borodino said that Murat "charged forward impetuously in the midst of his squadrons like the boiling Achilles."

Both triumph and tragedy occurred at the Beresina River during the French retreat from Russia. Napoleon's fighting strength was able to cross on hastily built bridges; however, thousands of weak and wounded fell prey to three converging Russian armies and the Cossacks.

By the time it left Russia, only a weakened remnant of Napoleon's army remained. The Emperor himself departed the retreat after the Beresina, because he needed to raise new forces in France.

Chief Joseph of the Nez Perce earned widespread respect for his bravery and eloquence. Though not a great warrior, he stayed with his people through a difficult retreat, and only surrendered to protect the tribe's innocents, women and feeble.

One-armed General Oliver O. Howard commanded a corps at Gettysburg and a wing of Sherman's army in its March to the Sea. After the Civil War he was named head of the Freedmen's Bureau.
Howard's devout Christianity eventually put him at odds with the Nez Perce, and he pursued them relentlessly across 1,500 miles.

Ollokot, Joseph's younger brother, led the warriors of their band in battle. Soldiers who afterward claimed to have seen Joseph displaying combat leadership had more likely seen his brother.
Ollokot was killed at the Bearpaw by 2nd Cavalry troopers, or perhaps by a Cheyenne.

The defile at Canyon Creek. The 7th Cavalry was unable to overcome the Nez Perce rear guard at this spot, thus allowing the great retreat of men, women and children to continue north.

The Bearpaw Mountain Battlefield in Montana. When Nelson Miles launched a surprise attack on the Nez Perce, much of the tribe streamed north to Canada, but many warriors clambered up the rocks to hold the soldiers off.

A French Char B, knocked out during the 1940 Battle of France. It not only sported a 37mm gun in the turret but a 75mm artillery piece in the hull. The tank stands as evidence that the French had not totally clarified their armored doctrine by the start of World War II.

A German Panzer III with a low-velocity 50mm gun. The Germans stressed mobility and air-ground coordination. They also grouped their tanks into divisions, their divisions into corps, and most of their corps into one huge group that found a vulnerable spot in French defenses.

French soldiers relaxing in May 1940. The presence of the German, to center right near the bicycles, indicates that these men are prisoners.

British dead, next to a destroyed or abandoned vehicle, on the beach at Dunkirk.

The British Expeditionary Force at Dunkirk was assisted by several days of bad weather, during which its men could line up for rescue without fear of Luftwaffe intervention.

A year after invading France, Hitler decided to launch the Wehrmacht against the Soviet Union, a far larger, and, as it turned out, stronger target.

First Panzer Army traversed the southern steppes of the Soviet Union, only to
encounter a geographic obstacle of mammoth proportions. Still, the Germans'
main objective was the Russian oil fields, many of which were located
north of the Caucasus range.

German 4th Mountain Division climbs the peaks, along with mules carrying
supplies and artillery. The Alpine troops were unable to hold their positions
against Soviets attacking from the other side.

Two ill-fated American generals in Korea. Eighth Army commander General Walton Walker (left) successfully held the Pusan Perimeter but then the Chinese caught him by surprise. He died in a car crash north of Seoul during the retreat. General William Dean of the U.S. 24th Division fell captive to the Communists at Taejon in the South after engaging in personal combat with North Korean tanks.

American troops advance through a Korean village during the UN drive beyond the 38th parallel.

In a shocking failure of intelligence, the United States did not realize that
hundreds of thousands of Chinese troops had entered North Korea,
determined to stop the northward U.S. onslaught.

A Chinese photo from the Fall of 1950 depicts Americans surrendering to the
People's Liberation Army.

U.S. armor heads south to Pyongyang, hoping to regroup. On the right, South Koreans head north to form a blocking position against pursuing Communists.

The road from Kunu-ri, after the U.S. 2nd Division ran a gauntlet of Chinese troops holding the heights on either side.

A North Vietnamese regiment marches south along Route 1 in April 1975, destination: Saigon. American Air Force pilots would have enjoyed such a scene during the earlier years of the war.

NVA tanks park in front of Independence Palace in Saigon, while civilian employees wait patiently, as prisoners, on the ground.

The last big clash of the Vietnam War took place at An Loc, northeast of Saigon. ARVN troops temporarily held fast, even as refugees looked for a means to escape.

A scene at the U.S. Embassy in Saigon during the last hours. Due to South Vietnam's precipitate collapse, the U.S. was unable to evacuate every civilian who wanted to leave.

After the U.S. Marines shut down the Embassy evacuation, Air America choppers continued to fly into Saigon to bring people off rooftops. Their efforts, too, concluded once the North Vietnamese had overrun the city.

these hedgehog positions and eventually receded with heavy casualties.

In early 1942, Stalin and his military council, STAVKA, despite their failure to inflict a Napoleonic debacle on Hitler, felt that they henceforth held the whip hand in the war. They had been taken by surprise, yet had absorbed the mightiest blow the Germans could deliver. The countryside before Moscow was littered with tens of thousands of tons of destroyed German ordnance as well as thousands of hastily dug graves. The prisoners taken were wasted and frostbitten, clothed in motley or in ragged remnants of summer uniforms—a testament to how badly the German High Command had misjudged its task. Stalin may not have realized that his war industry was already outproducing Germany's, but he did know that he had Britain for an ally, as well as, since December 10, the United States of America, and both were committed to supplying him with materiel. In population his country outnumbered Germany by more than two to one, and the weakened German armies now sat at the end of long, tenuous supply lines deep inside his territory.

For the Germans, the failure in the fall of 1941 and the crises of the following winter forced a drastic shake-up in the Army. That winter, and in the year to follow, so many commanders were replaced, from army group to division, that the purge resembled Stalin's infamous politically motivated one prior to the war. Emphasis was placed on younger commanders and those who were more attuned to the principles of National Socialism. In addition, Hitler fired the Army's commander-in-chief, Walter Brauchitsch, and assumed that role himself. It was unclear to him whether his generals, thoroughly schooled in military "operations," knew what was at stake. This war was not a matter of defeating opposing armies on the battlefield—maneuvering one's forces in order to claim an advantage or a victory. It was a matter of national survival. And the man with the broadest outlook on Germany's larger interests—especially its need for natural resources—needed absolute control over the military effort. On the other side, Stalin already held that role.

At a STAVKA conference late in March, to which senior field generals were invited, Stalin rejected the advice of his Chief of Staff, Boris Shaposhnikov, and Marshal Georgi Zhukov to adopt a defensive posture across most of the front. "Are we supposed to sit on our hands and wait for the Germans to attack first?" he asked. Instead, the Soviets would launch a number of attacks at various points to keep the Germans off balance. It was assumed that a renewed German effort would take place in the Moscow sector, so the heaviest concentration of forces would be arrayed before the capital.

Almost simultaneously, on March 28, Hitler held a conference in which he outlined the German plan for 1942. This time their offensive would be to the south with the ultimate objective of capturing the Caucasus oil producing region between the Black and Caspian seas. Flank cover would be provided by extending the front forward along the Don River and at one point to the Volga River, thereby cutting a vital Soviet north–south supply artery. At this point Hitler was fighting for economic resources, with the assumption that these would be so well defended that opportunities would also arise to destroy major enemy forces in the field. Reinhard Gehlen, who would become head of Germany's intelligence agency, Foreign Armies East, wrote: "Even if the coming operation was insufficent to destroy the Red Army or bring about an early collapse, the physical occupation of the vital Caucusus region and the blocking of the Volga as a Soviet waterway would cause untold harm to the enemy's economy." More important, Germany had already exhausted two-thirds of its oil stockpiled for the invasion in 1941. Denying Caucasus oil to the Soviets was desirable; winning those fields for Germany was imperative. Ironically, both totalitarian dictators, Hitler and Stalin, believed the other was near the end of his tether at the beginning of spring 1942. Stalin struck first and thus was the first to be proved wrong.

On May 12, 1942, five Soviet armies totaling 750,000 men burst out of their Donetz River bridgeheads on either side of the city of Kharkov in the southern sector. The attack was intended to hit the Germans in a vulnerable spot while their primary forces gathered before Moscow; Soviet intelligence had not realized that the south was exactly where the new German main effort was about to be launched. The attack, through sheer weight of numbers, pushed in the German front for 30 miles; however, First Panzer Army under General Paul von Kleist and Sixth Army under General Friedrich Paulus, both having been readied for an offensive of their own, remained poised on either shoulder of the bulge. On May 17 these armies attacked at the root of the penetration and joined hands. Over 200,000 Red Army troops surrendered in the cauldron; 73,000 more were killed or wounded.

On May 18, the Soviets suffered another blow when Erich von Manstein's Eleventh Army finished liquidating their position on the Kerch peninsula in eastern Crimea, a front into which Stalin had been pouring troops all year. This time 170,000 prisoners were taken. Manstein then turned his attention to the Soviet fortress city of Sevastopol in the southwestern Crimea, which fell by July 4 with the complete loss of another Soviet army of 100,000 men. This victory was especially welcome to the Germans, because now that Eleventh Army had obliterated all its opponents, it was available to reinforce other parts of the front.

If Army Group Center's failure before Moscow the previous winter was an irretrievable disaster, as it is often considered in hindsight, this was not apparent judging by German successes in the first half of 1942. Aside from the Red Army continuing to stumble into catastrophes, in North Africa General Erwin Rommel had shattered the British front in Libya and rushed on to capture 30,000 Empire troops in the fortress of Tobruk. By the end of June he was pursuing the remnants of the British Army back to Egypt. That month, an Allied convoy called PQ-17, heading for the Soviet port of Murmansk, was annihilated by Luftwaffe and naval action. With the United States' entry into the war, German U-boats were enjoying spectacular success off the American coast. Placing their priority on sinking tankers, the submarines were limited in their destructive power only by the number of torpedoes they could carry across the Atlantic. Throughout the spring Imperial Japan, which had come into the war on the side of Germany, achieved fantastic success in the Far East, ensuring that much of the Anglo-Americans' military potential would be tied down indefinitely far from Europe.

In Russia, the great German summer offensive of 1942 began on June 28. Case Blau, or Operation Blue, called for an assault by the northern wing of Field Marshal von Bock's Army Group South against the city of Voronezh on the Don. Then the offensive would turn south along that river and continue to follow it east. Below Kharkov, the Don made a huge bend until it came within 45 miles of the Volga, where sat the industrial city of Stalingrad. The Don then bent west again until it reached the city of Rostov, located at the northeast tip of the Sea of Azov. While the northern wing of von Bock's army group would cut the Volga and neutralize Stalingrad, the southern wing, comprised of First Panzer Army and

Seventeenth Army, would seize Rostov. Mobile divisions from the south would coordinate with the northern prong to encircle and destroy all Soviet forces in the Don bend. Once the defending elements of the Red Army were liquidated and Stalingrad was secured, the major objective of the campaign would be fulfilled by the southern wing conquering the Caucasus region.

The offensive got off to a clumsy start when a carefully planned pincer movement by Fourth Panzer Army and Sixth Army halfway to Voronezh succeeded in trapping nothing but thin air. In contrast to previous Soviet tactics, Marshal Timoshenko had pulled back his forces before they could be encircled. Von Bock then committed Fourth Panzer Army to street fighting in Voronezh itself. Hitler had made it clear that the city was not a major objective of the campaign and could be left to following infantry divisions; however, von Bock mistakenly thought he could win a quick victory against large Soviet forces there. On July 7 von Bock was asked to resign and Army Group South was split into two: the northern wing was designated Army Group B under General von Weichs and the southern wing Army Group A under Field Marshal von List.

On July 9, First Panzer Army and Seventeenth Army launched their end of the offensive from the Mius River, with Rostov as their first objective. As opposed to Voronezh, this city was integral to the plan because its bridges over the Don made it a gateway to the Caucasus region from the west. After First Panzer Army had been forced to retreat from the city in the fall of 1941, the Soviets had erected multiple defense lines with minefields and anti-tank obstacles. Nevertheless, the panzer divisions crashed once again into the city and engaged in bitter street fighting. Rostov was heavily garrisoned with NKVD troops—politically indoctrinated security forces, not unlike Stalin's version of the SS. For several days the Germans attempted to root out Soviet strongpoints, sometimes just blasting down buildings atop their defenders' heads. German Colonel Alfred Reinhardt recorded: "The defenders would not allow themselves to be taken alive . . . when they had been bypassed unnoticed, or wounded, they would still fire from behind cover until they were themselves killed. Our own wounded had to be placed in armored troop carriers and guarded—otherwise we would find them beaten or stabbed to death."

By July 23rd, 22nd Panzer Division had linked with 13th Panzer Division in the center of Rostov near a huge highway bridge over the

Don. Within 48 hours the city was cleared of last-ditch holdouts. The SS Viking Division, comprised of Scandinavian and other northern European volunteers as well as Germans, had meanwhile seized a ford six miles north and two infantry divisions had been passed over the river. On the 26th, reconnaissance troops of 13th Panzer and Brandenburg special forces seized the main bridge intact. The route to the Caucasus was open.

Army Group B, meanwhile, had failed to execute another encirclement battle in the vicinity of Millerovo on its belated drive south from Voronezh. Although some 18,000 prisoners were taken, the bag was puny by the standards the Germans were used to in Russia. To Hitler, the evaporation of Soviet resistance in front of Fourth Panzer Army proved that the Soviet Union had been so weakened in the previous year's battles that it could no longer offer serious resistance. The original idea that First Panzer Army should move northeast to link up with Fourth Panzer Army in order to trap Soviet forces in the Don bend was no longer considered practicable. Army Group B was just snowplowing meager enemy forces before it. First Panzer Army's left flank was already attacking in the Don bend, but there were no blocking troops to prevent the Soviets from escaping—there was no great encirclement of Soviet divisions to be had.

In the Kremlin on July 13, the Red Army Chief of Staff, Shaposhnikov, had impressed on Stalin that in view of the losses already suffered that spring, not to mention those of the previous year, the Soviet Union couldn't afford to cement its armies into place before full-blooded, good-weather German offensives. At the expense of temporarily ceding some of Mother Russia to the invaders, the Red Army would be better advised to preserve its strength until the Germans had once again overextended themselves. Perhaps humbled by the disasters of early spring at Kharkov and in the Crimea, Stalin agreed that the simplistic principle of "holding ground" should no longer dictate strategy.

To view the Red Army's retreat during summer 1942 as part of a master plan on the part of STAVKA, however, would be a mistake. As opposed to the battlefields of 1941 on the approaches to Kiev, Moscow and Leningrad, the 1942 campaign in the south was taking place primarily on vast empty steppes. This was excellent tank country, favoring mobile operations by an attacker. At the same time, unlike in European Russia, there were few centers of population worth holding; the Donetz industrial region had already once been traversed by the Germans and thence written off by the Soviets, much of the industrial equipment having been transferred to the east. At Voronezh, Rostov and—as the Germans would soon learn—Stalingrad, Soviet resistance was unrelenting. On the steppes themselves, the Red Army fought back where it could. The difference between 1942 and the initial onset of Barbarossa was not that Stalin decided to employ retreat as a strategy, but that he had recognized it as a viable tactic to employ in a long campaign. Still, the Soviet high command worried that the Germans would have the strength to reach their objectives, and were adamant that withdrawal should not turn into collapse. Stalin's Order No. 227 commissioned the NKVD to execute on the spot any Red Army soldier who was not doing his duty. Marshal Zhukov wrote:

> Due to our forced retreat, the enemy gained control of the rich regions of the Don and the Donetz Basin. We were faced with the direct threat of an enemy breakthrough to the Volga and into the northern Caucasus, and the loss of the Kuban Plain and of all communications with the Caucasus, a key economic region that was supplying oil to both the army and industry. At that point the Supreme Commander in Chief issued his Order No. 227, which set in motion severe measures to combat panic mongers and violators of discipline and condemned "defeatest" tendencies. Order No. 227 was backed up by intensive political agitation and other measures on the part of the Party's Central Committee.

On July 16, Hitler moved into new headquarters, codenamed "Werewolf," near the town of Vinnitsa in Ukraine in order to be closer to events. He immediately came to the conclusion that those aspects of Operation Blue aimed at achieving large-scale encirclements of Soviet armies should be discontinued, and instead the drive for the oil fields should begin. He pulled Fourth Panzer Army away from the advance to the Volga and ordered it to head due south to assist First Panzer Army in smashing through to the Caucasus. The powerful Sixth Army was left by itself to advance east, to seize the city of Stalingrad. It was one of the worst German decisions of the war.

First Panzer Army's commander, von Kleist, commented: "[Fourth Panzer Army] could have taken Stalingrad without a fight at the end of

July, but was diverted south to help me in crossing the Don. I did not need its aid, and it merely congested the roads I was using." The sudden concentration of two panzer armies also disrupted German logistics, which were tenuous enough given the single rail line available through Rostov. Sixth Army was denied fuel so that the southernmost armies could be supplied, and in mid-July its spearhead, XIVth Panzer Corps, mystified the Russians by sitting idle on the steppes for two weeks just as the road to Stalingrad seemed open. The tanks had run out of gas. Five Soviet armies were gathering to contest the approaches to Stalingrad, but by the time the Germans realized their mistake and ordered Fourth Panzer Army to head back north, Stalingrad was no longer ripe for the taking. Instead, it required a massive, concerted German effort.

On July 23, Hitler issued "Führer Directive No. 45," which described the alterations he had decided to make in Operation Blue. Whereas the original General Staff plan called for the Volga to be reached at Stalingrad as a prerequisite for the drive to the Caucasus, now the emphasis was switched to Army Group A in the south. Point one of the directive called for the annihilation of Soviet forces escaping to the south of Rostov. Point two ordered the seizure of the entire east coast of the Black Sea, which would eliminate the Russian Black Sea Fleet. Point three directed First Panzer Army to advance on Grozny and thence to the Caspian Sea, cutting the military highways through the Caucasus Mountains en route, and then to occupy Baku near the Iranian border.

Army Group B, including Sixth Army, was acknowledged in point four: "It will, in addition to organizing the defense of the Don line, advance against Stalingrad, smash the enemy grouping which is being built up there, occupy the city itself, and block the strip of land between Don and Volga." Then Sixth Army was ordered to dispatch "fast formations" to Astrakhan, at the mouth of the Volga on the Caspian Sea.

As originally planned, the German summer offensive of 1942 was designed to create a gigantic "balcony" into the southern Soviet Union. Now, protecting the northern flank of the attacking armies had been made a secondary concern to extending the incursion. It was a gamble based on the rationale that Stalin lacked the strength to resist—or counterattack—the Germans' main thrust. The risk may have been necessary because the longer the war continued, the weaker the Germans would become relative to their enemies, particularly in fuel supply. In 1941

Hitler had badly misjudged the political fragility of Stalin's regime; in 1942 there was no misjudging the tangible importance of oil to both sides. As long as the Wehrmacht could take the oil fields, in Hitler's view, the "learned gentlemen" of the General Staff could sort out the flank problem. The Germans had already become acquainted with "Generals Mud" and "Winter"; the often inept Red Army—now no doubt scraping the bottom of the barrel for personnel—could spring no more surprises. Prior to Operation Blue, Hitler had declared, "If I do not get the oil of Maykop and Grozny, then I must end the war." On the day Directive No. 45 was announced, however, Franz Halder, Chief of the German General Staff, recorded in his diary: "His persistent underestimation of the enemy's potential is gradually taking on grotesque forms and is beginning to be dangerous." The statements of both men were correct.

Due to the paucity of Russian forces willing to stand against the panzer armies in Operation Blue, and reports of panic and weak resistance among Soviet troops when confronted, the German High Command decided to withdraw significant forces from the southern sector. Manstein had assumed that his Eleventh Army in the Crimea would be employed to cross over from the Kerch peninsula to the Kuban and thus gain a jump on the advance into the Caucasus; at the least it could stand by as an operational reserve. Instead, he received orders to transfer the bulk of his army north to the Leningrad front. One division was dropped off en route to reinforce the German position before Moscow; another was dispatched to Crete, where, as Manstein wrote, "It was to lay more or less idle for the rest of the war." The High Command also pulled the SS division Leibstandarte Adolf Hitler out of the southern front at this time, sending it to France.

From its bridgeheads along a 100-mile stretch of the Don, Army Group A, with First Panzer Army, flanked by Seventeenth Army on its right and Fourth Panzer Army on its left, proceeded to advance on the Caucasus. On July 28, 23rd Panzer Division dodged an ambush by a Soviet armored corps and was able to get behind the enemy. After a close-range tank battle, 77 T-34s were knocked out. July 31 is when the Germans pulled Fourth Panzer Army out of the battle, less its XLth Panzer Corps, so it could help Sixth Army at Stalingrad. Von Kleist had complained of the army's arrival when he was still at the Don, but now he was upset that it was withdrawn just when he needed it on his left.

The Soviets had combined the retreating armies of their now-obsolete Southern Front with local forces to form the North Caucasus Front, consisting of the 56th, 18th, 12th and 37th Armies under Marshal Budenny. The Germans lined up SS Viking of Seventeenth Army, 13th Panzer and 16th Motorized of III Panzer Corps under von Kleist, and 3rd and 23rd Panzer of Fourth Panzer Army's XLth Corps. On August 1, Geyr von Schweppenburg's XLth Corps was placed under the command of First Panzer Army, as soon would be Viking.

On August 2, 3rd Panzer and 16th Motorized forced the Manych River, thus crossing the traditional boundary between Europe and Asia. The Wehrmacht in the Caucasus subsequently found itself in a land less familiar to Western armies than it had been to medieval conquerors such as Tamurlaine. Camels wandered the steppes and much of the population bowed three times a day to Mecca. In the 1940s the region was known as Transcaucasia, but modern observers know it better as the independent states of Azerbaijan, Georgia and Armenia and, north of the mountains, the disputed republic of Chechnya, which the Russian Army to this day struggles to control. In 1942, many of the Cossacks and Muslims in this region greeted the Germans as liberators and volunteered to help fight the Red Army. Since the German Army did not hold the region for long, these people didn't have time to be disillusioned by the Nazi political apparatus that generally trailed in the wake of the panzers. They were, however, severely punished by Stalin with massive deportations to Siberia once the front had moved on.

In the west, 17th Army attacked due south from Rostov toward Krasnodar in the Kuban steppe. In the center, von Mackenson's III Panzer Corps took the rail center of Krapotkin and then the industrial town of Armavir, capturing 50 planes on its airfield. Reconnaissance elements of 16th Motorized Division then discovered the rail line north of Armavir to be lined for twenty miles with Soviet freight trains that had been jammed up by Luftwaffe attacks. To the east, 3rd Panzer seized the city of Voroshilovsk (now Stavropol) but its partner, 23rd Panzer, was lagging behind due to a shortage of fuel.

By August 7, 13th Panzer and 16th Motorized had arrived at Maykop, the first of the region's three great oil centers. SS Viking came over from the Kuban steppe to lend its assistance. On the night before the attack, Brandenburg commandoes dressed in NKVD uniforms infiltrated the city to try to save oil tanks from destruction and to sow what chaos they could. The city was already in turmoil with demolitions and hasty evacuations, even as more retreating Soviet columns kept pouring in from the north. The Brandenburgers added their own explosions to the mix and for a while took over the city's main telegraph office, where they informed everyone the city had fallen. The Soviets literally headed for the hills, to the south, but they did blow up the oil tanks and refinery facilities. The Germans also learned that although Maykop was an oil "center," the fields lay farther to the south; by the time these too were seized they had been utterly destroyed. After dispersing the Soviet 45th Rifle Brigade, which had been left behind to defend the city, 16th Motorized inherited 400 American trucks, all with low mileage since they had just been delivered via the Allied supply route through Iran. On the army group's left, XLth Panzer Corps got within 70 miles of the Caspian Sea to cut the pipeline connection between Baku and Astrakhan. If the Germans had yet to capture working oil fields for themselves, they had at least begun to deny their use to the Soviets.

Across hundreds of miles of flat steppe the panzers had advanced south against scattered opposition. They had not only outrun their supporting infantry but also thousands of Red Army troops trying to escape on foot. A motorcyle patrol of SS Viking was sent to the rear to reconnoiter a large gap between their division and 13th Panzer to the east. Each time they came to an intersection they had to wait for Soviet troops to march by in the dark; finally they opened fire on a group, which scattered, and the cyclists effected a link with their neighbor. Bypassed Red Army troops presented a serious hazard to German supply columns following behind the spearheads; after each jump in the German advance, divisional commanders had to send strength backwards to protect their convoys from ad hoc ambushes.

On August 11, Viking plunged again into the Soviet rear, where enemy movement was hardly less chaotic than behind German lines. From a radio intercept, Viking learned that the 17th Kuban Cossack Cavalry Corps was assembling for a counterattack at a town northeast of Maykop. The division's panzer battalion, with Finnish volunteers from its "Nordland" regiment on top, broke into the assembly area and scattered the Cossacks, leaving riderless horses galloping in every direction. The rest of Viking, advancing on the left, ran into a Russian march column of 400 men, which quickly surrendered. Although the German High Command may have been frustrated at the Red Army's tactics of retreat, frontline troops far preferred the enemy to run. Before a head-on Viking attack against the town of Belorechenskaya, a detachment of Brandenburgers wearing Russian uniforms and riding Russian vehicles raced into the town feigning panic, shouting "Tanks! Tanks!" The defense turned into a stampede that not even the town's commissars could prevent. The Brandenburgers shot up a Soviet artillery battalion that was preparing to take position, and then seized a railroad bridge over the Belaya River, where they discarded their Russian overcoats and waited for reinforcements.

By now, however, the panzers had encountered a greater obstacle than the Red Army: the Caucasus Mountains. This range—higher than the Alps and considerably less charted—stretched across the entire neck of land from the Black Sea to the Caspian. It was cut by several "military highways," dating from the time of Peter the Great and earlier, which had not been constructed with tanks in mind. The range sloped toward the southeast, away from the Germans' strength. After Maykop, the second great Caucasus oil center, Grozny, sat in the foothills of the mountains near the Caspian side, behind the wide Terek River; Baku was beyond the mountains by the Caspian near Iran, even farther south.

The Germans though had brought along specially trained mountain divisions to force the passes of the Caucasus. On August 5, 3rd Panzer Division reached the foothills and opened the way for 1st Mountain Division to begin scaling the heights. 4th Mountain and 2nd Romanian Mountain Division would follow, although to von List's chagrin the Italian Alpine Corps hadn't been attached to Army Group A; it was serving as regular infantry on the northern Don. On August 21, a combined patrol of expert climbers from the 1st and 4th Mountain Divisions scaled 18,500-foot-high Mount Elbrus, the highest peak west of the Himalayas, and planted the German battle flag on top as well as their divisional flags. The achievement was celebrated and broadcast worldwide by the Nazi propaganda ministry; however, according to some reports, Adolf Hitler was less than amused.

Two days after the scaling of Mount Elbrus, 300 miles to the north, 16th Panzer Division fought its way to the Volga near Stalingrad. The division was immediately counterattacked on its left and right and cut off from behind. 3rd Motorized Division fought its way through to 16th

Panzer and was then itself cut off by Soviet counterattacks. For a week the two divisions held out against fierce tank and infantry assaults, sitting beneath artillery fire and barrages of Katyusha rockets until the rest of Sixth Army caught up. General Paulus then began the arduous process of hammering his way into the city, which stretched for 25 miles along the river. Fourth Panzer Army, having been pulled from the Caucasus minus XLth Panzer Corps, advanced against Stalingrad from the south. As American Soviet expert David Glantz commented on the 1942 offensive in general, "The Soviet armies had learned a little more thoroughly the art of withdrawal, and they managed to pull back the bulk of their forces fairly successfully. The only exception was the close approaches to Stalingrad where the Soviet 64th Army, 62nd Army, 1st Tank Army and 4th Tank Army were severely chewed up in German attacks." While most Soviet formations fell back on either side of Stalingrad in the face of the attacks, Soviet 62nd Army was assigned to stay in the city to contest its possession by the Germans.

In the Caucasus, Geyr von Schweppenburg's 3rd and 23rd Panzer Divisions attacked east through the foothills of the mountains with the ultimate objective of Grozny. They cut the Baku-Rostov oil pipeline but found all the pumping stations had been set afire. Due to lack of fuel the panzer divisions became so strung out they halted at the city of Pyatigorsk to let all their elements catch up. On August 14 they tried to cross the Baksan River to the south, but the Soviets blew up a dam; on the 18th severe thunderstorms worsened the problem. The Germans were by now receiving artillery fire from the mountains and the Soviets had slipped a corps of Cossacks into the steppes on their northeastern flank. The Soviet Air Force, which by this time included Hurricanes and Aerocobras, dominated the skies. The Luftwaffe's JG 52 kept moving its airfields closer in order to keep Me-109s in the air over the battlefront, but it was suffering from a severe shortage of planes. The Germans bounced away from the Baksan and continued east until they were north of the city of Mosdok on the Terek River. Across the fast-flowing Terek lay Grozny and then— if wishes were horses—Baku.

By August 22, 13th Panzer Division had arrived in the area from Maykop and was subordinated to XLth Panzer Corps. The 111th and 370th Infantry Divisions finally arrived after marching all the way from Rostov on foot. 16th Motorized Division, on the other hand, was pulled from the Caucasus to station itself in the middle of the Kalmuk steppes, halfway between Grozny and Stalingrad, because the Germans had no idea what was going on in the huge gap that had by now opened between Army Groups A and B.

The Germans were steadily paying the price for splitting the main focus of their offensive. The drive for the oil fields should have had as a prerequisite the neutralization of Stalingrad and the northern flank; instead, some kind of maelstrom was roaring on the Volga, sucking in German airpower, manpower and supplies. In the Caucasus, for First Panzer Army's von Kleist, it was now or never for a breakthrough to Grozny. For weeks, given the vast spaces of Russia, advances could be made only by panzer regiments that happened to have fuel at the time. Sister regiments would follow after they had been reached by a supply column, but by that time the leading regiment would have dried up. One concentrated effort, however, might gain all the fuel the army, and the Third Reich, needed.

On August 23, 3rd Panzer advanced on the city of Mosdok on the Terek; 13th Panzer sought a bridgehead at Isherskaya 18 miles east; 23rd Panzer raced even farther east to cross at Naurskaya. Outside Mosdok, which was held by the Soviet 4th and 8th Guards Rifle Divisions, three separate armored trains rolled out to engage German tanks in point-blank duels. One after another they were blown up. That night the Soviets reinforced the town with three airborne brigades and the Germans were forced out of the northern outskirts. What was left of JG 52 tried to hold off swarms of aircraft while German artillery bombarded the town and Soviet guns on the other side of the river blasted the attackers. On the 25th, 3rd Panzer finally took the city as the Soviets withdrew across the Terek, blowing the bridges behind them. The steep, strongly defended far bank, however, discouraged a German attempt to cross the river.

13th Panzer had meanwhile taken Isherskaya and panzergrenadiers prepared to cross the Terek in assault boats in three waves. The first wave got across thanks to surprise and artillery preparation; however, the second was badly shot up. By the time the third wave had finished only 5 of 36 assault boats were still afloat. At this point German engineers determined the site was poor for building a bridge, but the troops already on the far side were ordered to hold their ground. Elements of 13th Panzer moved on to help 23rd Panzer take Naurskaya but the Soviets had recrossed the Terek themselves to the east and were too strong for the fuel-starved Germans. Marking the farthest point of the German attack, tanks of 23rd Panzer got to a rail junction just 15 miles north of Grozny, from which they could stand astride the transport route between Iran, the east Caucasus and the northern Soviet Union; but they were unable to hold the position.

At Mosdok, 111th Infantry Division succeeded in establishing a bridgehead across the Terek on September 2; the tenuous toehold at Isherskaya was abandoned and the panzers were called upon to regroup around Mosdok to regather their strength. A 1,500-man unit of experts called the "Technical Oil Brigade," with trainloads of equipment, meanwhile waited on the steppe for the seizure of Grozny. Too late, they realized they should already have been attempting to retrieve the flow of oil from the destroyed derricks around Maykop.

On the Black Sea coast, Seventeenth Army was also meeting increased Soviet resistance, with the additional problem that mounting full-scale attacks across the Caucasus Mountains had proved to be impractical. The farther the mountain troops penetrated among the peaks, the more difficult it became to supply them; small Soviet units dominated entire passes and roamed the German rear setting up ambushes and pinning the mountain troops down. On the coast road, the Soviet naval base at Novorossisk was under attack, and would fall on September 10, but there was little hope of Seventeenth Army continuing to Tuapse and thence to Batumi by the Turkish border.

As summer turned to fall, 16th Motorized sat by itself in the middle of the huge Kalmuk steppe, sending reconnaissance patrols in every direction and occasionally blowing up northern-bound Soviet trains. The Kalmuks, descendants of the Mongol hordes of Genghis Khan, were helpful, but Soviet activity based on Astrakhan increased. In one incident, cadets from the Russian NCO school at that city wiped out a German outpost after its leading elements approached the troops of 16th Motorized in captured German vehicles and uniforms.

At his headquarters in the Ukraine, Hitler had already begun to sense that the main German effort of 1942 was failing. Far from a decisive blow against the Soviet Union, the offensive so far had succeeded in conquering nothing but 1,200 miles of grass and hills, at relatively little cost to the enemy. Von Kleist of First Panzer Army, who was supposed to have captured the Caucasus oil fields, was bitterly complaining about lack of fuel. Von List, who commanded Army Group A, had visited headquarters with an incomprehensible map, upon which he did not even seem to know where his units were. Halder, the Army's Chief of the General Staff, did nothing but formulate his arguments with negative provisos and constant warnings about Soviet strength, so that his recommendations could never be proven wrong. Hitler had risen to the mere rank of corporal during his own term of military service; however, he now had the responsibility of forcing the generals and field marshals of Germany to follow his directions. The oil fields of the Caucasus had to be secured; Soviet resistance at Stalingrad had to be liquidated. The war had already gone on too long for the resources Germany possessed. By the following year the Reich would only have grown weaker, and its enemies, through unlimited access to oil and natural resources, stronger.

On September 7, Hitler's most loyal general, Alfred Jodl, head of OKW, returned from a visit to the Caucasus to report that First Panzer Army did not have the strength to fulfill its mission as outlined in Führer Directive No. 45. In the heated argument that followed, Hitler accused Jodl of being a dupe of the army's generals, who Hitler was convinced were conspiring against him. Jodl fired back that von List, von Kleist and Konrad, who commanded the mountain troops aimed at the Black Sea coast, had only followed Hitler's orders to the letter, which was exactly why the entire offensive in the Caucasus had resulted in failure. The argument became ugly and Jodl later regretted having been so blunt with Hitler, even if his arguments were correct. "One should never," he reflected, "try to point out to a dictator where he has gone wrong since this will shake his self-confidence, the main pillar upon which his personality and his actions are based."

For his part, Hitler, who had always dined with his staff or visiting generals, began to take his meals alone. He also ordered a team of stenographers to fly in from Berlin to henceforth record every word spoken at military conferences. The next day he relieved von List, the commander of Army Group A, and took command of the army group in the Caucasus himself. On September 23 he fired Halder, the bespectacled Chief of the German General Staff, and replaced him with Kurt Zeitzler, a younger and more dynamic personality. Field Marshal Wilhelm Keitel took Zeitzler aside upon his appointment and warned him not to throw blame, casualty statistics or insoluble difficulties onto Hitler's lap, saying, "You have to spare the nerves of this man." Zeitzler disagreed: "If a man starts a war he must have the nerve to hear the consequences."

The new chief of the Red Army's General Staff, A.M. Vasilevsky, termed August 23, when German Sixth Army reached the Volga, "an unforgettably tragic day." At that point, the Soviets did not know whether their armies could hold the Germans at Stalingrad, or if the enemy indeed had the strength to seize and occupy the Caucasus. It was not until early September that STAVKA realized the German spearheads had worn themselves out. Just as in 1941, the enemy had overestimated its ability to force a decision by a sudden, massive attack. On September 13, Vasilevsky and Zhukov presented Stalin with a plan for a counteroffensive. While Soviet field marshals Timoshenko and Budenny had pulled back their main forces in the face of the German onslaught, the Soviet Union had created no fewer than ten new armies in the hinterland. These could soon be directed against the great German bulge that had been extended into the country in the south.

Throughout September, First Panzer Army tried to expand its bridgehead at Mosdok against the 9th and 44th Soviet Armies and in the face of swarms of enemy aircraft. During the advance from Rostov, Wolfram von Richthofen's 4th Air Fleet had provided good support for the panzers, with squadrons of Stukas on call to pulverize enemy concentrations. Sometimes, when columns were disoriented on the steppe, fighter planes would circle an objective to guide, or warn, the troops on the ground of what lay ahead. Now, however, Richthofen's aircraft were devoted to Stalingrad, that bleeding sore to the north where Sixth Army continued to face unceasing opposition. Near Mosdok in the south, the men of First Panzer Army were often forced onto the defensive to fight back Soviet counterattacks. Then they would stab in a new direction only to find the Russians waiting again in depth. By October 3, 13th Panzer had blasted its way through to the Elkhotovo Pass that led to Grozny, but by then its regiments had run out of strength. This battle featured a successful foray by two Soviet armored trains that shot up a number of Germans at the tip of the advance and then escaped back to their station.

On September 26, SS Viking was passed into the bridgehead and aimed at the center of the Soviet line toward the town of Sagopshin, astride another pass to Grozny. Viking's route was through a valley crisscrossed by Soviet fortifications and flanked by enemy infantry and guns on the heights. The prospects for Viking looked dim; nevertheless, on the morning of the 26th SS General Felix Steiner received a message from von Kleist: "The entire army is watching your division. You have the mission of advancing the army's attack toward Grozny. I expect to be with your lead attack elements this evening at 1800 hours near Sagopshin."

The Scandinavian "Nordland" Regiment was supposed to clear the flanking heights but they ran straight into a four-battalion Soviet attack coming the other way. The fighting in the wooded hills went back and forth, and in the afternoon Viking's panzer battalion, regardless of its flanks, charged up the middle. The lead company was cut off and by nightfall its five remaining tanks had to set up a circle for defense. One of them was destroyed during the night by a Molotov cocktail. The 2nd Company had it no better. According to its commander, "A great drama was being played out [near Sagopshin]. I could follow it all over my company radio, although I was also in a difficult situation. Mines were thrown in front of our tanks. The Russians climbed onto our tanks and threw hand grenades into the hatches. We had to keep watch over each other and fire our machine guns. . . . By the fall of darkness we rolled behind a small earthen wall and established a hasty defense in a cornfield. We were all alone."

On the 27th both sides launched minor attacks and probes. The next morning Viking's panzer battalion's commander, Mühlenkamp, decided to attack during the early morning fog so that he could get underneath the range of the enemy's guns and tanks. The panzers rolled across several trenches and then, as Mühlenkamp described:

Earlier than anticipated, at 0700 hours, the sun broke through; the fog was suddenly washed away. We were in the middle of the Russian defensive positions, between long rifle trenches and defensive nests, which were all well occupied. Through the hatch opening, I looked into the trenches which we were overrunning. The enemy was firing machine guns and machine pistols into our hatches and optical sights, throwing hand grenades. On a wide front at about 800 meters distance on our right numerous T-34 tanks were offering to duel. The first shot hit right behind my turret, the engine blew up, the turret lifted up slightly, the backrest from my seat was destroyed. I was thrown forward over the gun and yelled "abandon ship!"

Mühlenkamp would have two more panzers shot out from under him that day as 80 Soviet and 40 German tanks hammered each other in a ferocious melee. The second element of Viking's panzer battalion moved up on the left and found a clearing through a minefield, but its accompanying grenadiers were badly shot up. On day after following day, Viking attempted to twist itself into the Soviet defense belt, each yard's progress measured in dead and wounded. The division's "Germania" Regiment came into the battle on October 4, but it could only replace the losses already suffered. On October 20, Steiner reported to von Kleist that to continue the attack was senseless. Kleist agreed, but also mentioned that Viking had taken the pressure off III Panzer Corps. It would be their turn next.

While First Panzer Army planned its next attack at the extreme end of German supply capability, some 300 miles due north Sixth Army continued to weed out last-ditch Soviet resistance along the Volga at Stalingrad. The battle for this city had somehow taken on a life of its own, apart from the general strategic picture, as if a duel of willpower was taking place between the dictators of Nazi Germany and Communist Russia. In retrospect, it would have been wiser for the Germans to deploy the twenty divisions of Sixth Army along the Don bend rather than extend them in a concentrated fashion beyond the Don to the Volga. The waterway could still be closed off by the Luftwaffe and by the mere threat of a powerful German army only 45 miles away. On the other hand, Stalingrad and the Volga were "prestige" prizes, and once the German public had been informed that Sixth Army had achieved their conquest, it was difficult to voluntarily retreat on behalf of caution, or fear.

Sixth Army had a preponderance of force in Stalingrad itself and was also responsible for the neck of land between the tip of the Don bend and the Volga. Along that bend, the Germans had called upon their allies to hold the front with the river as protection. To the right of Sixth Army, Fourth Romanian Army held the line; to the left of Stalingrad, in a long stretch north all the way to Voronezh, stood Third Romanian Army, Eighth Italian Army and Second Hungarian Army. (It had been considered prudent to separate the Romanians and Hungarians.) The Germans still did not have an exalted view of the Red Army's offensive capability; at that point in the war, any one German division could take on three of the enemy's. Still, there was nervousness among the German leadership about their allies' capability.

In the Caucasus, von Mackenson submitted a plan to von Kleist for his 13th and 23rd Panzer Divisions to advance to the east. He also needed the 2nd Romanian Mountain Division to advance up the Georgian Military Highway to block Soviet reinforcements from coming through the mountains. On October 25 the panzer attack began, making excellent progress toward its first objective, the town of Ordzhonikidze. But as the days wore on, in a microcosm of the entire war in the East, the Germans grew weaker with every mile gained; Soviet resistance seemed to grow. On November 6 the Red Army launched a counteroffensive that got behind 13th Panzer, cutting it off. Subsidiary attacks pushed back 23rd Panzer and the Romanians; when 13th Panzer sent out reconnaissance to make contact with its neighbors, they discovered nothing but Soviet troops and tanks. The flanking divisions were held down by their own problems and could offer no assistance. 13th Panzer grouped itself around the town of Gisel, and on November 10 determined to break out back to the lines.

Fortunately, elements of Viking had arrived in the area and its officers arranged with von Mackenson to help rescue the trapped division. That night, 13th Panzer began its breakout to the north, tanks in the van, followed by vehicles of wounded and then grenadiers, while Viking fought its way forward to hold open a corridor. Throughout the night the division forded a marshy stream to safety; however, when daylight came the Soviets on the surrounding heights realized what was going on. The ford was plastered with fire and the retreat turned into a rout. Viking's panzer battalion, with a Luftwaffe flak battery in tow, probed into the rear of 13th Panzer and found Soviet tanks assembling to pursue. The 88s of the flak battery leveled their sights and destroyed the concentration of Russian armor. By the morning of November 12, the pocket had been evacuated by personnel, though over 500 of 13th Panzer's vehicles were left behind. Viking served as rearguard until all of 13th Panzer's effectives and wounded had been taken out. During the next few days, heavy rains made operations impossible for both sides, which was just as well for First Panzer Army. By now it had no further plans for additional attacks in the Caucasus.

As another brutal Russian winter approached, the German armies in the south had been stopped short of their objectives by a Red Army that was no longer trading space for time. The Germans held Ukraine and the Donetz industrial region; they'd cut the Volga transport artery and severed the Caucasus oil-producing region from the rest of the Soviet Union. But the next move was up to Stalin. Field Marshal Keitel, in a memoir written while awaiting his execution after Nuremberg, captured the precariousness of the German position.

> There was fierce and unprofitable fighting among the northern spurs of the Caucasus mountains, there was uncertainty along the weakened front in the steppes between the mountains and Stalingrad, there was very heavy fighting in and about Stalingrad itself and the gravest possible danger to our allies holding the front along the river Don. The uneasy question overshadowing everything was: where are the Russians going to launch their counteroffensive? Where were their strategic reserves?

On November 19, the great Soviet counteroffensive of 1942 commenced, with an overwhelming attack that shattered the Romanian Third Army above Stalingrad. Two days later, another hammerblow Soviet offensive struck the Fourth Romanian Army below the city on the lower Don. Red Army T-34s and T-70s from both prongs of the offensive rushed to link hands at the tip of the Don bend, at Kalach, to sever German Sixth Army from the rest of the Axis front. General Paulus had to pull his divisions in from the Don and form a pocket. He was surrounded at the eastern tip of the Axis front, his flanking armies destroyed or in full retreat.

Hitler, as commander-in-chief of the German Army, still refused to believe that the Soviets could muster the strength in that second year of the war to seriously threaten his most powerful armies. As opposed to the winter of 1941–42, this time the Germans were well prepared for winter operations. If Stalin could fling men into the battle that was one thing; to defeat the cream of the Wehrmacht was another. The fact that the Russian offensive had been aimed against the Romanian sectors was proof enough that the concentrated German divisions shouldn't panic. His order of the previous year to "stand and fight" had been vindicated and now, when

Germany was more urgently in need of Caucasus oil and Ukrainian coal and ore than ever, there could be only one order: "Hold fast."

General Paulus, in keeping with General Staff principles of operational warfare, requested permission to withdraw after he learned his Sixth Army had been cut off. Hitler, sensing that if Germany could not succeed in this campaign it could not hope to win the war anyway, refused. Sixth Army must hold Stalingrad; First Panzer Army had to maintain its position in the Caucasus. If retreat were sanctioned, Germany would never again get so close to the raw materials it needed in order to successfully prosecute the war. If the Wehrmacht and its commanders could keep their nerve in this crisis, perhaps the Soviet Union would have finally shot its bolt. In order to keep Sixth Army alive, the Luftwaffe would temporarily supply it by air; meanwhile, the Soviets could not possibly be as strong as their successes against the Romanian sectors indicated.

In Moscow, Stalin followed the progress of the Soviet offensive on a gigantic map in the room in which he met with STAVKA. The thrust of Russian diplomacy to the Anglo-American powers the past year had been to urge for a second front in the West in order to draw pressure off the Soviet Union. The German High Command had been privy to various cables and had become increasingly convinced that the Soviet Union could not hold out much longer against the Wehrmacht. Given the clouds of postwar bravado, it's unclear whether, to Roosevelt and Churchill, the perception of Soviet weakness or strength would have better hastened a second front in Europe. Stalin evidently thought that by his feigning weakness the Anglo-Americans would adopt a greater sense of urgency; he might not have considered that the prospect of a Russian collapse would interminably postpone any major offensives from the West. As it turned out, the mere rumor of an imminent second front in France was enough to tie up a number of Germany's best divisions. And while the British and American armies sought to pin down Rommel in North Africa in late 1942, Stalin had meanwhile assembled a gigantic force that could trap Germany's entire Army Group South against the Sea of Azov at Rostov.

During the fall, Erich von Manstein had been unable to execute his orders to capture Leningrad because the Soviets had launched an offensive south of Lake Ladoga on the eastern flank of the German encirclement. Manstein's Eleventh Army took its neighboring army under wing and then crushed the offensive, leaving some 40,000 Russians dead on the field or captured. On November 23 Manstein received orders to take command of a newly created Army Group Don near the Black Sea, where, as he was informed, Operation Blue had completely fallen apart.

When Manstein set up his headquarters at Novoverchask near Rostov, he learned that First Panzer Army and Seventeenth Army had run into a brick wall in the Caucasus and had been unable to secure the Russian oil-producing region. Far worse, Sixth Army had just been cut off from the rest of the Wehrmacht at its far-flung battlefield around Stalingrad at the Volga. The real problem, however, was that Soviet armored corps were pouring across the Don to the west of both Army Groups A and B and threatening to cut off the entire German position in southern Russia.

His first assessment was that First Panzer Army, even if it lacked the strength to conquer the Caucasus, could hold its own there on a static front. The Russian armies in the south could be supplied by British Mark III tanks, American Stuarts, Ford trucks and Spam rations, but Stalin himself could not transfer great strength there. Otherwise, Manstein faced a devil's choice: restore the integrity of the German front, or concentrate east of the Don River, in still another gamble, to try to get through to Stalingrad. Upon its encirclement, Hitler had declared Stalingrad to be a fortress and announced that it would be supplied by air. Manstein, however, judged that the Germans at Stalingrad were doomed unless they could be reconnected with the rest of the front. As a career officer raised and educated in the centuries-old Prussian military tradition, he had only one option: save Sixth Army.

In the Don bend, 22nd Panzer Division was supposed to have stiffened up the Third Romanian Army; however, it had lost half its tanks to mice. Being short of fuel, the division had buried its tanks in the ground to protect them from cold, only to find their electrical insulation eaten away once they were dug out. Herman Balck's better-equipped 11th Panzer Division, on the other hand, was in the Don bend working in tandem with the 336th Infantry. Balck was rushing back and forth, knocking out the spearheads of one Soviet armored corps after another. His tactic was to pin the Russians down in front and then pass his attacking strength into their rear. On one day, Balck's panzers managed to approach a large marching column of T-34s from behind and shoot up 43 tanks before the Russians realized that their "second wave" was the enemy. A few hours later the panzers fell in behind another column and claimed 22 more, all at no loss to themselves. Much as the T-34 has been revered as a superior weapon in World War II, most of the Russian tanks weren't equipped with radios, and the tank commanders had to double as gunners. When fellow machines began brewing up all around them, the typical T-34 commander had no idea where the shots were coming from. By contrast, all German tanks had radios and a commander whose primary job was to see what was going on.

The situation in the Don bend consisted of many fleeing Romanians, sundry German supply troops pressed into frontline combat, a couple of isolated panzer divisions, and some half-million onrushing Soviet troops. Luftwaffe infantry units were in the theater but these tended to break apart upon contact. Hermann Goering's wish to create his own private army, as Heinrich Himmler had done with the Waffen SS, resulted in a waste of manpower that could more profitably have been allocated to regular formations. German Army commanders invariably refer to Luftwaffe divisions as ranking in effectiveness somewhere between the Italian Army and Russian "Hiwis." The Waffen SS, on the other hand, turned out to be a successful experiment in creating valuable fighting formations. Though ridiculed during the campaigns in Poland and France, the SS gained experience, and during the first crises in Russia came into their own. Aside from their access to top equipment, thanks to Himmler, SS volunteers tended to be young, unmarried men—in all wars the best fighters.

Manstein quickly built up his main forces east of the Don for the relief of Stalingrad. Heinrich Hoth's Fourth Panzer Army would be the rescue vehicle, with 23rd Panzer brought up from the Caucasus, 17th Panzer drawn in from the Don bend, and the newly arrived 6th Panzer, which, unlike its neighbors, was at full strength. On December 12, from Kotelnikovo, Fourth Panzer Army began its drive on Stalingrad.

In Moscow, Red Army Chief of Staff Vasilevsky had already found his plans disrupted. He had intended that his offensives through the Romanian-held fronts on either side of Stalingrad should be followed by another attack, Operation Saturn, designed to propel the powerful Second Guards Army straight to Rostov to cut off German Army Group

A in the Caucasus. His mistake was that he thought only 80–90,000 German troops were concentrated at Stalingrad; instead he found he had encircled over 200,000 men—20 entire German divisions. This was the toughest nut on the entire southern front; further, the Germans were continuing to commit their main strength toward Stalingrad to rescue the surrounded army. At all costs, the German rescue effort had to be headed off. Second Guards Army was redirected to oppose Manstein's relief force. And the Soviets had still another card to play.

Fourth Panzer Army, with thousands of trucks full of food and supplies trailing in its wake, was making good progress toward Stalingrad. As the center division, 6th Panzer was proving too strong for the Soviet blocking forces piling in front of the advance. On December 19, however, STAVKA sprang another offensive against the Don bend, this time against the Italian-held sector of the front. The Italians, who were hopelessly outnumbered and outgunned, caved in. Another flood of Soviet tanks and troops began pouring into the Don bend. By that time the troops in Stalingrad were able to witness flashes of gunfire to their south and say to each other, "Manstein's coming." Little did they know that von Manstein, whose forces had gotten within 30 miles of the city, was unable to continue his rescue mission. In defiance of Hitler's orders, he had been pleading with Paulus to break out from the pocket and join him, or at least attack from the north to relieve some of the defensive pressure against Fourth Panzer Army. Paulus refused because of his orders from Hitler to hold fast, and he also claimed he didn't have enough fuel for his remaining motorized forces to move.

It has been opined that Paulus' predecessor as commander of Sixth Army, von Reichenau, as well as a number of other German generals, would never have let the army succumb to such a catastrophe on the Volga. It's beyond doubt, in fact, that Paulus should have made every effort to attack south from the pocket to link with Manstein's relief force, since Sixth Army would never have another chance to escape annihilation. Instead, the defenders of Stalingrad enjoyed a relatively quiet week while the Russians threw every spare division they could find against Manstein. On the other hand, Paulus had advanced to Stalingrad in the earnest belief that the rest of the German front would not collapse behind him. In mid-winter, when Manstein was suddenly imploring him (outside official channels) to abandon his lines and break out across the snowy steppe in the midst of superior enemy forces, leaving his guns, wounded and most of his infantry behind, he was not overly eager to comply. His men had fought off Soviet counterattacks from the north for three months; they had constantly advanced against infernal resistance in the city itself while under fire from enemy artillery concentrations across the Volga. It was the job of Sixth Army to hold Stalingrad, as it had been ordered to do; it was the job of the High Command to restore the integrity of the rest of the front.

After the Italian Eighth Army collapsed on the upper Don, Manstein had no choice but to detach 6th Panzer Division from the relief force into the Don bend to protect Rostov and other German supply routes to the west. Sixth Army at Stalingrad was doomed to annihilation. Nevertheless, the fact remained that Paulus' army was the strongest knot of German forces in the southern theater. It was a magnet for Soviet offensive efforts and by January was tying down seven armies comprising 90 rifle divisions and armored brigades. If these were let loose to join the other 170 Red Army formations pouring through gaps in the front, the remaining German armies in the south would share the dismal fate of Stalingrad's defenders. If Sixth Army did not hang on, a million more German troops would be cut off from the west.

To Manstein, the vital next step was for the High Command to order First Panzer Army in the Caucasus to retreat. As long as it stayed on the Terek River line, over 300 miles southeast of Rostov, the Germans' main strength in the south had to be kept east of the Don. Fourth Panzer Army couldn't withdraw west because it needed to protect First Panzer Army's back and hold open its escape route. Stalin seemed to recognize the Germans' peril and was urging Soviet armored columns ever deeper into the Don bend without regard to their flanks or supply. If they could get to Rostov, not just one but five German armies would be destroyed— the entire German front in southern Russia. The war would be won.

Adolf Hitler was the one man who still failed to realize, or admit, to the magnitude of the Soviet counteroffensive. He continued to hesitate on the question of whether the rest of the Germans, aside from Sixth Army, should retreat. Manstein wrote: "Hitler was still not disposed to give up the Caucasus region. He still thought he could somehow maintain a front south of the Don which would at least safeguard his possession of the Maykop oil fields."

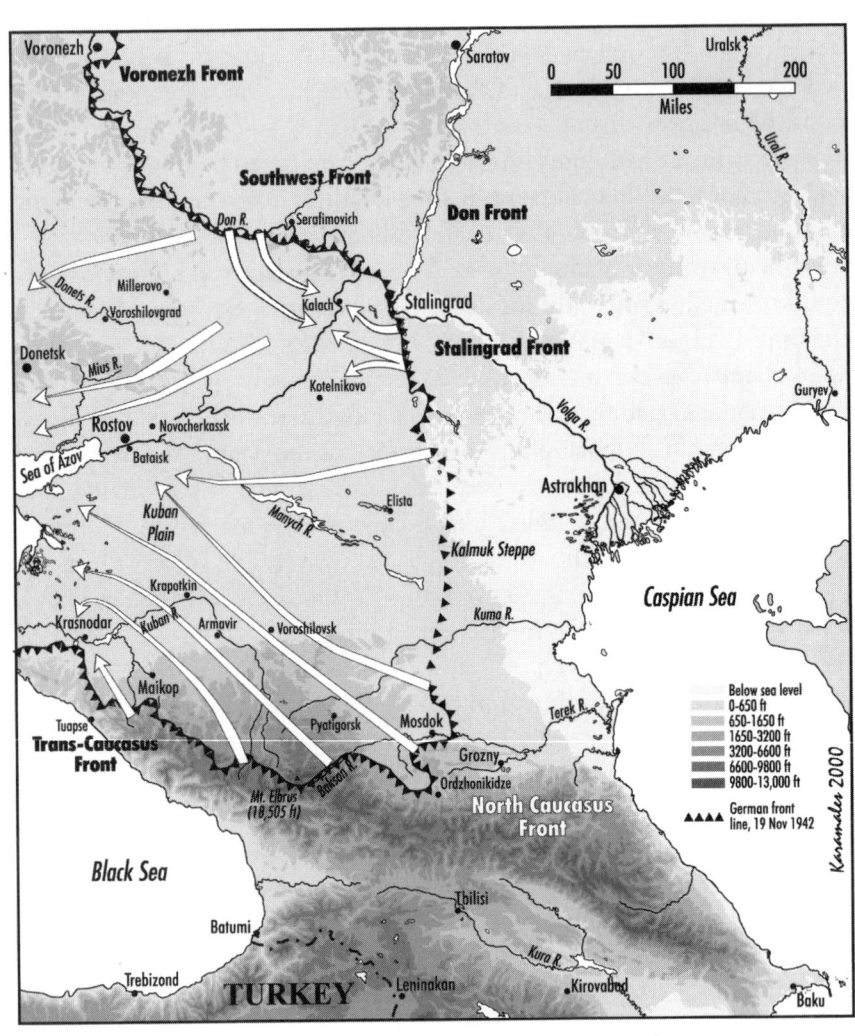

Voronezh
Voronezh Front
Saratov
Uralsk
0   50   100   200
Miles
Southwest Front
Don R.   Serafimovich
Don Front
Donets R.
Millerovo
Voroshilovgrad
Kalach
Stalingrad
Donetsk
Mius R.
Stalingrad Front
Rostov   Novocherkassk
Kotelnikovo
Guryev
Sea of Azov
Bataisk
Elista
Astrakhan
Kuban
Plain
Manych R.
Kalmuk Steppe
Volga R.
Krapotkin
Kuma R.
Caspian Sea
Krasnodar
Kuban R.
Armavir   Voroshilovsk
Maikop
Pyatigorsk
Mosdok
Terek R.
Tuapse
Below sea level
0-650 ft
Trans-Caucasus
Front
Baksan R.
Grozny
650-1650 ft
1650-3200 ft
3200-6600 ft
6600-9800 ft
9800-13,000 ft
Ordzhonikidze
Mt. Elbrus
(18,505 ft)
North Caucasus
Front
German front
line, 19 Nov 1942
Black Sea
Tbilisi
Karandler 2000
Batumi
Kura R.
Trebizond
TURKEY   Leninakan
Kirovabad
Baku
Ural R.

In the Don bend, Soviet tanks overran Tatsinskaya airfield on December 24, from which the Luftwaffe had been flying supplies into Stalingrad. When the Germans counterattacked four days later they discovered the horrifying irony that hundreds of wounded Sixth Army soldiers had been butchered after being flown to "safety." Across thousands of square miles, a few German formations stood like rocks in a stream of rapids against the multi-pronged Soviet offensive, but the Russians had free rein across much of the steppe. A Cossack division intercepted a troop train coming from the west and killed the German recruits with sabres. 22nd Panzer became so shot-up in its attempt to stand behind the Romanian Third Army that it had to be disbanded.

On December 29 the German High Command finally ordered First Panzer Army in the foothills of the Caucasus to withdraw from the Terek to a line on the Kuban River. Throughout the month the Red Army had pinned down von Mackenson's troops with constant attacks, which not only complicated the problem of retreat but also made First Panzer Army reluctant to lend assistance to Manstein. Nevertheless, SS Viking had been pulled out of the line and arrived in Fourth Panzer Army's sector by the end of December, as did 16th Motorized Division, which was more chased back from its isolated outpost in the middle of the Kalmuk steppe by new Soviet forces streaming out of Astrakhan. The Soviets were approaching Rostov from the north and the east; at one point they were 30 miles away from the bridges, while Germans in the Caucasus were 350 miles to the southeast, not yet authorized to flee.

To the soldiers of First Panzer Army, the order to retreat was unwelcome. Not only had they suffered badly to get to the Terek River, they had also fortified their positions against the cold and the Red Army, and they had no wish to start marching now that winter had come. Nevertheless, withdrawal was better than being killed or captured. The army's retreat from the Terek line consisted of a series of jumps, fights and counterattacks against closely pursuing Soviets. On January 3, 3rd Panzer was the last division to exit Mosdok, blowing its bridges and destroying industry in the city as it left. During their own retreat the Russians had attempted to destroy everything that could be of use to the Germans; now, to the extent that bridges and other facilities had been restored, they were blown up again as the invaders pulled back.

Retreating German infantry sometimes found Soviet tanks waiting for them at the end of their day's march. On January 6, 3rd Panzer had to save 111th Infantry from a blocking line of T-34s. The next day 50th Infantry lost an entire regiment that was cut off by the enemy pursuit. The German combat troops had no problem with ammunition as they retreated past a series of their own supply dumps, but the rear logistics were confused. According to a Soviet account of actions in the sector of 13th Panzer: "A tank battalion under the leadership of Captain Petrov and additional rifle companies advanced on January 11 . . . crossed the Kuma to the eastern edge of Mineralnie Vody and then further to the railroad station, where two military trains stood. First the two locomotives were set afire. Another four trains coming from Georgievsk were blocked and had to halt. Two other trains with tanks and ammunition were intercepted in front of the station."

On reaching the Kuban River, 3rd Panzer destroyed all its vehicles that had engine or other problems; nevertheless, the division's repair units made prodigious efforts to retrieve tanks. The commander of a Mark IV wrote:

> We had to drive to the rear area because of a damaged fuel pump. The snow was knee deep and the temperature was minus 20 degrees. The tank treads looked like polished chrome. The road was icy and the tank skidded. The crew jumped out. The tank landed in a frozen trench and we tried to back it out. The ice broke. The Panzer IV sunk deeper. The water was already coming through the floor plates. The motor stalled!
>
> Then it was night. Long columns passed by; they were becoming ever thinner. We were freezing from the cold. Should we blow it up or wait for the repair team, that was the question. Then a message came over the radio: "Hold on—we're coming!"
>
> We were all alone. There were engine sounds—were they ours or Russians? It was Technical Inspector Bärwinkel with two 18-ton prime movers. He told us: "There was no question of blowing it up!" Each prime mover had a Panzer III in tow. One of the prime movers used a Panzer III as a block. They righted our Panzer IV by turning it over twice. Water, ice and oil covered the inside of our tank. We took our places and followed the prime movers back to the regiment.

While Hitler continued to hold on to any shred of hope that he could keep First Panzer Army in the Caucasus, he ordered Manstein in the north to form some kind of front in the face of Soviet attacks. Manstein recalled: "If Hitler thought he could order us, in the face of that preponderance of forces and with such an expanse of territory to cover, to make the army hold some 'line' or other, or else to obtain his approval before undertaking any withdrawal, he was seriously mistaken. As an obstacle, a hard-and-fast line was likely to prove about as effective as a cobweb in Fourth Panzer Army's situation." On January 7 Soviet armor got within 12 miles of Manstein's army group headquarters. The Germans fetched some tanks out of a workshop and were able to fight them off. Farther north, on January 15, the Soviets launched a new offensive that destroyed the Hungarian Second Army and the Italian Alpine Corps, which had been holding out on its right flank, widening the hole in the Axis front to over 200 miles. Russian armor was now advancing on the Dnieper bridges, which, if taken, would add the Crimea to the Soviet bag.

British correspondent Alexander Werth toured the Soviet front south of Stalingrad at this time and reported a landscape of carnage: destroyed villages, trains and military equipment along with hundreds of bodies of men and horses. Marshal Malinovsky, Soviet Minister of Defense for the southern theater, told him:

> For the first time the Germans are showing signs of great bewilderment. Trying to fill in gaps, they are throwing their troops about from one place to another—which shows that they are short of reserves. Many of their troops are retreating west in a disorderly way, and abandoning enormous masses of equipment. Such troops are an easy target for our aircraft. Most of the satellite troops have been knocked out altogether.

Fourth Panzer Army, still in the open steppes, had the Soviet 51st and 2nd Guards Armies to its north and 28th Army attacking from the east, trying to get between it and the forces in the Caucasus. The 17th and 23rd Panzer Divisions, along with 16th Motorized and SS Viking, attempted to keep in touch with each other while Red Army tanks and infantry flowed in and among their positions. By now the rivers were

frozen over, and roads had little significance since nearly all of the flat terrain was equally passable. An attack could come from any direction, at
any time. The tension of those days is captured by Oberst Flügel of
Viking, who had successfully headed off a Soviet tank thrust during the
day, and after darkness was trying to find his neighbor. Flügel wrote:

> Darkness fell and we had no further opportunity to establish con-
> tact with the 17th Panzer Division. Unfortunately, we also had
> wounded. Since we had lost contact with our battalion, we broke
> contact with the enemy and drove in two columns, with cannon
> and machine guns aimed in all directions . . . Orientation was
> almost impossible. After a short distance we drove past a Russian
> camp where Russians were sitting around a fire. . . . The 17th
> Panzer rearguard and the pursuing Russian tanks were firing at
> each other. It was difficult for us to make contact with the friend-
> ly formations because we could be fired upon as an enemy.
>
> We withdrew from the vicinity of the combatants and came
> upon some houses. We noticed some movement, but was it
> friendly people or Russians? . . . we could not determine whether
> the headlights were yellow Russian or white German. . . . We con-
> tinued to drive. Our withdrawal became a problem because of a
> lack of fuel. However, because of the cold, we could not stop the
> tank engines. I drove in the lead in order to determine from the
> tank tracks what we would have to deal with. Finally, at 0500
> hours, Landser jumped from their positions and yelled:
> "Password?" which I did not know. However I soon was able to
> establish that this was the security of the 17th Panzer Division. I
> drove with the company to the nearest troop surgeon and turned
> over my wounded. . . . We quickly refueled and stocked up on
> ammunition. We had just finished supplying the tanks when
> Russian T-34s drove into my halted company. They fired in all
> directions; two of my tanks were hit. In the meantime we took up
> the battle. Of the five T-34s, two were hit at close range. The rest
> retreated. We rolled to the west at the break of day.

To the south, on January 18, 13th Panzer arrived at Armavir and 3rd
Panzer at Voroshilovsk, which had been First Panzer Army's initial objec-

tives on its drive into the Caucasus the previous summer. Both divisions destroyed their artillery that could no longer be pulled back by prime movers. Wilhelm Tieke, a veteran of the campaign, wrote, "The exertions of the grenadiers were unimaginable. Blown vehicles and destroyed equipment, burned-out ration stores and euthanized horses littered the routes. By the time they reached Rostov the 3rd Panzer Division had lost over half of its trucks and tanks."

Hitler's quandary was whether to comply with Manstein's repeated requests to pass First Panzer Army through Rostov, or withdraw it into the Kuban peninsula, which could serve as a powerful bridgehead for a renewed effort against the Russian oil fields in the spring of 1943. On January 24 he compromised, ordering 13th Panzer into the Kuban and the rest of First Panzer Army to Rostov. The decision was made without a moment to spare, because the German divisions holding open the escape corridor were close to being overwhelmed. Mühlenkamp, the commander of SS Viking's panzer battalion, wrote:

> During the day the panzers fought in all directions. At night they marched toward Rostov in the grim cold. . . . We often drove by the stars, like sailors. Yesterday I had to go into a house to re-orient myself, also in the hope of finding some warmth because I then would have to spend the entire night standing in the turret in the minus-35-degree cold. The extreme temperature difference temporarily knocked me unconscious. The constant strain on the nerves also had something to do with this. . . . We followed the oil pipeline back to Bataisk. Our dead comrades, whom we could no longer bury, accompanied us to Rostov. In one of the tanks lay Büscher . . . He was shot in his tank, killed immediately. After a short time he was frozen stiff and it was impossible to extract him from the tank. We had to break his arm before we could remove him, but we did not want to leave him behind.

On February 2, 1943, the last German holdouts in the city of Stalingrad laid down their arms. Theirs was a unique battle in modern history in that over 200,000 men were abandoned by their high command, forbidden to retreat. By Christmas it had become clear that the army could not be rescued, but it fought on as long as it could, tying down over half a million Soviet troops until the other German armies in southern Russia had gotten to the west. Some 25,000 Sixth Army wounded were flown out of the pocket before the airfields were overrun. The rest of the trapped soldiers could only die or be taken prisoner, which turned out to be much the same choice. The next day 3rd Panzer Division, as rearguard, crossed the bridges over the Don at Rostov, and on February 7 a unit of 16th Motorized became the last element of First Panzer Army to retreat from the Caucasus.

Manstein placed the divisions of First Panzer Army on the left of his army group and then finally withdrew the rest of his thin armored screen around Rostov to the west of the Don. Adolf Hitler's dreams of gaining the oil fields of the Caucasus were shattered, as was his entire idea of forcing a victory against the Soviet Union through offensive action. In a discussion at Army Group Don's headquarters on February 5, Hitler humbly confessed to Manstein that the disaster at Stalingrad had been solely his fault. Soviet tanks approached to within 18 miles of the headquarters while the men were talking. The field marshal received temporary permission to conduct further operations in the south of the Soviet Union as he saw fit.

The Red Army had meanwhile retaken the Caucasus, including Mosdok, Maykop, Armavir, Krasnodar and all the other cities. They had also retaken Stalingrad, Voronezh, Rostov and Kharkov on the steppes. Stalin continued to urge the victorious Soviet armies onward as the Germans retreated day upon day before the irresistible onslaught. The entire rear infrastructure of the German southern front was in full retreat, endless columns of vehicles fleeing to the west. The panzer divisions were no longer standing in the Don bend but joining the massive exodus. David Glantz has commented: "Soviet intelligence detected large-scale German redeployment of armored forces westward from Rostov toward the Dnieper but steadfastly interpreted those movements as a German withdrawal to new defensive positions along the Dnieper River. Consistently, the STAVKA and the Front commands clung to their optimistic view as they spurred their advancing forces on, even as Soviet lower-level commanders began to suspect and fear the worst."

Manstein, given freedom of maneuver, had finally been able to consolidate his forces for the purpose of operations, as opposed to warfare aimed at gaining territory. He now had First Panzer Army as well as

Fourth Panzer Army available for a counteroffensive, plus a powerful new SS Panzer corps that had arrived near Kharkov, to the north of the flow of onrushing Soviet columns. The permission he had received from Hitler was to pull back the front in order to lure Soviet armor into a new battle in which he would hold the initiative. Soviet commanders at the forefront of the Russian offensive reported back to STAVKA that they were running short of fuel and ammunition; they feared a trap. Stalin, however, sensed an opportunity to destroy the German invaders once and for all, concluding the entire war in one brilliant campaign. He did not realize that the continued movement of German armor was now designed to place them in a position to strike back.

Manstein's counteroffensive in mid-February 1943 would restore the German front, leaving tens of thousands of Soviet dead on the field as well as hundreds of destroyed tanks. The steppes between the Donetz and Dnieper became littered by burning or abandoned Russian vehicles beside the treadmarks of German panzers and halftracks once more driving east. Masses of Soviet infantry attempted to outrun the German armor, but in regimental and even divisional strength failed and were forced into captivity. By the time Manstein's counterstroke ended, the SS Panzer Corps had retaken Kharkov and First Panzer Army was back at the lines it had held in the spring of 1942. Despite strategic ineptitude on the part of its high command, left to its own devices the Wehrmacht had proven that on the battlefield the Red Army was not yet at the point where it could win the war by itself. Ultimately, the Soviet Union would need a major second front to be opened by the Anglo-Americans. Still, the Germans' only net gain for the year's fighting was the Kuban bridgehead, which presented at least a theoretical threat to the Caucasus. Their only net loss on the southern front was a large Soviet-held bulge of territory surrounding the city of Kursk.

Strangely, it was only at this point in the conflict that the Germans declared a "total" mobilization of their war industry. The Russians had already done this, as had the British and Americans. Early in 1943 the Third Reich followed suit. Germany had always counted on a short war against the Soviet Union, but now the Nazis realized that the only development that would shorten the war would be their own collapse. The second front in France was over a year away, but the Germans had already been forced by the Russians onto the strategic defensive.

Stalingrad is the battle most commonly considered to be the turning point of World War II, but that assessment is imprecise. The primary objective of the Germans' 1942 campaign was seizing the Caucasus oil fields, not the city on the Volga. When Sixth Army, having been ordered not to withdraw, was wiped out by Soviet forces, the broader ramification was that the German position in the Caucasus became untenable. It was when First Panzer Army was forced to retreat from the Soviet oil-producing region that Hitler realized all possibility of gaining the resources necessary to successfully continue the war was lost. Henceforth his hopes would be pinned on "wonder weapons" or a political break between the Soviets and the Anglo-Americans. The true turning point of the war was the German failure, in 1942, to gain and hold the Caucasus.

By the spring of 1945, when Russian soldiers were probing the ground above the subterranean bunker in Berlin where Hitler had shot himself, the Allies had proved beyond a doubt that their industrial capacity could dwarf that of the Third Reich. Germany suffered a number of material shortages as the conflict continued, but its primary, crippling need during the final two years of the war was for oil.

Chapter 6

EIGHTH ARMY
IN KOREA

By September 2, 1945, when Japanese leaders appeared before General MacArthur in Tokyo Bay to surrender, America's war machine had grown to incredibly vast proportions. Aside from twelve million men under arms, it consisted of history's largest navy and air force, plus countless tanks, armored vehicles and guns. Ironically, however, in view of the astonishing quantity of lethal weapons produced by the "arsenal of democracy," the Japanese surrender was prompted by just two bombs: "Little Boy" and "Fat Man." As if it weren't difficult enough for America's enemies to combat gigantic fleets of conventional machines, U.S. scientists had come up with secret weapons so hideous they could destroy entire cities in one blinding flash. Emperor Hirohito might even have felt a sense of relief after Hiroshima and Nagasaki, because Japan suddenly had a face-saving excuse to get out of the war. Not even Samurai could be expected to stand against such destructive might.

Of course, in the postwar world, the sudden revelation of American power carried with it new responsibilities. The young, fresh-faced dough-boys who had rescued the Western allies at the end of World War I had been able to return to their continent unbothered by additional commitments. After World War II, the situation had changed in that former enemies and allies alike stared in awe at the New World democracy that had been forced to reveal itself as the strongest military power in history. The nation that once felt protected by oceans in its isolated hemisphere was now seen able to project its might east or west, anywhere around the globe. The result was that not only the Americas, or Europe, but the entire world had become a sphere of interest to the United States.

221

Unfortunately, one other superpower had emerged from the smoke and ashes of World War II, and, as fate would have it, its political-economic system was antithetical to everything Americans believed in. The amity enjoyed between the United States and the Soviet Union lasted little longer than the handshake shared by their troops at the Elbe when Nazi Germany was about to be destroyed. After that, events moved fast, and the United States was on the defensive. An "Iron Curtain" clamped down across Europe, behind which Soviet-occupied nations were forced to become allied with Moscow. Poland, Czechoslovakia, Romania, Hungary and other countries became part of the Communist bloc, as well as the eastern part of Germany. In each newly established "workers' paradise" the men in charge were hand-picked by the Kremlin. The Soviets attempted to strangle the Allied zones of Berlin in 1948, but a massive Anglo-American airlift foiled the plan. Greece was up for grabs, as was Iran. The Americans sponsored a junta and a shah, respectively, to forestall insurgencies. In Washington it had become clear that Moscow was attempting to expand its power, through the agency of Communism, by any means possible.

As 1949 drew to a close, the West was shaken by dangerous new developments. The Soviet Union exploded an atomic bomb, thereby closing its military inferiority gap with the United States. In addition, mainland China fell to Communist arms. A decade earlier, Japan had been the counterforce to China in the East as was Germany to Russia in the West. Now these states had been neutered and their strategic role taken on by America, which had troops in both nations. While U.S. and Soviet armies glared at each other across the Iron Curtain in Europe, the east coast of Asia had become a Communist domain with only two exceptions: in the south France was quixotically fighting both the tide of history and a formidable Communist guerrilla army to maintain its hegemony over Vietnam; and the United States had taken responsibility for the southern half of the Korean peninsula.

The success of nuclear weapons in forcing a quick end to World War II put a seal on the American conviction that strategic airpower would be the primary means of waging future conflicts. Even before Hiroshima, Allied strategic bombing had wrought devastation on the Axis powers, and larger, faster aircraft like the B-29 were still coming on line as the conflict entered its final stage. The predictions of Douhet, Mitchell and other prophets of military dominance from the sky had gradually been vindicated and, with the atomic bomb, had evidently reached fruition. If the Soviet Union were to attack American forces, retribution would be immediate and massive; in fact, prohibitive.

The conclusion that warfare had entered a new age, however, was only correct in the event the superpowers were to fight each other. As it turned out, the concept dubbed MAD—Mutual Assured Destruction—proved successful in preventing nuclear war. But no one anticipated the Cold War, in which the superpowers would battle each other with spies, propagandists, proxy armies or, indirectly, their own conventional forces at tangential points of contact around the globe. Since total war had been rendered infeasible in the nuclear age, "limited war" emerged as a substitute. And, once again, the prophets of airpower supremacy had proven premature: even during the atomic age, the only true results would be won by soldiers on the ground.

Aside from the Damocles sword of nuclear holocaust, the Cold War was unique in American experience because it involved an ideological as well as a military struggle, and in politico-economic terms the United States did not hold the advantage. The Communist ideal was that all people should be considered equal, socially and economically, governed by fellow citizens devoted to ensuring unprejudiced benefits for everyone. "From each according to his ability and to each according to his need" was the motto, and its logic was difficult to refute. Though intended for Western industrial countries by its inventor, Karl Marx, the philosophy worked best in practice with poor agrarian societies, with the added incentive that it was directly contrary to the systems of former colonial powers. Europe's degradation in World War II unleashed a generation of young leaders in former colonies looking to Communism as the quickest means to social and economic stability. Transplanting the American system, a faith in individualism based on perpetual economic opportunity, was difficult in poor countries with no democratic tradition. A weak nation could become stronger more quickly by dismissing the concept of individualism and uniting all its resources—particularly human ones—under absolutist guidance by the well-intentioned state.

America had already experienced one Red Scare, in 1919 after the Russian Revolution. The prosperous Roaring Twenties had shoved such worries aside, but during the Great Depression of the 1930s the Soviet model gained new adherents. After World War II American fear of Communism, which had by then taken on the aspect of serious military threat as well as undesirable socio-economic system, became almost paranoiac. Politicians built careers on finding Reds in Washington, Los Alamos or Hollywood. Unfortunately, some spies and traitors, as well as many sympathizers, did exist, which only increased public perception that the United States was under some kind of insidious attack.

During the all-too-brief period of relaxation at the end of World War II, Winston Churchill behaved like a skunk at the garden party with constant warnings to his ally about Soviet designs. His fears stemmed from an institutional British grasp of world power struggles based on three centuries of experience; however, the United States, new to world leadership, refused to take heed. By the time Americans realized "Uncle Joe" Stalin did not have their best interests at heart it was already too late.

The Soviets had been invited—actually, implored—by the U.S. to open a second front against the Japanese as soon as the Red Army could spare forces from the war in Europe. American intelligence believed the Japanese had a million men in northern China and Korea, including most of their heavy and mobile forces. The war was in its last month before the Americans realized their atomic bombs would render Soviet assistance unnecessary. The Russians nevertheless hurriedly shipped forces east and performed a textbook blitzkrieg against the Japanese. Aside from gains in China, which the Soviets later gave up, after destroying or confiscating everything of use, they occupied the northern half of the Korean peninsula down to the 38th parallel. They then set up a Communist government in their occupied territory so that Korea, a Japanese colony from 1910 to 1945, was divided into two countries: North Korea, quickly molded into a Stalinist state, and South Korea, occupied by the Americans and which was supposed to be a democracy.

The Americans sponsored Syngman Rhee as the leader of South Korea, even though he had spent most of his life in the United States. A brilliant intellectual, the 73-year-old Rhee was autocratic in his methods, ruthless toward political opponents, and many of his fellow right-wing bureaucrats were corrupt. Nevertheless, he was a fierce patriot, wholly devoted to a free, independent Korea. He saw the Soviet-occupied zone as still in thrall to a foreign power; worse, enslaved by Communist evil. When the Americans created the South Korean army, they took care not to supply it with tanks, planes or heavy artillery because they were afraid Rhee would try to invade the North.

The Soviets had a Korean of their own to place in power in their zone, a man half Rhee's age named Kim Il Sung. Although it is difficult to separate truth from legend in Kim's background (one story has him fighting with the Red Army at Stalingrad), the Soviets found him to be a talented young soldier and a dedicated Communist, if a bit immodest. Upon establishing his seat of government in the city of Pyongyang, he relinquished the rank of "Comrade" and assumed the title "Exalted Leader." In contrast to the half-hearted U.S. arming of the South, the Soviets supplied the North with tanks, self-propelled guns, heavy long-range artillery and an air force of some 200 Yak fighters and Ilyushin ground-attack aircraft. The North also gained a preponderance in nomenclature, the South's official name, Republic of Korea (ROK), outweighed by the People's Democratic Republic of Korea (the North).

In 1949, after the Chinese Communists won their war on the mainland, America discontinued aid to Chiang Kai-shek's Nationalists, who had fled to China's island province of Taiwan. While the Communists prepared to invade Taiwan, 30,000 Koreans who had fought under Mao Tse-tung returned home, providing Kim Il Sung's army with a hard core of veterans. The North thus benefited both from Soviet expertise in how to wage offensive operations with superior forces and from Chinese knowledge of how to fight at a material disadvantage. Returnees from the Chinese civil war combined with domestic mobilization gave North Korea an army of 135,000 men. South Korea had 98,000, but a third of these were domestic security troops, devoted more to implementing Syngman Rhee's will than to national defense.

To Kim Il Sung, the elimination of the U.S. puppet regime in the South was a logical next step in the worldwide struggle against decrepit Western capitalism. The Soviet Union had won the Great Patriotic War against Europe's fascists. China had pushed Chiang Kai-shek's corrupt, American-supported minions off the continent. Now it was time for Kim to follow suit, erasing the artificial separation the United States had created between the Korean people.

Meanwhile, the Soviet attitude toward Kim's designs was curiously ambiguous. On the one hand, Stalin had no wish to fight the United States directly; neither did he desire a client regime to provoke the

American behemoth, causing it to remobilize and work itself into an anti-Communist bloodlust. On the other hand, Kim claimed that he could conquer the southern half of the peninsula in five days, faster than America could mobilize, presenting the United States with a fait accompli. Having trained and armed the North Koreans for such an action, the Soviets could hedge their bets by responding with a shrug if Kim's ambition fell short, implying through their disinterest that American ire should be aimed not at themselves but at Communist China.

China, too, sanctioned Kim's plans, allowing Soviet arms shipments to cross its Manchurian rail network prior to the North's invasion of the South. Having spent a quarter of a century achieving their own people's revolution, the Red Chinese were not ones to stand in the way of further progress in Korea—especially if it could be done quickly and without provoking massive violence from the West. Prior to its colonization by Japan, Korea had for centuries been considered a "little brother" to China. America replacing Japan as the dominant power in southern Korea was considered unwelcome by the new regime in Peking (now Beijing), and probably dangerous.

In Washington DC, policy toward Korea was not so much ambiguous as nonexistent. Behind its new arsenal of war-winning bombs and in view of its roaring domestic economy, the United States had demobilized after World War II with shocking speed. Matthew Ridgeway summed up the prevailing attitude toward Korea, such as it existed:

> By 1949, we were completely committed to the theory that the next war involving the United States would be a global war, in which Korea would be of relatively minor importance and, in any event, indefensible. The concept of "limited warfare" never entered our councils. We had faith in the United Nations. And the atomic bomb created for us a kind of psychological Maginot Line that helped us rationalize our national urge to get the boys home, the armies demobilized, the swords sheathed, and the soldiers, sailors and airmen out of uniform.

By 1950 the U.S. Army had shrunk from 100 to only 10 divisions, four of which were amusing themselves on occupation duty in Japan. The strategic air force was at a pitch of readiness, but the other services had been stripped of men and funds. In 1949 the U.S. Army pulled out of South Korea, leaving behind only some advisers. In January 1950, moreover, Secretary of State Dean Acheson gave a speech, reflecting the views of Douglas MacArthur and other military leaders, in which he described the American defense perimeter in Asia as a line extending from Alaska through Japan and Okinawa to the Philippines. Notably excluded from his speech was any mention of "indefensible" Korea. To Kim Il Sung, and to his benefactors and moral supporters, Acheson's words sounded very much like a go-ahead.

On the rainy Sunday morning of June 25, 1950, after an artillery bombardment by over 1,600 guns, the North Korean Army burst across the 38th parallel. Four of the seven divisions in the initial assault were targeted on Seoul in the west, two more were aimed down the center of the peninsula, and the last, aided by amphibious landings, advanced along the east coast. The South Koreans were caught by surprise. The heavier Communist artillery outranged the ROK guns, and had been targeted in advance. Outnumbered at points of impact, the ROK army also lacked anti-tank guns. Russian-built T-34s (late models with 85mm high-velocity cannon) broke through with ease, leaving destruction in their wake and panic on all sides.

Only in the center, where the tankless NK 2nd Division ran into the ROK 6th Division, were the North Koreans stopped. The 6th Division had been active in combating guerrillas in the south and in launching its own probes to the north prior to the invasion, and it had canceled all leaves that weekend. The North Koreans were forced to backtrack their 7th Division, spearheaded by 30 tanks, to the scene, at which point the ROKs had to retreat. Above Seoul, the ROK 1st Division also made a good showing, but its flanking units collapsed and it was ordered to withdraw. The commander of the division, like many ROK officers, had served under the Japanese in World War II, as opposed to many on the other side who had learned their trade under Mao.

By June 28 the North Koreans were at the outskirts of Seoul, from which the government had already fled. American civilians and advisers were being hastily evacuated on C-54 transports while jet fighters soared in from bases in Japan to provide escort. At the nearby port of Inchon, 680 Western civilians were crammed aboard a Norwegian fertilizer boat.

Thousands of Korean civilians and ROK soldiers filled the roads out of Seoul, seeking safety across the Han River to the south. The main highway bridge was prematurely blown while packed with refugees, sending at least 500 to their deaths and cutting off the escape of ROK forces still fighting in the capital's suburbs.

In the United States the decision to defend South Korea was made with remarkable quickness and virtual unanimity at the higher levels of government. Informed of the attack while spending the weekend at his home in Missouri, President Harry Truman came to the conclusion that the invasion was not about Korea so much as it comprised a challenge by worldwide Communism to the Free World's courage and resolve. He instinctively decided to fight back. The holy word that guided Western leadership at this time was "Munich." It was an article of faith in Washington that if the Allies had only stood firm before Hitler's initial aggression, World War II would have been unnecessary. Although the logic of how Nazi Germany would have been rendered harmless if Allied diplomats had refused to cede the Sudetenland in 1938 is difficult to follow, the "Munich" principle—firmness at an early stage of aggression— was not unsound. The more difficult aspect of the principle was the part that called for peaceful nations to be forever prepared for armed confrontation. In 1950, initial caution was required in case Korea was a feint to draw off U.S. forces prior to a larger attack by the Soviet Union elsewhere, but once this possibility proved groundless American reaction was clear. Koreans were not just attacking other Koreans; worldwide Communism was on the move.

Fortunately, at the time of North Korea's invasion, the dozens of former colonial possessions that would eventually come to dominate the United Nations had not yet emerged, and that organization was still in sync with Western policy. The United States called an emergency session of the UN Security Council at which North Korea's aggression was condemned. Two days later the Council authorized all "necessary assistance" to South Korea. The Soviet Union could have vetoed either resolution, but was boycotting the UN over its refusal to recognize Red China. The Truman administration had already decided to authorize U.S. air and naval action against the invaders, but now its effort to save South Korea could take place under the flag of the United Nations: the world community.

On June 29, the new "UN" commander, General Douglas MacArthur, flew to Korea to assess the situation firsthand. While coming in his plane was buzzed by a Yak-9 fighter, forcing everyone aboard to duck, but MacArthur went straight to a window to watch his escorting Mustangs chase it off. Since 1945, MacArthur had made a transition from all-conquering war hero to benevolent dictator of Japan, ruling that nation from his Tokyo headquarters called the Dai Ichi Building (the former home of an insurance company). Now he was back at war and the 70-year-old general felt invigorated. The American party rode north from the airfield in two jeeps and then MacArthur led his staff up a steep hill that overlooked the Han River and the smoking city of Seoul. Occasional mortar rounds burst nearby as MacArthur witnessed the hordes of fugitives, the "dreadful backwash of a defeated and dispersed army." On returning to headquarters he wired the Joint Chiefs of Staff that sea and airpower would not be enough to stop the Communists: he would need U.S. troops, at the time estimated as two full divisions.

Even as its resolve to defend South Korea became manifest, the Truman administration made another decision to project American power. After all, Korea was only one of the East Asian nations threatened by Communist guns: another was Taiwan. Though Chiang Kai-shek's island refuge, like South Korea, had been omitted from U.S. strategic defense planning, the Americans were determined not to send the same ambiguous signal to Peking that they had mistakenly sent to Pyongyang. The U.S. Seventh Fleet was ordered to sail for the strait that separated Taiwan from the mainland. Mao Tse-tung's growing invasion fleet of junks, sampans and steamers would be blocked by modern carriers and battleships. To the Red Chinese, America's re-intervention in their civil war, combined with U.S. forces descending on Korea, caused alarm. Until then, Kim Il Sung's gamble had not been a matter of deep concern. Now, as the United States roused itself to fight what it suspected was the first step in a worldwide Communist offensive, Red China saw itself assailed by a two-pronged American attack.

In Seoul, where North Korean troops had been greeted by a fair number of red flags, the Communists promptly began rounding up their political enemies, thousands of whom were later found in mass graves. Syngman Rhee's police had acted similarly against leftists before leaving the city; however, as in other respects, Rhee's regime simply wasn't as efficient as Kim's. To complete the chaos, U.S. fighter planes roamed the front, but without ground control could only guess where to unload their ordnance. Many ROK units as well as columns of fleeing refugees were hit from the air.

After capturing the capital, the North Koreans paused to reorganize. Three additional divisions were coming down from the north to reinforce the initial assault. Meanwhile, the ROK army had shrunk from 98,000 men to 54,000 after a week's fighting—even worse than it sounds because most of the losses were front-line troops. Assembled on the American system, the ROK army was heavy on supply and logistics while the North Koreans, based on the Soviet model, were more lean.

On July 3 the Communists renewed their offensive, their ultimate objective the port city of Pusan at the southeastern tip of the peninsula. The mountains of South Korea funneled drives into predictable routes, but opposition was so weak that, despite disparate air attacks, the invasion was becoming a walkover. The NK 4th Division was thus surprised on the morning of July 5 to find an enemy unit standing firm along a ridgeline straddling the Seoul-Pusan highway.

The North Koreans led with their unstoppable T-34s, which waded unscathed through a barrage of artillery fire. On reaching the pass through the heights, defending troops in the ditches fired with bazookas at the rear of the tanks but with no effect. Finally the first two tanks were stopped by a howitzer just beyond the pass firing armor-piercing shell over open sights. The crew of one tank jumped out of their damaged vehicle, and a North Korean tanker killed a defending soldier with his burp gun before the crew itself was mown down by small arms. A third tank destroyed the howitzer and proceeded through the pass, followed by eight others, guns blazing, bazooka shells and machine-gun fire bouncing off their armor. By now the tankers realized that their enemies weren't South Korean troops—they were Americans.

Task Force Smith, a heroic little force of two infantry companies and an artillery battery, had been flown in from Japan and trucked as far north as possible, without flank support, to hold an entire North Korean division. The U.S. 24th Division had been based in Honshu, the closest Japanese island to Korea, and Smith's companies were from its 21st Regiment. Even if the soldiers had been "softened" by occupation duty in Japan, unprepared for combat as well as unequipped to kill T-34s, the

American high command backed them with the highest confidence. After the North Korean tanks got through the pass, they shot up the American transport and then dueled with the artillery battery as they proceeded south. The Americans had given all six of their anti-tank shells to the howitzer up front, and though the rest of the howitzers fired at the tanks with high-explosive shells at pointblank range, a total of 33 tanks passed through the American lines.

The 400 U.S. infantrymen dug in along the ridgeline listened to the battle taking place in their rear, while a more immediate problem emerged to their front. The rest of the NK division was coming down the road in a gigantic column stretching to the horizon. Thousands of NK infantry in mustard-brown uniforms disembarked from trucks and spread out on the plain. Mortar shells began falling on the ridge as the Koreans advanced; more worrisome, many of the Communist troops were splitting off to left and right to flank the American positions. Soon machine-gun fire came in from heights on both sides. For three hours the Americans held their ground until Charles Brad Smith, the task force commander, ordered a retreat before his men were surrounded. With the tanks on the road behind them, the men had to scramble across fields and rice paddies in between encircling NK troops. The seriously wounded were left behind. The artillery battery abandoned its guns and escaped in trucks that the T-34s hadn't seen. By evening, 250 men had gotten back to a line held by the newly arrived U.S. 34th Infantry Regiment at the town of Pyongtaek.

The next day, the same North Korean division hit the 1st Battalion of the 34th Infantry and broke it even faster than Task Force Smith. Having been warned that the North Koreans were not just "natives" but cold, trained troops, the Americans fled before a similar flanking assault. General William Dean, commander of the 24th Division, fired the commander of the 34th and replaced him with an experienced World War II vet, Colonel Robert Martin. Two days later Martin attempted to hold the town of Chonan but was killed while firing a bazooka at a tank at close range. Only 178 men of his 3rd Battalion escaped from the battle. North Korean troops seeped in and behind the isolated American positions while the T-34s blasted everything in their path.

The 24th was still the only division MacArthur had been able to put on the peninsula, and for the next week its three regiments alternated standing and fleeing before the North Korean onslaught. On many occasions the troops were seized by "bug-out fever," abandoning their equipment when NK forces were discovered to their rear. On a positive note, air-to-ground liaison teams had arrived to direct air strikes. On July 9, U.S. jet fighters ducked in beneath a low overcast to find a column of NK tanks and trucks stalled before a wrecked bridge. Every available plane from Japan was vectored onto the target, leaving a mile-long stretch of blazing vehicles. After this incident the North Koreans remembered the lesson learned by America's enemies in World War II—camouflage by day, move by night.

By July 13 the 24th Division had deployed behind the Kum River, the last major obstacle above the important communications center of Taejon. The river was low during that dry, hot summer, however, and the North Korean infantry was able to cross at will. The NK 4th Division attacked the 34th Regiment, tanks enfilading the front while small groups infiltrated between the American companies. A large flanking movement from the west overran the regiment's artillery battalion, capturing all ten howitzers while inflicting 136 casualties. The 34th's reserve battalion failed to execute a counterattack and the regiment abandoned the river.

To the north on the following day the NK 3rd Division hit the 19th Regiment. Here the Americans held out longer, but by the time the order to withdraw was given, the enemy had set up a strong fireblock along the highway to their rear. Divisional reserve attempted to break the NK position from the south with two tanks and four AA halftracks, but the antiaircraft vehicles were destroyed and the U.S. tanks retreated after using up their ammunition. Above the fireblock the 19th Regiment's vehicles were piled up, the men trying to keep their nerve as North Korean attackers closed in from both sides. A few jeeps and trucks were able to run the gauntlet but by dark on July 16 the remainder of the regiment took to the hills to make their way south. The Americans destroyed their vehicles before abandoning them so that the highway resembled a ribbon of fire. The regiment lost 650 men.

After the North Koreans broke the Kum River Line, the 24th Division was ordered to protect the city of Taejon, astride the Pusan highway. General Dean deployed his 19th and 34th regiments, holding the 21st in reserve. But on July 19 the NK 3rd Division somehow penetrated the American lines. Tanks got into the city to find U.S. defenders armed with new 3.5-inch bazookas firing a shaped shell that could penetrate any armor. Dean put himself in charge of a bazooka team and went stalking T-34s. At one point he climbed to the second floor of a building and found an enemy tank below him. One round set it afire, and two more quieted the agony of its crew. Unfortunately, while the general was out hunting tanks, NK infantry had surrounded the city and U.S. troops on the perimeter were fleeing in panic. North Korean snipers set up shop in buildings, firing at anything that moved in the streets. The Americans formed convoys to escape, infantry atop the vehicles blasting away at upper windows, but their lead vehicles were often destroyed, the enemy shooting first at the drivers. Some Americans got lost in the labyrinthine streets and drove into dead ends where they were easy prey. Outside the city, NK ambush positions swept the roads with machine-gun fire, and at one of these General Dean was forced into a ditch. He set off on foot but became detached from his party and was subsequently captured.

Since the American Army in 1950 was laced with World War II veterans, unaccustomed to retreat, the performance of the 24th Division was severely criticized at the time. However, thanks to that same hard core of veterans, the beleaguered regiments never lost their cohesion. The worst that can be said for the 24th was that it was repeatedly beaten in battle, resulting in a certain number of stragglers. But there was no wholesale collapse. After each setback the division would reform and fight again, its young soldiers unsupported against ruthlessly aggressive enemy assault troops. The North Koreans did not help their own cause by callously murdering American captives. Sometimes U.S. soldiers were found with their hands bound behind them and a bullet in the head. Kim Il Sung admonished his troops for killing prisoners, knowing full well that the practice hardened, rather than softened, resistance.

By mid-July two more American divisions had arrived from Japan, the 25th and the 1st Cavalry. Thrust into a losing battle, the inexperienced troops were further spooked by reports that the masses of civilians streaming south included disguised enemy soldiers. In one incident, Cavalrymen gunned down scores of South Korean civilians who approached the UN lines. In combat, the two new divisions were no more successful than the 24th, repeatedly breaking before NK onslaughts. The all-black 24th Regiment of the 25th came under particular scrutiny, but on a front where "bug-out fever" had become epidemic the attention was unfair. At this time the U.S. Army decided that segregated black units were anachronisms and resolved to integrate the U.S. armed forces in the future.

Aside from the perils of holding a line, the Americans also had trouble with counterattacks. A regiment of green troops, the 29th, was half-destroyed attempting to reach the village of Hadong in the far south. It fell into an NK ambush and 313 U.S. bodies were later counted on the field; a hundred others had surrendered.

In the first months of the war, the North Koreans made full use of their psychological advantage. Once they had proven they could make the South Koreans and Americans run, NK attacks were pressed home with lethal determination, in full confidence that the enemy was only waiting to break. The string of NK successes created a corresponding sense of helplessness among UN forces. The most unnerving thing about the North Koreans was that their battlefield tactics ran counter to those of the Americans. While U.S. commanders tried to avoid letting their units become isolated and cut off from supply, the North Koreans purposely sought to place isolated units throughout the UN rear. "Front line" lost its meaning when at daybreak the Americans would find key heights held by NK companies that had infiltrated in the dark, and ambushes and roadblocks astride their line of supply.

The Communists could have conquered the peninsula had their strategy been equal to the tactical skill of their fighting troops. With America in the war it was only a matter of weeks before the North Koreans would be outnumbered and enormously outgunned, yet they failed to mass their forces, particularly armor, at a decisive point to shatter the vulnerable UN front. Instead, they pressed on all sectors, their weaker forces against weak opposition, their main strength against the UN's main strength, dozens of local successes disguising the lack of a true war-winning strategy. After the capture of Seoul the NK 6th Division, comprised of veterans from Mao Tse-tung's war, had been dispatched to secure the southwestern corner of the country instead of cutting along the undefended coast to seize Pusan in the east—a move that would have doomed the UN army. The 6th Division joyrided through town after undefended town, establishing Communist rule, while the United States poured every available reinforcement through Pusan into the southeast. When the 6th Division finally rejoined the offensive, its route to Pusan was blocked; the strategic opportunity to flank the UN forces had been thrown away.

General Walton Walker was named commander of U.S. Eighth Army, which established headquarters at the city of Taegu. At that point U.S. forces had fallen back over 100 miles and had yet to inflict a significant setback on the enemy. On July 27, MacArthur flew in and informed Walker, "There will be no Dunkirk in this command. To retire to Pusan will be unacceptable." Walker, a short bulldog of a man who had commanded a corps under George Patton in Europe, took MacArthur's order to heart. Addressing his staff, he announced that the days of headlong retreat were over. From now on, he said, "this army fights where it stands."

Of course, "stand and fight" orders are not meant to be taken literally. Eighth Army staff identified a defense line based on the Naktong River that extended 80 miles north to south and 50 miles across in the southeast corner of the peninsula. This tiny Pusan Perimeter would be the last-ditch enclave of the UN forces. The NK 6th Division had disappeared from UN situation maps but then re-emerged, racing from the west along the coast for Pusan. Walker took advantage of his short interior lines to switch the U.S. 25th Division from the north-central front to the south to plug the gap. By July 31 the Americans had dug in behind the Naktong and their line began to flesh out with reinforcements, not just from Japan but from the United States. The 2nd Infantry Division arrived, as did the 5th Regimental Combat Team, a battalion of upgunned Sherman tanks and a brigade of Marines. The Marines had a battalion of M-26 Pershing tanks (superior to the T-34) and their own offshore air support from the carriers *Sicily* and *Badoeng Strait*. The British 27 Brigade was on its way from Hong Kong. According to historian Max Hastings, its brigadier received a cheerful greeting at Pusan from an American officer: "Glad you British have arrived—you're the real experts at retreating."

The Battle of the Pusan Perimeter raged through August as the North Koreans probed, infiltrated and penetrated the UN front, only to be met by counterattacks, air attacks and increasing concentrations of artillery fire. In the "Bowling Alley," a long, steep valley, T-34s and Pershings dueled every night for a week while opposing infantry clashed on the heights. In the south the Americans attempted a counteroffensive, but the NK 6th Division got behind the attackers and overran two artillery battalions. Marines made good headway, but Army units became lost or stalled, and the entire force withdrew to its start-line. Toward the end of the month North Korean pressure eased, but then exploded again on

August 31 with simultaneous attacks along the entire front. The Perimeter sagged in the north and was broken along the west; Walker considered abandoning the Naktong line to retreat to one last bastion, just slightly larger than the final Dunkirk perimeter, surrounding Pusan. Fortunately the North Korean Army, now dependent on thousands of untrained conscripts, was unable to achieve a decisive breakthrough. In desperate fighting the UN was able to hold, sealing off the NK bridgeheads that were too large to push back.

Although the UN was still on the defensive in its small corner of the peninsula, Kim Il Sung had cause for worry. His attacking troops were now outnumbered by the defenders—half-American, now being joined by the British—who also had more tanks and guns and complete control of the air. Successes could not be exploited, and though his troops retained their psychological edge, their ranks were thinning, tank and artillery strength diminishing. Kim had no means of preventing UN reinforcement and supply, while his own supply columns were assailed by vicious swarms of aircraft. At one point successive waves of B-29 bombers obliterated a three-by-twelve-mile sector of the NK front. Fortunately no troops had been present, but many witnessed the utter devastation caused by UN saturation bombing. The campaign was already slipping when Kim suddenly received news that it had fallen apart.

On September 15 the U.S. Marines came ashore at Inchon, 200 miles northwest of the Pusan Perimeter, only 20 miles from Seoul. The 1st Marine Division was backed by the 7th Infantry Division and thousands of ROK troops, together designated X Corps. An armada of 270 American, British, Australian and Dutch ships provided fire support for the landings as waves of carrier-based aircraft blasted the port, the city and any possible route of reinforcement. The 1,500 NK troops that were available to contest the landings were quickly overwhelmed. The United Nations had turned the war on its head.

Inchon was MacArthur's masterpiece, an idea that had come to him in June when he first looked out over the Han River at the blazing South Korean capital. At first no one believed in his plan and he had been forced to summon all his powers of persuasion to convince the U.S. Navy and Joint Chiefs of Staff to go along. Some called it a "5,000-to-one" shot. To the amazement of all who had doubted MacArthur's genius, however, his intuition proved correct. Eighth Army's front, the scene of so much initial disappointment, then resolve and heroism, became a mere sideshow to MacArthur's strategic counterstroke: placing an entire UN army far in the enemy's rear.

Inchon was so far from the Pusan Perimeter, in fact, that the North Korean Army did not immediately realize it was outflanked. Ridgeway said, "The enemy, although we bombarded his lines with news of Inchon, for some days just did not seem to know he had been beaten up north." Walker had been asked to launch an offensive on the day of Inchon but asked for a day's delay so that the news could take effect. On the 16th cloudy skies grounded his air support, but Eighth Army nevertheless attacked out of the Pusan Perimeter. The North Koreans turned out to be as resolute on defense as offense and hung on to their positions tenaciously. For days Eighth Army tried to blast NK troops off hilltops, while infantry suffered heavy casualties in direct attacks. Walker had trouble getting his armor and heavy guns across the Naktong until some enterprising 1st Cavalry troops found an underwater bridge built by the North Koreans. On September 19, the southernmost NK divisions began to withdraw. By the 23rd the entire front was giving way, and within hours the North Korean Army suddenly disengaged in a mad scramble to escape north. American motorized infantry and mobile firepower had entered its element. Thousands of formerly ferocious North Koreans surrendered, sometimes simply standing at the roadside to stare at UN columns roaring by. NK divisions lost their cohesion as individuals and small groups attempted to make their way home.

On September 27, a column of the 1st Cavalry Division joined hands with elements of the 7th Infantry south of Seoul, uniting the two parts of the UN front. In Seoul the NK 18th Division battled the 1st Marine Division in fierce street fighting, but for public relations reasons (three months to the day since the invasion) MacArthur announced the recapture of the capital on September 25. On the 28th it became fact; the Communist invaders had been beaten and evicted.

Inchon ranks with the most brilliant military maneuvers in history. More than a counterstroke affecting a battle, or even a campaign, it had cut the legs from under an entire nation's war effort, reducing the tough, aggressive North Korean Army into a diminished, disorganized stream of ill-equipped fugitives. MacArthur's triumph was so complete, in fact, that it required a quick reappraisal of America's political objectives. Prior to

Inchon, the Truman administration would have been delighted to see Walker's Eighth Army fight its way back to the 38th parallel. After Inchon, not only was the border secured but Kim Il Sung's domain stood defenseless against a further counteroffensive. The North's goal had been to reunify the Korean peninsula by force, backed by their supporters in the Kremlin. Now the peninsula could indeed be unified, but by the South, backed by its supporters in the Dai Ichi Building. In his Tokyo headquarters, said by visitors to resemble a 17th-century European court, MacArthur basked in new waves of adulation. To America's Supreme Commander in the Pacific, the only goal in war was total victory. The Communists would learn that naked aggression against the West would not only fail but carry with it dire consequences.

However, while the Cold War concept of "limited war" was no part of MacArthur's thinking, it weighed heavily in Washington. To the Truman administration the true conflict was East–West on a global scale, not simply North–South in Korea. The problem was how to fight a small Communist state without drawing the larger Communist powers into World War III. While U.S. forces fought in barren, primitive Korea, the larger prizes were Europe and the Mideast, where the United States would be hard-pressed to defend against vast numbers of Red Army tanks, artillery and aircraft. In the event of another worldwide conflagration, the United States could most certainly destroy Russian cities with atomic bombs, but only at the sacrifice of Paris, London and perhaps U.S. cities to Stalin's own nuclear resources. Aside from worst-case global scenarios, enlarging the conflict in Korea would be impractical. The country bordered both the Soviet Union and Red China, while the United States was 5,000 miles away. Strategically, North Korean real estate had no value to the West and occupation of those cold, underpopulated wastes would be detrimental if it meant having to fight a major war.

In moral terms, on the other hand, it was obvious to the American public that Kim Il Sung should pay for his failed aggression with his regime. The South Koreans, unsurprisingly, were even more adamant. In Seoul, Taejon and other places, UN troops had found thousands of executed civilians in shallow graves—hideous testaments to the North's criminal nature. A reunified peninsula free from Communist rule, thenceforward guided by peaceful democratic principles, would be an additional reward for the suffering and sacrifices made before. In domestic political terms during those heady days of victory, it would have been far more difficult for the Truman administration to declare North Korea inviolable than to let MacArthur continue to win for the cause.

A minor problem was that the UN had made no decision about invading the North, and the Soviets had returned to the Security Council to veto any new resolutions. U.S. Secretary of Defense George C. Marshall finessed the issue by wiring MacArthur, "We want you to feel unhampered tactically and strategically to proceed north of the 38th parallel." The UN preferred, he said, "not to be confronted with necessity of a vote on passage of 38th parallel, rather to find you have found it militarily necessary to do so." Marshall also made it clear that offensive operations in the North were contingent on the Soviets and Chinese staying out of the conflict. Unfortunately, MacArthur interpreted Marshall's careful words as a mandate to invade the entire North in a hell-bent style. Subsequent directives from Washington that only ROK troops should be allowed to approach the Chinese and Soviet borders were ignored.

The UN invasion of North Korea began on October 9, 1950, although ROK troops, urged on by Syngman Rhee, had been advancing up the east coast of the peninsula since September 30. In the west, the U.S. 1st Cavalry Division spearheaded the assault on Pyongyang, which fell on October 19. The North Koreans had lost the strength to put up serious resistance and Kim Il Sung's government fled to the town of Kanggye in the far north, a rugged mountainous area that had served as a guerrilla base during the Japanese occupation. The U.S. 187th Airborne Regiment was dropped north of Pyongyang to try to intercept Kim's government, as well as American POWs who might be in transit, but was only able to trap the North Korean rearguard, soon crushed between the paratroopers and the British 27 Brigade. Men of the 1st Cavalry were later sickened to find over 70 American prisoners in a railroad tunnel, killed by the North Koreans during the chaos of their retreat.

MacArthur, the grandmaster of amphibious strategy, went to the well once too often with his deployment of X Corps during the invasion of the North. Instead of letting it drive through weak opposition from the vicinity of Seoul, he funneled the 1st Marine Division (at Inchon) and the 7th Infantry Division (at Pusan) back onto ships so they could land at the North Korean port of Wonsan on the east coast opposite Pyongyang. The logistics of the move threw the flow of UN supplies into reverse at the expense of Eighth Army as the two divisions executed their retrograde movement, commandeering shipping and port facilities. By the time the Marines arrived off Wonsan, ROK troops had already taken the city and had proceeded beyond; to make matters worse, Wonsan harbor was blocked by 3,000 Soviet-built mines. The Marines sailed back and forth for six days waiting for the mines to be cleared, while 7th Infantry was diverted to a different port. The operation crossed the line from ill-advised to ludicrous when the Marines finally came ashore at Wonsan to find Bob Hope had staged a USO show in the city the previous evening.

Once X Corps was back in the line of battle its mission was to advance due north in the eastern part of the peninsula parallel to Eighth Army in the west. North Korea was divided down the middle by the Taebek mountain range, making a division of UN forces sensible. In addition, X Corps was commanded by Edward (Ned) Almond, MacArthur's loyal ex-chief of staff, while Eighth Army's Walker had been a Patton man. Rather than sublimate X Corps to Eighth Army, both wings of the UN advance would be coordinated from Tokyo.

On October 15, MacArthur responded to a summons from President Truman to attend a meeting at Wake Island in the Pacific, midway between Hawaii and Japan. U.S. intelligence was confident that the Soviets were not gearing up for battle. Truculent signals, however, were coming from Peking, and Chinese troop movements were underway to Manchuria, just above the Korean border. Foreign Minister Chou En-lai had bluntly stated to Indian Ambassador Sadar Pannikar that if the United States attacked above the 38th parallel China would enter the war. Pannikar was known to be a Communist sympathizer, however, and his reports were considered suspect. MacArthur told Truman that the time to worry about Chinese intervention was over. If they had come in when the UN was hanging on by its fingernails at Pusan, their intervention could have been decisive. Now, MacArthur said, "We are no longer fearful of their intervention. We no longer stand hat in hand. . . . Now that we have bases for our air force in Korea, if the Chinese tried to get down to Pyongyang there would be the greatest slaughter." His statement turned out to be correct, though not in the sense he intended.

To the man who had led the Allied effort against Japan—an industrialized nation with the world's most fanatic fighters—the prospect of facing Chinese on the peninsula did not seem daunting. The American military had become familiar with the Chinese during World War II and considered that country's military establishment inherently second-rate: peasant troops with hand-me-down weapons, completely lacking an air force, navy or any semblance of sophisticated military doctrine. If anything the Communists, skilled only in guerilla warfare, would be more ragged and ill-trained than the Nationalists. Truman was delighted to accept MacArthur's assurance that the war was as good as won. The President and his all-conquering general shook hands and posed for photographers. The meeting would prove additionally useful to the President in the off-year elections just three weeks away. As usual on the domestic political scene, Truman needed all the help he could get.

In Korea the American catchphrase was "Over by Thanksgiving, home by Christmas." The North Koreans could still spring an isolated firefight but many UN units were unopposed in their advance. The ROK Capital Division was racing all the way north along the east coast to the Soviet border. The Eighth Army was targeted on the Yalu River, which formed the border between Korea and China. A reconnaissance platoon of the ROK 6th Division drove into the town of Chosan on the Yalu and its American advsior, Major Harry Fleming, saw scattered North Korean troops fleeing across the river. On the other side stretched the endless wastes of Manchuria. Fleming couldn't have known that he would be the only American in Eighth Army to reach the Yalu. The ROK 7th Regiment, to which he was attached, had miraculously advanced through the staging ground of four enemy armies. Unknown to General Walker, General MacArthur or the U.S. Joint Chiefs of Staff, the Chinese were already in Korea, deployed in depth to 50 miles south of the Yalu.

On October 25, Red China launched an offensive. From all sides of the ROK 1st and 6th Divisions at the tip of Eighth Army's advance, enemy troops sprang from hidden positions in the hills. Machine guns and mortars fired into the startled South Koreans while stealthy columns of Chinese troops cut between their positions. A regiment of 6th Division found its retreat cut off and fell apart. Two more ROK regiments counterattacked only to run into a Chinese buzzsaw. On the 29th the 7th Regiment was ordered to quickly withdraw south from the Yalu; however, it's luck had expired. The regiment was crushed, survived only by terrorized stragglers who eventually made it back to UN lines. The badly wounded U.S. adviser, Major Fleming, was captured by the Chinese.

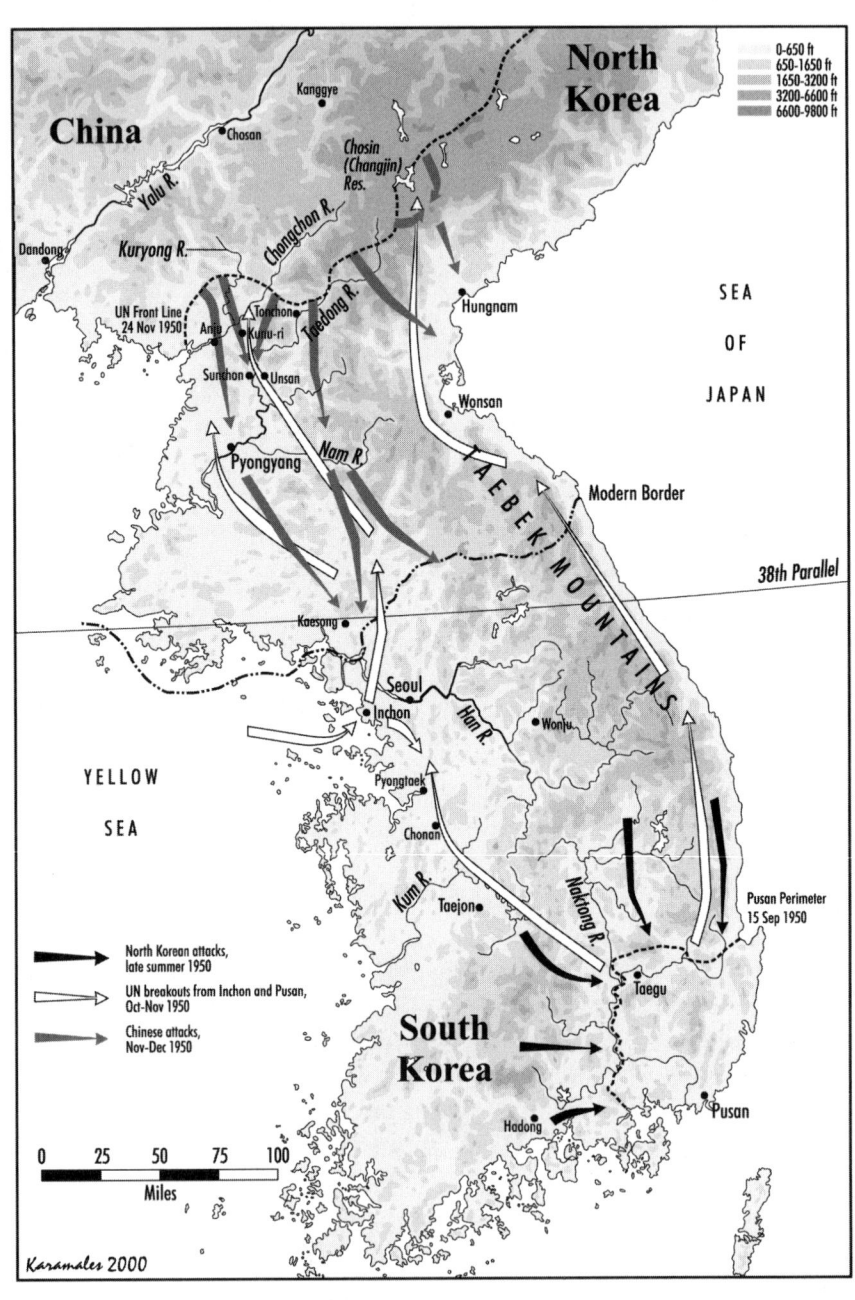

China

North Korea

0-650 ft
650-1650 ft
1650-3200 ft
3200-6600 ft
6600-9800 ft

Kanggye

Chosan

Yalu R.

Dandong

Kuryong R.

Chongchon R.

Chosin (Changjin) Res.

UN Front Line 24 Nov 1950

Tongchon

Anju • Kunu-ri

Taedong R.

Hungnam

SEA

OF

JAPAN

Sunchon • Unsan

Wonsan

Nam R.

Pyongyang

TAEBEK MOUNTAINS

Modern Border

38th Parallel

Kaesong

Seoul

Inchon

Han R.

Wonju

YELLOW

SEA

Pyongtaek

Chonan

Kum R.

Taejon

Naktong R.

Pusan Perimeter 15 Sep 1950

North Korean attacks, late summer 1950

UN breakouts from Inchon and Pusan, Oct-Nov 1950

Chinese attacks, Nov-Dec 1950

Taegu

South Korea

Hadong

Pusan

0   25   50   75   100
Miles

Karamales 2000

The first Chinese prisoner taken by the ROKs was hustled up the chain of command like some kind of exotic animal, freely explaining to everyone en route his army's order of battle. It took time before Americans realized that even the lowliest soldier in the Chinese People's Liberation Army (PLA) was made privy to strategy and its details. Meanwhile, Eighth Army and higher headquarters in Tokyo tried to figure out what was going on. ROK 1st Division was still holding together, backed by independent battalions of U.S. armor and artillery. Its commander, a veteran of the Japanese Army, inspected the enemy dead on his front and reported to corps headquarters. This wasn't a last-ditch resistance by North Koreans stiffened by volunteers; the enemy on his front were Red Chinese regulars.

The U.S. 1st Cavalry Division was ordered north from Eighth Army reserve, its 8th Regiment taking up positions around the town of Unsan that dominated an important road junction. During the day the Americans stared at huge, ominous clouds of smoke to the north. The Chinese were setting forest fires to conceal their advance from UN planes. On the black night of November 1, U.S. companies in the line heard bugles and whistles announcing the attack of unseen infantry. Flares went up. Katyusha rockets slammed into 1st Battalion's positions, exploding an ammunition truck in what turned out to be a rare use of Chinese artillery. Suddenly the Chinese were inside the 1st Battalion's perimeter. Outposts could see enemy troops infiltrating the gap between the 1st and 2nd Battalions, the latter also coming under attack. Enemy soldiers were moving along ridgelines headed south. The U.S. companies fell back, sometimes grappling hand-to-hand with their attackers. Reports arrived that the ROK 1st Division on the right had finally vacated its positions.

Just before midnight 8th Cavalry received orders to withdraw. Some vehicles tried to get through the town of Unsan but Chinese were already there. The main rendezvous was a crossroads south of town upon which fleeing U.S. and ROK troops, vehicles and Chinese soldiers all converged. The Americans set up strongpoints around their tanks while battalion columns roared south down the road to the town of Ipsok. Unfortunately, the Chinese knocked out a truck, causing its towed howitzer to jack-knife across the road. A tank tried to clear the obstruction but Chinese sappers threw a satchel charge into its treads, disabling it and further blocking the route. Chinese were seen advancing across the fields toward the road; fire was coming in from all directions. As the pile-up of vehicles increased at the roadblock so did panic among the 8th Cavalry troopers. Everyone had to make a break for it, across the hills to the safety of Ipsok. About half the vehicles and guns of the 1st and 2nd Battalions made it out.

The real disaster, however, was waiting to hit 8th Cavalry's 3rd Battalion. Positioned along the Nammyon River southwest of Unsan, the men of Major Robert Ormond's 3rd Battalion had heard the sounds of battle echoing in their rear but had otherwise been unmolested. At three in the morning, a platoon of ROK troops marched in good order across a bridge into their perimeter from the south, the guard barely nodding as they passed by. The newcomers went straight to the battalion command post, where suddenly they began throwing grenades into the trucks and a satchel charge against the door of the CP. They were Chinese commandos and their attack was the signal for an assault on the entire perimeter.

Many Americans on the line fled into the darkness but others held out, clustered in isolated groups. The next day, UN planes raced in to blast the surrounding hills; a helicopter tried to descend to take off wounded but was chased away by machine-gun fire. The 5th Cavalry Regiment tried to break through to the 8th's beleaguered 3rd Battalion from the south, but Chinese held them off, inflicting heavy casualties. General Gay, commanding the 1st Cavalry Division, was forced to call off the relief attack, leaving 3rd Battalion to its fate. On the second night, Chinese overran the American battalion's command post, by then crowded with wounded. Those who could walk were taken prisoner. The next day, weather kept the air force out of the battle while Chinese continued to squeeze the perimeter, pouring in barrages of mortar fire. At daylight on November 4 the 200 remaining able-bodied men of 3rd Battalion decided to break out. They got through the Chinese lines and wandered through the hills. The following afternoon they were surrounded by PLA troops and most were killed or captured; a few made it back to U.S. lines, which were then only a mile or two away.

While these actions were taking place, UN forces in the eastern half of the Korean peninsula were also encountering Chinese. On the road to the Chosin Reservoir, the ROK 3rd Division had taken prisoners from the Chinese 42nd Army on October 29. The ROKs were replaced by the 7th Regiment of U.S. Marines, which continued to plow its way forward. Farther east, too, the U.S. 7th Infantry Division found Chinese. But on

November 6, just when the UN high command was on the verge of reassessing its entire war strategy, everything became still again.

Just as suddenly as it had risen from the ground the Chinese Army in Korea disappeared. Eighth Army, as well as X Corps on the other side of the mountains, once more found themselves with no major enemy to fight, little opposition to a further advance. The hills were quiet; UN air reconnaissance couldn't detect any movement. The UN command wondered what had just hit them, and why; for the Chinese it was a period for tactical reassessment, and to see if their warning shots had had the desired effect.

While the United States believed that its task was to confront monolithic world Communism—a global boa constrictor that happened to raise its head in Korea—China's motives were based as much on their national history as on current ideology. MacArthur called their first attack "one of the most offensive acts of international lawlessness of historic record," but to Red China the actions were purely defensive. They were not only defending North Korea, just as America had defended the South, but also defending their border.

Just one year after the Kuomintang had been vanquished, America had taken Chiang Kai-shek back under its wing. As if that weren't enough, U.S. armies under the warmonger MacArthur were advancing through Korea, the staging ground for Japan's invasion in 1936. Mao still had five million men under arms. If the revolution was not yet finished, if the fight needed to be continued a little longer against an American counteroffensive, the people could pay the price. For centuries the Middle Kingdom had suffered exploitation and degradation at the hands of Western powers, but those days had ended. This American army raging up to the Yalu would discover a different China standing against them. Of course, one benefit of testing the waters with a limited offensive was to see whether the United States would use its atomic bombs or otherwise react in a berserk fashion. That scenario had not developed. Evidently Korea alone would be the forum in which China and America would match their strength.

To General Walker of Eighth Army, the near-destruction of a regiment of 1st Cavalry was disturbing, but had not turned the war around. The Chinese had announced that their fighters in Korea were volunteers, and perhaps he had taken their best shot. (In retrospect China's "volun-

teer" subterfuge, designed to deflect the wrath of America's nuclear arsenal, seems hardly less contrived than Truman's "UN police action" rationale.) In explaining away the presence of Chinese troops Walker said,
"After all, there's a lot of Mexicans in Texas."

General MacArthur alternated between dismissing Chinese intervention as too little and too late, and wiring apocalyptic visions to
Washington, in which he foresaw the utter destruction of his forces unless
he was allowed complete freedom to unleash his air power. His point was
that the UN armies were so close to the Yalu—only two days' march—
that air interdiction in that narrow corridor would be ineffective. He
needed to destroy the Yalu bridges and hit the Chinese staging grounds
on the other side of the border. The Joint Chiefs of Staff were persuaded
enough by the urgency of MacArthur's demands that they permitted targeting the bridges. On the first day of attacks, November 8, Mig-15 jet
interceptors rose from Chinese airfields. In history's first air battle
between jets, a U.S. F-80 Starfighter shot down a Mig. For a month U.S.
bombers hit the bridges, finally destroying the Korean halves of several,
but by that time the river had frozen over.

The larger problem was that the UN commander—a man whose
accomplishments, genius and prestige caused him to tower over his military peers—wanted to fight a larger war than desired by his political leaders. Douglas MacArthur would never shrink from a challenge by Mao
Tse-tung; in fact, he relished it. The American general knew exactly what
the U.S. armed forces were capable of. How sure was Mao that his peasant hordes could survive so immense a torrent of American firepower that
his war against Chiang Kai-shek would look like a mere skirmish?

MacArthur's bellicosity aside, the Truman administration would
clearly have opposed an advance to the Yalu if it had sensed Red China's
determination. It was absolutely not in the interest of the U.S., or the
UN, to sacrifice thousands of lives for bleak, cold territory on the Chinese
border. Ironically, the incredible stealth of China's field armies blinded
both the politicians and the UN military to the situation. China's persistent warnings, through diplomatic channels and over its national radio,
seemed like empty threats when air reconnaissance could not detect as
much as an overturned leaf between Eighth Army and the Yalu. The
Chinese only moved at night, and during the day employed expert camouflage. If an aircraft approached, masses of men would be ordered to

stand completely still, their brown uniforms merging with the color of the ground. UN signals intelligence was foiled by the fact that the Chinese didn't rely on radios. The PLA, of course, did not have a massive "tail" of heavy weapons and supplies. It had no tanks and practically no artillery, its men armed with old Japanese rifles or American-made arms captured from the Nationalists. It was a gigantic guerrilla army, trained in the tactics of secrecy, surprise and traveling light. It could appear out of nowhere, and disappear just as rapidly.

On November 20 Chinese 66th Army issued a pamphlet called "Primary Conclusions of Battle Experiences at Unsan," detailing the lessons learned from fighting the U.S. 1st Cavalry Division. As reported by Colonel Roy Appleman in his official U.S. Army history, the Chinese had immediate respect for U.S. weapons:

> The coordinated action of mortars and tanks is an important factor. . . . Their firing instruments are highly powerful. . . . Their artillery is very active. . . . Aircraft strafing and bombing of our transportation have become a great hazard to us . . . their transportation system is great. . . . Their infantry rate of fire is great and the long range of fire is still greater.

Still, it wasn't all bad news for the Chinese. Describing the American infantry when cut off by penetration attacks, the pamphlet said they

> abandon all their heavy weapons, leaving them all over the place, and play opossum. . . . Their infantrymen are weak, afraid to die, and haven't the courage to attack or defend. They depend on their planes, tanks and artillery. At the same time, they are afraid of our firepower. They will cringe when, if on the advance they hear firing. They are afraid to advance farther. . . . They specialize in day fighting. They are not familiar with night fighting or hand to hand combat. . . . If defeated, they have no orderly formation. Without the use of their mortars, they become completely lost . . . they become dazed and completely demoralized. . . . At Unsan they were surrounded for several days yet they did nothing. They are afraid when the rear is cut off. When transportation comes to a standstill, the infantry loses the will to fight.

It came as an enormous relief that what the Chinese called their "First Phase Offensive" had not brought down the revenge of nuclear weapons or any other attacks on Chinese soil. American planes were being so cautious they were only hitting the Korean sides of the Yalu bridges. Chinese troops continued to pour over the Yalu at night. Soon there were at least 150,000 men hidden away just miles from Eighth Army; 120,000 more were on the east side of the peninsula in the vicinity of X Corps. The big question was: Would the Americans keep coming?

MacArthur, for one, would not be denied his victory. Afterward, observers likened his attitude to George Custer's on the eve of the Little Big Horn. The American Joint Chiefs of Staff were paralyzed by his prestige and his reputation for unerring judgment. The Truman administration and its CIA offered no guidance. MacArthur controlled the show. On November 23, Thanksgiving, the American logistics network in Korea performed a prodigious feat in delivering lavish turkey dinners to every U.S. soldier in Korea. On November 24, the "End the War" offensive began.

The final lunge was intended to place UN troops along the northern border of Korea. Eighth Army was deployed along the Chongchon River in the West, which ran parallel to and some 50 miles south of the Yalu. In view of China's October surprise, General Walker resolved to keep his army close in hand with measured advances by mutually supporting formations. His I Corps on the left consisted of the U.S. 24th Division (along the coast) and the ROK 1st Division. In the center IX Corps was comprised of the U.S. 25th and 2nd Divisions, backed by the 5,000-man Turkish Brigade. On the right was the reassembled ROK II Corps of three divisions, the 6th, 7th and 8th. The U.S. 1st Cavalry Division and British 27 Brigade were held in reserve.

On the east side of Korea, over the Taebek range, ROK I Corps deployed along the coast while Ned Almond's X Corps drove north inland. Almond's U.S. 7th Division would advance on the right with the newly arrived U.S. 3rd Division in reserve. The cutting edge of X Corps was the 1st Marine Division, ordered to advance along the western side of the Chosin Reservoir, a body of water created to provide hydroelectric power. (In Korean the reservoir was called Changjin, but Americans, relying on their Japanese-made maps, called it Chosin.) The Marines would reach the edge of the reservoir and then veer west to join Eighth Army near the Yalu in a pincer movement to destroy any final Communist opposition.

During the pause following their brief, mysterious First Phase Offensive, the Chinese moved 42nd Army from the vicinity of the Chosin Reservoir to the Eighth Army front. There it joined its parent XIIIth Army Group. The Chinese IXth Army Group came over the Yalu to await the advance of X Corps in the east. Overall command of People's Liberation Army forces in Korea was held by Peng D'Uhai, a veteran of the Long March. Eighth Army estimated Chinese forces in Korea—volunteers, specialist units, a token force of the PLA—at 40–70,000, to be faced along with up to 80,000 scattered North Koreans. Incredibly, despite aircraft crisscrossing the territory south of the Yalu, UN intelligence failed to detect the magnitude of the enemy buildup. Having no air reconnaissance, the Chinese used ground patrols to pinpoint UN positions, while the topography of North Korea, allowing for only a few roads through separated mountain passes, informed them exactly where to prepare their traps.

On the 24th, Eighth Army surged north from the line of the Chongchon River. On the first day the only opposition was on the army's right, where ROK II Corps had been in contact with enemy forces for a week. The next day the methodical advance continued, even as local commanders began coming across disturbing signs. Korean civilians reported many Chinese in the hills, and troops saw that people had fled their villages, not unlike birds before a storm. The 38th Regiment of the 2nd Division found an abandoned mine packed with over 500 tons of American-made ammunition.

The 25th Division formed a task force of all arms to spearhead its advance. That afternoon, two former American captives approached from the north, saying that 24 more wounded men were waiting up the road. The men were survivors of 8th Cavalry's debacle earlier in the month, released by the Chinese for unknown reasons. Soon Task Force Wilson reached the scene of 3rd Battalion's annihilation, where American bodies still lay on the ground, many in sleeping bags.

Chinese outpost units fell back during the day, but by late afternoon U.S. troops had run up against isolated strongpoints that the enemy meant to keep. A Ranger company acting as spearhead of 25th Division's task force was badly bloodied while taking its final objective of the day,

Hill 205. (Hills were named according to their height in meters.) A company of U.S. 2nd Division to the right made it halfway up Hill 1229 against Chinese skirmishers, even as they sensed the presence of stronger enemy troops above. B Company of the 2nd Division's 9th Infantry got to the north side of Hill 219, which turned out to be the apex of Eighth Army's advance. At the end of the day the UN forces halted, some men assigned to stay awake on company perimeters while the rest bedded down. Meanwhile, the bleak territory to their north came alive with thousands upon thousands of marching troops. With the coming of darkness on November 25, the Chinese Second Phase Offensive began.

The first massive blow fell on ROK II Corps, Eighth Army's right flank. The Chinese managed to find the boundary between its 7th and 8th Divisions, overran several companies and poured additional men through the gap. Reserve ROK battalions were hesitant to counterattack in the dark and instead set up for defense. Chinese 38th Army made the breakthrough while 42nd Army attacked along the ROK right, cutting its lines of communication in the rear.

On the first night of their offensive, the enemy seemed more intent on penetration than frontal attacks, but isolated UN units that held important ground found themselves assailed. On the hill that dominated the road to Ipsok, 25th Division's Rangers came under attack from all sides. Supporting U.S. tanks and artillery blasted the slopes but the company was overrun. Of 83 Rangers only 21 made it off the hill, many of them wounded. A mile to the west, Company E of the Task Force had set up in a Korean graveyard on Hill 207. Brief bursts of firing were heard on the lower slopes, followed by periods of quiet. Flanking squads were being wiped out by advancing Chinese. The 3rd Platoon frantically tried to call in artillery fire but the airwaves were jammed with too many other emergency requests. The men heard a babble of voices at the bottom of the hill and then there was silence. Suddenly, hand grenades began exploding at the summit and Chinese troops rushed the platoon from only yards away. A U.S. machine gun tried to fire but a Chinese armed with a Thompson submachine gun ran over and shot the firing team. When the platoon fled, the Chinese remained among the gravestones shouting, "Come on back, G.I. Afraid, G.I."

Chinese 40th Army, considered one of the PLA's elite formations, was targeted on U.S. 2nd Division, which had advanced up the main road that led to Kanggye in the north. On the division's left, K and L Companies of the 9th Infantry were in position near a dry creek bed when outpost troops heard the sound of hundreds of tramping feet. Columns of Chinese were trotting by in the dark, officers on horseback urging the troops on. K Company might have gotten through the night unscathed, except a soldier fired his rifle just as the last of the enemy was nearly out of sight. Two companies of Chinese peeled back and K Company was destroyed. Just to the south, L Company had lit numerous bonfires to comfort its men against the cold. At one point two South Koreans ran into the perimeter and told an officer, "You in great danger." An attempt was made to put out the fires but the company's position had already been given away. The sudden disappearance of K and L Companies on the night of November 25 was a mystery to 2nd Division until, days later, a few survivors made it back to Eighth Army lines to tell what had happened.

One after another, 2nd Division's units suffered surprise attacks. At Hill 219, the farthest point of Eighth Army's advance, 82 of B Company's 116 men were lost. In the center of the division, a regiment of Chinese forded the Chongchon River and overran the 61st Artillery Battalion. Pfc. Jimmy Marks of A Battery recalled, "All up and down the river valley, all hell had broken loose. Tracers and explosions, left and right. Flares would explode, giving too much light, then flutter down and extinguish themselves in frozen corn stubble. The Chinese blew bugles and whistles and shouted American profanity. I thought their bugles were playing, 'Silent Night, Holy Night.' Between shots and explosions I could hear the wounded crying for help."

Farther south, another PLA regiment crossed the river and overran a platoon of 23rd Infantry, but a counterattack with tanks retook the position. The Chinese took their pants and shoes off to cross the river and the Americans were astonished to find many of their attackers naked from the waist down. As more and more Chinese came across the river, the 23rd Infantry took to shooting them while they were trying to put their shoes on. U.S. tanks put down a curtain of fire on the riverbank. A 1st Battalion counterattack regained the 61st Artillery positions where they found many Chinese huddled in foxholes for warmth. At dawn the Americans counted 400 enemy bodies in the area and 102 prisoners.

On November 26, Eighth Army had cause for concern if not alarm.

U.S. 24th Division on the left along the Yellow Sea had been unmolested, though its neighboring ROK 1st Division had come under heavy attack. 25th Division's Task Force was engaged in fierce fighting but all was quiet in the sectors of its flanking 35th and 24th Regiments. 2nd Division was bringing steady firepower to bear on the PLA units in its midst and American aircraft were pounding the surrounding hills. The biggest worry was ROK II Corps on the army's right, where it was hard to tell what was going on. Part of the problem was that ROK commanders were reluctant to inform the Americans they were having difficulties. In fact, beginning on the night of November 25, two PLA armies were dicing the South Koreans into small pieces.

The crossroads town of Tonchon was reported held by the South Koreans but it had already been seized by the Chinese. American advisers attached to the ROKs called for planes to take them out, but Eighth Army canceled the evacuations. A T-6 flying over Tonchon reported dozens, then hundreds of destroyed U.S. vehicles around the town. The U.S. advisers around Tonchon were killed or captured by the Chinese. The ROKs had more holes in their front than they had reserves to plug them; Chinese columns were dogtrotting at will through the rear of the corps. That night the Chinese launched a violent head-on attack that, combined with the enemy presence in the South Korean rear, caused II Corps to collapse. Only the 3rd Regiment of ROK 7th Division on the left escaped intact by retreating into the lines of its neighbor, U.S. 2nd Division. The remaining troops of ROK 7th and 8th Divisions either became casualties or fugitives heading south. ROK 6th Division in Corps reserve was still trying to patch itself together from the blows it had suffered near the Yalu weeks earlier. It too fell back as the Chinese advanced 50 miles, veering west to cut off the remainder of Eighth Army.

That night the U.S. 2nd Division faced more assaults throughout its sector while the rest of the 25th Division came under attack. On more than one occasion a man approached an American-held hill and asked in good English, "How many men you got up there?" Sometimes the G.I.s would naïvely answer, other times they would fire. On the hill held by F Company of the 27th Infantry the question was taken up by dozens of voices, first to their front, then left and then right: "How many men you got up there?" in a chant coming from all sides. A 38th Infantry platoon witnessed perhaps the strangest prelude to an attack. Three Chinese calmly approached the American ridgeline and halted. They raised flutes to their lips and began playing beautiful music. Soon other Chinese came out of hiding and began dancing around the flutists. The performance under bright moonlight finally began to wear on the Americans' nerves and they opened fire, causing the Chinese to vanish. Unfortunately, the next thing the U.S. platoon heard was bugle calls, and the Chinese opened up with machine guns and mortars. The platoon got off the ridge with 10 survivors, five of them wounded.

During the night, Task Force Wilson and the 25th Division's 27th Infantry were afraid for the safety of their many wounded and decided to move their medical collection stations to the rear. The column was wiped out by a Chinese ambush before it had gone a mile, however, leaving a gruesome tangle of human and mechanical wreckage to be found in the morning. In 2nd Division's rear, PLA troops gained a height where they could fire down into the 38th Infantry's command post and medical station. 23rd Infantry's command post was briefly overrun; several officers were shot when a Chinese soldier ran through the headquarters tent firing a tommy gun.

One reason both Task Force Wilson and U.S. 2nd Division were finding so many Chinese in their rear was that the 25th Division's 24th Infantry, between the two formations, had become disorganized. No one, including its commander, seemed to know where the 24th's companies were, and Chinese were pouring through the area. As it turned out, 3rd Battalion in the regiment's center had halted its advance on the afternoon of the 25th, before the enemy offensive had begun. 1st Battalion to its left, on the other hand, had advanced too far into Chinese territory, losing contact with its neighbors. B Company of the 1st managed to escape detection by the Chinese but C Company stumbled into an enemy staging area and surrendered intact (prompting headlines in Peking). On the regiment's right, E and G companies of 2nd Battalion were chased back by Chinese and retreated into the lines of the 3rd Battalion of 2nd Division's 9th Infantry, coincidentally the only other segregated black unit in Eighth Army. F Company was sent to retrieve E and G companies but it ran into a Chinese roadblock. In view of the problems, 27th Infantry was brought up from 25th Division reserve to back up the porous sector. 24th Infantry's highly decorated commander, Colonel Don Corley, found himself left with only one battalion, the 3rd, because his

1st was put under the control of 27th Infantry and most of the 2nd spent the remainder of the battle fighting with 2nd Division's 9th Infantry.

The popular conception of Chinese infantry attacks is of massive human waves crashing successively against U.S. perimeters. This image stems partly from American movies, as well as from U.S. battles against the Japanese such as Guadalcanal. The reality in Korea was even more frightening. The Chinese were nightfighters and their method was to constantly probe, looking for weak spots in the American lines, preferably seams between platoons. They would quietly get as close as possible and their sudden charge would typically consist of 40 men. When the enemy sprang from only yards away in the dark, it was difficult to tell whether the attack was by 100 men, 40 or only a dozen. Machine-gunners would be lucky to get off a burst before the enemy had closed. Chinese favored the hand grenade for nightfighting because it could be delivered without giving away their position. Many PLA troops only carried grenades, which may explain why Americans reported so many fistfights when Chinese were overrunning their positions.

When the Chinese piled up and died by the score in front of American firing lines, the problem on their end was communication. On a company level the PLA used bugles for communication, and on a platoon level whistles. Once an objective was assigned, the Chinese often reinforced failure because local commanders lacked means to change, or even question, orders from higher headquarters. Whereas a U.S. company given an impossible assignment could radio to its battalion that an attack was unfeasible, a Chinese unit would have to persist until it was destroyed; succeeding troops would then take up the task. While the image of "Chinese hordes" endures among the public, the reality lies more in the question originally enunciated by a U.S. Marine: "How many hordes are there in a Chinese platoon?"

During the short daylight hours of November 27 the Chinese were seen to move in daylight for the first time, exploiting gaps in the UN front. The ROK debacle necessitated moving reserves to bolster the army's right, so the 5,000-man Turkish Brigade was dispatched down the Tokchon road. The previous day the Turks had reported a great victory, killing or capturing hundreds of charging Chinese, but when an Eighth Army inspector arrived on the scene he found the Turks had mistakenly slaughtered a group of fleeing ROKs. On the morning of the 27th the Turks' lead battalion was ambushed by Chinese and suffered heavy losses including all its vehicles. The 1st Cavalry Division was also dispatched from reserve to protect the right flank, followed by the British 27 Brigade.

That day, on the eastern side of the Korean peninsula, the 1st Marine Division was scheduled to spearhead X Corps' half of the "End the War" offensive. Instead, the Chinese IX Army Group sprang from concealment, executing its part of the Second Phase Offensive. The Marine Division was heavily outnumbered, its regiments under vicious attack from all sides; even worse, Chinese overran parts of the Marines' only supply road, alongside the Chosin Reservoir. The most powerful formation of the U.S. armed forces was threatened with destruction.

On Eighth Army's front, U.S. 24th Division on the far left was still not under pressure although Chinese 50th Army lay above it, waiting an order to advance. To the 24th's right, ROK 1st Division was in a desperate struggle to hold off Chinese 66th Army. Considered the best ROK division, the 1st had been equipped with a full complement of artillery, and its combat capability was considered equal to American units. To 1st ROK's right, U.S. 25th Division was in serious straits. The division's 35th Regiment, to the west of the Kuryong River, was falling back. The division's spearhead was Task Force Wilson to the right of the stream, but the powerful task force was reeling. On the afternoon of the 27th it was disbanded, its surviving components returned to their parent formations. 24th Regiment on the division's right seemed helpless to prevent large Chinese units from marching through its sector. Everything now depended on 2nd Division, the most exposed of the U.S. formations, and which, after ROK II Corps' collapse, had become the right flank of Eighth Army.

On the night of November 27, increasing numbers of PLA formations dissected the U.S. divisions from all directions. Flares and rounds of white phosphorous could reveal the approaching enemy, but just as often Americans would die before they realized the enemy was upon them. Tanks served as mobile pillboxes laying down machine-gun and cannon fire, but the tanks themselves became targets of Chinese mortars, bazookas and infantry armed with satchel charges or grenades. Roy Appleman recounted one U.S. counterattack:

> The tank drivers and assistant drivers ran the tanks with their hatches partly open so that they could see. One driver saw a face

and a hand appear at the partly opened hatch. He snapped it shut, and four fingers fell off inside the tank. But in other instances, Chinese who climbed on the tanks pulled off American wounded who had been placed there. The Chinese infantry were very aggressive in attacking these tanks. Often they climbed on them to throw grenades inside the hatch. They also made every effort to blind the crews by heavy concentrations of small-arms fire and to knock out periscopes, radio antennas and the deck machine gun.

That night all three divisions of Chinese 66th Army attacked the ROK 1st Division. Its 15th Regiment was surrounded and the command posts of both the 15th and 11th Regiments were overrun, the commanders killed or captured. American artillery supporting the ROKs was already packing up when the divisional commander, General Paik Sun Yup, arrived at the front line. Somehow he was able to rally his broken units and launch a successful counterattack. The ROKs held firm until dawn, when the enemy withdrew to escape retaliation from American planes. If ROK 1st Division on the left had collapsed, its neighboring U.S. 25th and 2nd Divisions would have been caught in a giant pincer.

On November 28, four days after the start of the UN "End the War" offensive, and two and a half days after the beginning of the Chinese Second Phase (Counter) Offensive, General MacArthur realized that all his premises had been mistaken. He wired the U.S. Joint Chiefs of Staff:

> The developments resulting from our assault movements have now assumed a clear definition. All hope of localization of the Korean conflict to enemy forces composed of NK troops with alien token elements can now be completely abandoned. The Chinese military forces are committed in North Korea in great and ever-increasing strength. No pretext of minor support under the guise of volunteerism or other subterfuge now has the slightest validity. We face an entirely new war.

Part of MacArthur's fear, which would approach panic in the coming days, was that if the Communists had truly thrown down the gauntlet, the U.S. could expect waves of Soviet-made aircraft to enter the battle.

Already, Migs had been in contact over the Yalu, and a Pyongyang airfield had been hit, destroying or damaging 11 Mustangs parked there. What if the Soviets as well as the Chinese came in with both feet? MacArthur demanded authority to hit back at the Yalu, Manchuria—even Peking—to stem the onslaught of Communist troops. When asked about use of atomic bombs, MacArthur replied, "I cannot comment at this time." The vagueness of his reply compelled British Prime Minister Clement Attlee and his staff to rush to Washington for an urgent conference with the Truman administration. Surely the Americans were not contemplating a worldwide holocaust over their difficulties against the PLA in North Korea? Truman calmly assured the British that the United States had no wish to further widen the war. The American view that nuclear weapons should remain a deterrent, or defensive, element in warfare rather than a tactical recourse in battle established a precedent that guided the super-powers through the end of the century. America remains the only country to have employed nuclear weapons in anger, against Japan, but is also the country that renounced their use in its first battlefield crisis, establishing an international concurrence not unlike that which followed the use of poison gas in World War I.

Eighth Army, meanwhile, realized that it needed to retreat. U.S. 25th Division was a shambles and ROK 1st Division was hanging on by its fingernails against hammerblow attacks. The advance infantry regiments of U.S. 2nd Division were down to a third of their strength. ROK II Corps, the army's right flank, had disintegrated. The only positive note was that during daylight hours American artillery and aircraft were killing hundreds, if not thousands, of Chinese who could not reply with counter-battery or AA fire. It was only after darkness that there seemed to be no stopping the PLA from reaching farther into UN lines, biting off whole companies in their nocturnal attacks. Every time a unit was overrun there would be some degree of shatter; although a company or battalion might reform or even counterattack to regain its lost position, a certain number of cut-off, disoriented or fleeing men would be flung off from the original impact. Isolated groups from a multitude of units were turning up as far to the rear as Pyongyang.

It cannot be said that Eighth Army was outgunned or even outnumbered by the enemy; however, it can be said that the UN troops were unprepared for the size and ferocity of the Chinese offensive. The Americans had thought they would be home by Christmas; instead, they faced a new, larger enemy. The speed of the Chinese penetrations had an effect as if a gigantic airborne army had landed behind and among the UN divisions. Anyone could be attacked at any time; there was no clear front. The advancing Chinese had acquired the psychological edge—and, to the officers and men of Eighth Army, the objective had become to get away.

As Eighth Army flowed back to Pyongyang, Laurence Keiser's U.S. 2nd Division held the key to preventing retreat from turning into utter rout. In preceding days the embattled Indianhead Division had seen Turks, ROKs and 24th Regiment troopers come streaming into its sector. The 2nd's 9th and 38th Infantry Regiments had been decimated in the fighting thus far, but its 23rd Regiment under Colonel Paul Freeman was still intact. 2nd Division had the Chinese 40th Army pressing against its front and the 38th and 42nd Armies on its right. Effectively the rearguard of the army, 2nd Division had fallen back through the town of Kunu-ri, where its artillery was firing away in all directions. By November 29 all units of Eighth Army were south of the Chongchon River except for the 24th Regiment's 3rd Battalion, which was out of contact somewhere to the north. The order came through for 2nd Division to withdraw the next morning.

There were two roads out of Kunu-ri: one went west, paralleling the Chongchon, to the town of Anju in I Corps' sector. It was easy to picture this road being cut by Chinese forces crossing the river in pursuit of the army's left wing. All of I Corps—U.S. 24th, 1st ROK, U.S. 25th (placed under I Corps after the Chinese offensive had begun)—was being funneled through Anju and thence south to Pyongyang. Nevertheless, I Corps commander Frank Milburn, a personal friend of General Keiser, offered use of this road to 2nd Division. Some noncombatant units and wounded went out that way on the 29th. The other road from Kunu-ri led due south to the town of Sunchon, held by British 27 Brigade and U.S. 1st Cavalry in 2nd Division's own IX Corps sector.

Some accounts of the battle have Keiser flipping a coin over whether to retreat through Anju or Sunchon; others hold that he opted for Sunchon because it was the shortest, or most direct, route to safety. In fact, IX Corps ordered Keiser to retreat through Sunchon rather than pass into I Corps' sector. The question then became which road was more open. Reports came in of Chinese astride the Anju road. These reports turned out to be false, but reconnaissance sent out to verify failed to report back to 2nd Division headquarters. On the morning of the 29th, meanwhile, a Turkish supply convoy had been ambushed while coming up the Sunchon road. Survivors said there was at least a PLA company overlooking the road, but, perhaps due to the language barrier, 2nd Division didn't learn that the Turks had actually been ambushed twice, two miles apart.

A detachment of military police was sent to investigate, but upon reaching the (second) Turk ambush site they were shot up with only a few survivors. The 2nd Division Reconnaissance Company was ordered to clear the Chinese fireblock; however, they weren't strong enough to do so and an infantry company was also sent in. A platoon of Shermans from the 72nd Tank Battalion was ordered down the road and drove without problems all the way to Sunchon. They estimated a battalion of Chinese near the road but radio communication was difficult among the hills and 2nd Division still didn't learn that, aside from the nearest fireblock, there was an ominous collection of dead Turks and destroyed vehicles farther down the road. Neither was General Keiser aware that during the night of the 29th more Chinese arrived from the sector vacated by ROK II Corps in the east, and the ambush position had expanded until it was nearly seven miles long.

The plan for 2nd Division's retreat on November 30 was that two battalions of 9th Infantry would clear the hills overlooking the Sunchon road on either side. The 38th Infantry would then spearhead the column, followed by the headquarters group (including MPs, signal, reconnaissance and supply troops), followed by the divisional artillery battalions, then the 2nd Engineer Battalion, and finally the 23rd Infantry as rearguard. British 27 Brigade would attack from the south to clear the road on that end, and meet 2nd Division on its way down.

November 30 began inauspiciously for 2nd Division's rearguard when two PLA companies approached from the north during the dark, early morning hours and assaulted Hill 201 on 23rd Infantry's perimeter. A score of Chinese crept to the top of the hill and shot twelve B Company men. Other Chinese opened fire on the regiment's supporting tanks and, when the tanks blasted back their shots went long, bursting amid B Company on the hill. Nineteen Americans ran to avoid the tank fire and were never seen again. B Company lost 20 dead and 70 wounded to the tank fire. A counterattack later reclaimed the hill, and after dawn aircraft came in to douse the fleeing Chinese with napalm. To the nervous 23rd Infantrymen, daylight also revealed more columns of Chinese approaching from the north, west and east.

Around 7:00 A.M., two battalions of 9th Infantry—diminished through casualties so that each numbered only 200 men—moved south to clear the hills on either side of the road. Within minutes they came under fire because the Chinese ambush position had by now gotten to within a mile of 2nd Division's CP. 72nd Battalion tanks laid down supporting fire from the road. The problem for the infantry was that the enemy was dug in along successive hills or ridges. They'd take the nearest only to receive fire from the next. Four tanks were ordered to traverse the length of the road to make contact with British 27 Brigade. Evidently the Chinese were content to let tanks go back and forth unmolested while they waited for more vulnerable prey. To the surprise of the American tankers they didn't meet the British on the road but only when they had driven nearly all the way to Sunchon. The British Middlesex Battalion had attacked the previous day, suffered 30 casualties, and was now in defensive positions with no further inclination to advance.

Back at 2nd Division, the ROK 3rd Regiment had shown up after a brief disappearance and was ordered to help clear the west side of the Chinese fireblock. Two companies of Turks appeared and were ordered to assist 9th Infantry on the east side. The ROKs attacked intrepidly, reaching the top of the nearest ridge. Unfortunately, friendly tank fire again took a hand as supporting armor fired short, killing some South Koreans just as the Chinese on the hill counterattacked. The ROKs reformed and attacked again, only to suffer more tank fire from behind just as they reached the summit. After the second mistake, the ROKs discontinued their attack.

By now it was after midday and General Keiser couldn't wait any longer for the flanking heights to be cleared. Colonel Freeman of 23rd Infantry reported that Chinese were coming across the Chongchon in droves. The division had to get out. At 1:30 the 38th Infantry, mounted on tanks, led the breakout down the Sunchon road. 9th Infantry troopers came running in from both sides where they had been trying to clear the hills. The 9th's 2nd Battalion hopped on tanks and other vehicles while the 3rd Battalion set up a perimeter on the road through which following units could pass. The lead tank fired straight ahead with its machine gun every time the column came to a curve, while the infantry riders fired with small arms to either side. After less than two miles the column halted; it had come to a roadblock the Chinese had constructed during the morning, consisting of a wrecked Sherman, a half-track and a truck. Everyone dismounted, taking cover in ditches while the lead tank, commanded by Lt. James Mace, pushed the vehicles off the road. When the column started up again Mace waited for his riders to reboard. Other tanks didn't have time to wait. Roy Appleman has commented:

> The tanks in the division column did not cover themselves with glory. They showed a marked tendency to run off after a stop and leave the infantry who had been riding on their decks stranded in the ditches, and some of them deliberately left their assigned mission during the breakout to seek their own safety.

Six miles down the road, Lt. Mace came to a pass where the road cut between 50-foot ridges close on either side. He surprised some Chinese who had been milling around in the pass, which stretched for half a mile. After getting through, he came to another roadblock of wrecked vehicles and other material the Chinese had piled on. He broke this barrier by driving through at full speed. The last obstacle was a steep bypass across a stream next to a blown bridge. The stream was three feet deep and the ascent on the far side difficult, but an American bulldozer had previously tried to level it out. Shortly after clearing the bypass, Mace met a tank from the Middlesex Battalion. American artillery attached to the British had been firing north, but the Middlesex itself had not received firm orders to attack.

Unfortunately the rest of 2nd Division hadn't been able to keep up with the pace of the lead tank platoon. The soft-skinned vehicles were taking machine-gun and small-arms fire, and mortars had begun to find the column. Soldiers took cover in the ditches while each obstacle was dealt with. Two regiments of Chinese from 113th Division, 38th Army, formed the initial fireblock during the afternoon. 2nd Division suppressed much of the enemy fire from the road, but a bigger help was provided by American aircraft plastering the hillsides. Many of the tactical ground control teams were by now dead or missing; however, pilots could see where 2nd Division was aiming its fire and reinforce that fire with their own ordnance. Some of the bombs hit close to the road, sending rock fragments into the column, and at the pass napalm dripped onto the road.

The 38th Infantry reached Sunchon during the afternoon with only 400 able-bodied men; 100 stragglers arrived during the night. Its 2nd Battalion had held off a rush of PLA troops just short of the pass. When the enemy poured onto the road, infantry jumped off tanks they had been riding and Chinese clambered on with grenades. The U.S. troops then shot the Chinese off. The tanks traversed to hold off the rest of the Chinese and aircraft zoomed in to punish the attackers.

The divisional provost marshal, Lt. Colonel Henry Becker, made good progress down the road at the head of 34 vehicles, but artillery began landing on his column, wounding 21 men and destroying 6 jeeps and trucks. It came from American 155mms south of Kunu-ri misplacing their shots. Just then, Chinese machine guns opened up. Becker ran back to find two tanks, which he ordered to suppress the enemy fire. As soon as he turned his back, however, they gunned their engines and continued south. A unit of Engineers came up and started climbing the ridge toward the enemy positions. Becker found an M-16 half-track with quad-50s farther back in the column and its fire helped get his unit through. As the afternoon wore on, the route became increasingly difficult as more and more wrecked vehicles and abandoned equipment littered the road and growing numbers of wounded called for help from the ditches.

When General Keiser's headquarters group began its withdrawal from Kunu-ri, there were barely enough vehicles for the men and wounded so there was no question of carrying out the dead. The command group was held up repeatedly by enemy fire and at one point Keiser's bodyguard was killed. The general laid his body in a roadside ditch in order to make room for a wounded man in his jeep. When the column stalled Keiser got out of his jeep, which eventually emerged with 7 bullet holes, and fired at the hills with his personal rifle, a 1903 Springfield. One problem was that during intervals in the column's progress men would take cover, not firing back at the enemy. Keiser walked up and down the column rallying the men, encouraging them to fire back.

At this stage American aircraft were also making things miserable for the Chinese. One man counted 7 spotter planes seeking out the location of one enemy mortar that was subsequently knocked out by an F-80. The intense air support, perhaps combined with a temporary Chinese ammunition shortage, caused enemy fire to die down for almost two hours in late afternoon. The division's command group made it through the pass to safety around 5:00, just as darkness was falling. After that, close air support would vanish and the reinforced Chinese would be in their element.

The first artillery unit to run the gauntlet was the 17th Field Artillery Battalion, which possessed 8-inch guns, the only ones in Eighth Army. The battalion started down the road at two o'clock but was held up by enemy fire and traffic jams. As dusk fell a truck stopped behind a jeep that had two men in it. After a while the driver realized the two men were dead. The 17th Battalion got through the pass and was approaching the bypass next to the blown bridge when one of the howitzers rolled over down a ravine, killing 8 ROK soldiers who had been riding on it. Another gun missed the bypass and was about to topple over the edge of the destroyed bridge, when its driver suddenly noticed the abyss in the dark. It took some time to back up the column to let the gun withdraw from the edge. At the bypass the column was stalled by three trucks abandoned in the stream. The artillerymen were clearing these obstacles when two tanks drove up from the south with their lights on. Chinese opened fire on the illuminated scene; a PLA soldier was killed just yards away from an American officer. The tanks doused their lights and the 17th Battalion got through. The bad news was that the Chinese had extended their position all the way to the bypass—the fireblock was now ten miles long.

By nightfall half of 2nd Division was through or at least halfway down the road to the safety of Sunchon. The road was a mess with abandoned vehicles and equipment, and, even worse, bodies lining the route interspersed with wounded. Pfc. Kenneth Ross of the 9th Infantry remembered:

> The Chinese hit us from both sides. The road was clogged with every imaginable type of vehicle—jeeps, trucks of all sizes, tanks and artillery. I was walking, sometimes I caught a ride. It was mass confusion on the road. The noise of the battle was tremendous. A mortar round landed nearby and the concussion blew out my eardrums. . . . A company would go up and try to clear the Chinese off a hill but would be blown away. The Chinese were

hitting us with machine guns and mortars. They would let the tanks go by, then close in behind them. They'd knock out the lead vehicle, then beat hell out of us while we tried to push the thing off the road or went around it.

Lt. Colonel Kelleher, commanding the 38th Infantry's 1st Battalion, proceeded slowly to the pass, having his men clear the road of corpses and check the roadside for survivors. After getting through, he described the scene in between the pass and the blown bridge:

For the next 500 yards the road was temporarily impassable because of the numerous burning vehicles and the pile-up of dead men, coupled with the rush of the wounded from the ditches, struggling to get aboard anything that rolled. When we checked to make a turnout, away from a blazing wreck, either there would be bodies in our way, or we would be almost borne down by wounded men who literally threw themselves upon us.

Once U.S. aircraft were out of the battle, the Chinese pressed increasingly against the pass and the lower end of the route where they could get closer to the road. Still, they had thus far been unable to prevent the division from fighting its way out. It was a tough battle but 2nd Division mobility and firepower, combined with U.S. air support in the afternoon, had kept the route open. Back at Kunu-ri, however, the divisional rearguard, the 23rd Infantry Regiment, had come under enormous pressure.

At Kunu-ri, 23rd Infantry's afternoon began with a mass of men running down from the north to the Chongchon River above the 23rd's perimeter. It was the long-lost K and L Companies of the 24th Infantry's 3rd Battalion, the last American troops remaining above the Chongchon. They had Chinese right on their heels. Col. Freeman ordered divisional artillery to lay a barrage on the Chinese and had a liaison plane drop a note to the companies telling them to make a break for 23rd Infantry's lines. The Chinese were so close that 23rd troopers could see them catching up to 3rd Battalion men, pushing them down to the ground. Nearly 200 24th Infantry troopers were able to escape. During the afternoon Freeman was taking mortar fire on his command post. Despite growing numbers of enemy on his front, he had the impression that many Chinese were passing him to the east and south to join the huge fireblock astride the Sunchon road. His reconnaissance probes to the west, meanwhile, informed him that the Anju road was clear.

Freeman asked for permission to place the 2nd Division's three remaining artillery battalions under his command, as infantry. His request was flatly refused. He also urgently requested permission to retreat down the Anju road with the artillery and all units that hadn't already started to Sunchon. 2nd Division chief of staff Colonel Gerald Epley denied this request. General Keiser was out of contact, but at 2:30 Freeman made the same request to assistant division commander General Joseph Bradley, who said he would consider it. Colonel Freeman wanted to get the 23rd Infantry going before dark. The way things were proceeding on the Sunchon road he would have to hold his rearguard position throughout the night, and he thought his command would be wiped out by Chinese converging from all directions.

At 3:30 Freeman radioed again for a decision but General Keiser still couldn't be reached. General Bradley called back a half hour later, and as assistant division commander gave his assent. Bradley later stated: "It was not what was desired but as presented to me by feeble and interrupted radio and Colonel Freeman's strong recommendation, I authorized him to use his best judgment in extricating himself from what he implied was immediate 'suicide' to comply with orders."

2nd Division's rearguard had received permission to withdraw in a different direction from the rest of the division. The question was: Now that Chinese troops were streaming toward the battle on the Sunchon road, who would be the rearguard? The 37th Artillery Battalion had already started south down the gauntlet. The 503rd Battalion was forming up to follow. Freeman asked Lt. Colonel Robert O'Donnell of the 38th Artillery if he wanted to come out through Anju. O'Donnell said he would take the Sunchon road. The 1st Battalion of 9th Infantry said it would continue to follow its own regiment per orders from its colonel. The only other independent unit was the 2nd Engineer Battalion, holding two hills near 2nd Division's former command post. The 2nd Engineers were determined to stay with the division—they would be the rearguard.

Colonel Freeman did have the 15th Artillery Battalion and elements of the 72nd Tanks and 82nd AA Battalion under his direct command,

and these units accompanied 23rd Infantry. As soon as he received permission to escape to the west, Freeman ordered 15th Artillery to fire off all its remaining ammunition at the Chinese. The 105mm howitzers shot off over 3,000 rounds in 22 minutes. The barrels were smoldering and the gunners, too, must have worn themselves out, but Chinese return fire slackened. The artillerymen then disabled their guns. The three battalions of 23rd Infantry came racing down from their perimeter positions, scrambling aboard any vehicles they could reach. Unit cohesion was less important than time as 23rd Infantrymen replayed the scene in *Drums Along the Mohawk*, racing away with the enemy on their heels. Freeman ordered the vehicles to put on their headlights against the growing darkness and shouted, "Get the hell out of here and don't stop!" The rear of the column caught Chinese small-arms fire but the regiment made it safely down the clear road to Anju.

For the remainder of 2nd Division's units on the Sunchon road, the disappearance of 23rd Infantry turned the retreat from difficult to disastrous. The 37th Artillery Battalion, which had already started south when Freeman left, got through the gauntlet losing only 8 of its 18 howitzers. A battery of the 503rd Battalion, however, was overrun by Chinese before it had gone a mile. The Chinese ran among the wreckage taking what they needed and then disappeared again, leaving a burning ammunition truck to illuminate the scene. Major John Fralish of the 503rd and Lt. Colonel O'Donnell, commander of the 38th Artillery, came up to the carnage, but a PLA machine gun just 40 yards away riddled O'Donnell with bullets. The remaining officers decided to bring up all the AA vehicles that were back in the column. The antiaircraft guns had yet to see a Chinese plane, but had proven to be devastating weapons against enemy infantry. Fralish found three M16s with quad .50 cal. guns and two M19s with dual 40mms. He pointed out the Chinese machine gun to one of the M19s and the six-man enemy crew was quickly obliterated under a torrent of fire. The column began to move.

The remainder of the 503rd Artillery made it through the pass but then saw large numbers of Chinese to their right, revealed by a burning village. The decision was made to charge the AA vehicles past the village followed by trucks loaded with the battalion's wounded. The M16s and M19s drove the Chinese to cover, but one was knocked out and another missed the bypass and went over the end of the destroyed bridge. The few jeeps and trucks that followed the AA guns were the last vehicles to escape the fireblock. By now, the bypass over the stream was clogged with wreckage and the area was under Chinese fire. During the afternoon, tracked 2nd Division vehicles had stayed on the south side of the stream to pull out trucks that were having trouble with the ascent, but after nightfall the bypass became an impassable junkyard. It's significant that not only had 2nd Division lost its own rearguard, but no other units came up from the south to help the division hold that vital last stage of its escape route.

During the night, additional PLA units continued to arrive from north and east, and Chinese soldiers came down to the road from the hills to stalk 2nd Division survivors. Entire stretches of the road were deserted except for abandoned equipment, debris and helpless wounded. The Chinese commando unit that had surprised 1st Cavalry's 3rd Battalion a month before found an M16 AA halftrack. Pushing aside the American dead in the vehicle, they tried to fire at an approaching tank but couldn't work the traverse. Soon all vehicle traffic on the road ceased, held up by wreckage or by Chinese firing teams along the route.

All of 503rd Artillery's 155mm howitzers were lost as the battalion's survivors took to the hills on foot to reach Sunchon. Farther back, 38th Artillery bumped up against the rear of 503rd, and while stalled some of its 105mms dueled with Chinese on the eastern hills. When all hope of further progress disappeared, 38th Artillery destroyed its guns and dispersed. Both artillery battalions suffered terrible casualties during the night, the few survivors straggling into Sunchon during the next day.

The last 2nd Division unit to leave Kunu-ri was the 2nd Combat Engineer Battalion, which had been deployed on two hills south and southwest of the former divisional CP. Half of the men never got started because after the division's original rearguard, 23rd Infantry, departed, the Engineers were attacked from all sides. Those men on the northernmost hill were wiped out before they could retreat. Those from the southernmost hill fell in behind the last artillery units on the road. Sometime during the night, when the retreating column shattered into small pieces, the survivors of the battalion left the road in small groups to make their way south. Less than one-third of the 2nd Engineers, 266 men, eventually reassembled at Sunchon.

Eighth Army later counted 2nd Division's casualties during the retreat from Kunu-ri as 4,037, not counting ROKs and Turks. Its overall losses since the Chinese offensive began on November 25 were 5,295. (The 25th suffered the second-highest casualties among U.S. divisions with 1,313.) The 2nd's artillerymen, who found themselves toward the rear of the retreat from Kunu-ri, suffered the most with 1,529 casualties; 9th Infantry had 1,474 and 38th Infantry 1,178 losses. The only major unit in the division to come out intact was 23rd Infantry, which had 545 casualties. Aside from personnel, three artillery battalions had lost all their guns, and 49 AA vehicles—the most useful weapons during the ordeal—were destroyed. Thirty percent of all other vehicles were lost. The division reassembled south of Sunchon, where it was declared combat ineffective, and was placed in Army reserve to recuperate and refit.

The stage of Eighth Army's retreat that the soldiers themselves called the "big bug-out" began after the army's fighting withdrawal from the Chongchon River line. Eighth Army fell back on Pyongyang but immediately decided it couldn't hold the North Korean capital. General Paik Sun Yup of ROK 1st Division recalled:

> Geysers of flame from fires around the city lighted the darkening sky. Pyongyang's fires proved to be funeral pyres for U.S. Army supply storage points. After the railroad link with South Korea had been restored, veritable mountains of U.S. supplies had appeared in Pyongyang. General Walker had ordered that these be destroyed to prevent them from falling into enemy hands. . . . At the Taedong River railroad station, I saw eighteen tanks on flatcars surrounded by oceans of flame. The tanks had crossed the broad Pacific Ocean to support our battle, but now at the end of their long journey they were being destroyed before they had fired a single shot.

Hundreds of thousands of tons of supplies and equipment were blown up with demolitions as everyone fled south to the 38th parallel. American aircraft were kept busy destroying abandoned vehicles to keep them out of enemy hands. The mechanized UN forces held the advantage during the bug-out because they could retreat much faster than the PLA could pursue. The Chinese plodded along behind, at the rate of six miles a day, enjoying a bonanza of U.S. rations and supplies, while completely unable to keep up with Eighth Army.

But while China's People's Liberation Army achieved a spectacular success against Eighth Army, the question remains how it was faring on the other side of the peninsula, against X Corps.

The Chinese Second Phase Offensive in the east got off to a good start: three battalions of U.S. 7th Division were all but wiped out on X Corps' east flank; a convoy of mixed units to the south was horribly bloodied in an ambush. Unfortunately, China's IXth Army Group in the east chose as its major prey the largest, most powerful division in the world: the U.S. 1st Marines.

At four main base camps and countless lonely foxholes along the Chosin Reservoir, the Marines held like rocks against the fury of nocturnal Chinese assaults. The division was isolated, surrounded and outnumbered, its battalions tied together by a single precarious supply road. At MacArthur's headquarters, fear of the 1st Marines' annihilation caused nightmares; X Corps commander Ned Almond suggested that they retreat at full speed, leaving their equipment behind. The Marines themselves, however, did not view their task as retreat so much as an order to "attack in another direction." By the time they marched away from the Chosin Reservoir, bringing out their dead, wounded and all of their serviceable equipment, Chinese IXth Army Group had been destroyed as an effective fighting force. Those Chinese who hadn't impaled themselves on Marine firepower were debilitated by frostbite and hunger and thereafter could only hide from the vengeance of the U.S. Air Force or the Marines' own "black Corsairs."

The Marines subsequently joined perhaps the most leisurely seaborne evacuation in history, concluding December 11, 1950, from the east Korean port of Hungnam. Aside from 105,000 UN troops, the ships brought off 98,000 refugees, 17,000 vehicles and 350,000 tons of cargo. Material not worth taking was destroyed. The only consolation to the feeble survivors of Chinese IXth Army Group, especially those few who dared come within sight of the last departing ship, was that their sacrifices had resulted in the retreat of all UN forces from North Korea. The Marines had fought through like a colossus, leaving nothing but a pitiful wreckage of PLA units behind. On the other hand, some Chinese might have reflected on the basic truth Churchill had uttered ten years earlier: "Wars are not won by evacuations." Even in the east the Chinese had seen the backs of their enemy.

That day, Kim Il Sung returned to Pyongyang to make a triumphant speech, after which he held a reception for Chinese and North Korean officers. The North Koreans were well turned out in Soviet-style uniforms decorated with medals and epaulettes; Mao's guerrilla veterans, who had done all the recent fighting, were more shabby in their worn, unmarked quilts. When Kim desired to review a formation of his allies, the Chinese called upon one of Chiang Kai-shek's former units which still knew some parade-ground moves. It took two weeks for the Chinese Army to assemble south of Pyongyang and receive a minimum of supply; then it advanced again, across the 38th parallel, continuing to drive Eighth Army before it.

Douglas MacArthur had gambled his army and lost, with fatal consequences for thousands of men. The unerring accuracy of hindsight informs us that if the UN had exploited its Inchon triumph by seizing a large, but less than obliterating, slice of North Korean territory, terminating on a defensible line across the peninsula, the tragedies to follow would have been avoided. The North Koreans would have been punished for their aggression, the South Koreans compensated for their suffering and China's People's Liberation Army would never have crossed the Yalu. Instead, MacArthur sought total victory, causing China to become alarmed at the sudden transformation of a defensive UN force into an aggressive American army rampaging up to the Chinese border. During the dark days of retreat, MacArthur alternated between fear of reliving on a larger scale his experience at Bataan, and demanding authority to broaden the war. His recommendations to open a second front from Taiwan, employ Chiang's troops in Korea, and devastate Chinese territory from the air with every available weapon, up to and including nuclear bombs, indicate that China's anxiety about America's greatest general may have been warranted.

On December 23, Eighth Army commander General Walton Walker was killed in a traffic accident north of Seoul. It may be unkind to remark that the "bulldog" of the Pusan Perimeter, whose energy and fighting spirit helped save the UN army early in the war, was not overly mourned after he had presided over the longest retreat in American history. The disorganization always apparent at Eighth Army headquarters was revealed most starkly when U.S. 2nd Division was left unaided to retreat through a vortex of three converging Chinese armies on the night of November

30. British 27 Brigade and U.S. 1st Cavalry had been available to pull the 2nd Division out, but were not ordered to do so. Walker's successor, Matthew Ridgeway, arrived in Korea to find the army mentally, if not physically, near collapse. He wrote:

> The men I met along the road . . . conveyed to me a conviction that this was a bewildered army, not sure of itself or its leaders, not sure what they were doing there, wondering when they would hear the whistle of that homebound transport. . . . Every command post I visited gave me the same sense of lost confidence and lack of spirit.

The Chinese launched their Third Phase Offensive on New Year's Eve, 1950. The UN command determined that Eighth Army wasn't capable of holding Seoul so the retreat continued, now less of a bug-out than a strategic withdrawal. The Fourth Phase Offensive, down the center of South Korea, was held. The Fifth Phase, in April 1951, was utterly crushed and the Chinese fled north, a revitalized Eighth Army retaking Seoul and advancing once more above the 38th parallel, destroying whole divisions of the PLA en route. By then MacArthur had been relieved of his command and Ridgeway was in control.

Unlike U.S. hawks who clamored for enlarging the war by bombing China, Ridgeway recognized that strategic results could not be won by expanding the war in the air. "It has always been tempting for men removed from the conflict to envision cheap and easy solutions through naval blockades and saturation bombing," he wrote, "but any man who has fought a war from close up must know that, vital as are the sea and air arms of our combat forces, only ground action can destroy the armed forces of the enemy."

In the age of nuclear weapons, Korea inaugurated the concept of "limited warfare" between superpowers. The Soviet Union soon felt confident to pour enormous quantities of weapons into the conflict in support of its allies. The Communists were unable to surpass, however, the soldierly courage and firepower America could provide on behalf of the UN. The war ended with both sides having experienced both victory and defeat, and with the border between North Korea and South Korea largely the same as when the fighting began. On taking command of Eighth

Army, Ridgeway felt it necessary to issue a memo to his troops entitled "Why Are We Here? What Are We Fighting For?" It read, in part:

> It is not a question of this or that Korean town or village. Real estate is, here, incidental. . . . The real issues are whether the power of Western civilization, as God has permitted it to flower in our own beloved lands, shall defy and defeat Communism. . . . this has long since ceased to be a fight for freedom for our Korean allies alone and for their national survival. It has become, and it continues to be, a fight for our own freedom, for our own survival, in an honorable, independent national existence.

His words were stirring, his convictions pure. Despite its initial retreat, America did indeed project enough force, half a world away, to protect a small enclave of Western values in the shadow of immense enemy military power. Not many years would pass, however, before a similar challenge arose, and many Americans would be found to have lost their passion for such crusades.

Chapter 7

THE FALL OF
SOUTH VIETNAM

If Korea has become known, quite justifiably, as "the forgotten war," Vietnam remains so famous a conflict that its memory has haunted the American public and Washington policymakers for over a quarter of a century, primarily in the guiding principle: "No more Vietnams."

The trauma resulted from America sacrificing 58,000 lives and untold amounts of treasure across an entire decade in pursuit of an unattainable political objective: the preservation of a non-Communist South Vietnam. In the process, America also ceded a significant amount of moral authority and an even greater degree of domestic social harmony. Vietnam took place during a tumultuous period in U.S. history, its effect not unlike a jagged knife releasing a stream of blood into the mix of turbulent social revolutions—racial and generational—that were simultaneously taking place.

A curious aspect of the Vietnam War is that, in view of subsequent events, nearly everyone who had a passionate opinion about the conflict at the time turned out to be wrong. Even the dispassionate common wisdom—that from America's point of view the war was a tragic waste—can be questioned, since the United States went on to win a spectacular victory in the Cold War, the greater conflict of which Vietnam was only a part. Who can be sure that America's militant willingness to "go anywhere" and "pay any price" to oppose Communism as in Vietnam (and Korea) did not ultimately influence the Soviet Union's decision to abandon its socio-economic system and its parallel competition for global hegemony with the West? Still, as befits the most confusing of wars, the questions of why the United States became so heavily involved, whether

it could have defeated the Vietnamese Communists, or whether it should
have cut its losses at an earlier date continue to evade clear answers.

The "why" of United States involvement in Vietnam is least mysterious. To the current generation of Americans, accustomed to viewing
nations such as Libya, Iraq and Serbia as dangerous enemies, the danger
of the Cold War between the superpowers, with its omnipresent threat of
global nuclear holocaust, may seem like a faraway concept. Nevertheless,
following the 1953 armistice in Korea the world only became a more dangerous place as the intensity of the struggle between East and West
increased. The stakes of the conflict were enormous and the United States
felt compelled, and inclined, to fight against what seemed like a rising
Red tide.

In 1954 the French, fighting to hold their former colony of Vietnam,
suffered a climactic defeat at the hands of Ho Chi Minh's Viet Minh at
the Battle of Dien Bien Phu. They were able to leave behind a fragile non-
Communist state in the south after the country was divided at the 17th
parallel. In Europe in 1956, the Soviet Army rolled into Budapest and
brutally crushed a Hungarian uprising. The following year, with the
launch of Sputnik, the Soviets revealed their lead over the United States
in space exploration, a term which to military planners was synonymous
with missile technology. In 1959 Cuba, just 90 miles off the American
coast, fell to Marxist revolutionaries. Three years later the Soviets
attempted to plant nuclear missiles on that island, but President John F.
Kennedy deployed the U.S. Navy, while secretly trading off U.S. weapons
based in Turkey, to coolly steer the superpowers back from the brink of
mutual destruction. In 1960 the Soviets shot down an American spy
plane over their territory and the following year erected the Berlin Wall,
punctuating this move with a test explosion of the largest bomb in history—50 megatons, or 2,500 times the explosive might of the weapon
that destroyed Hiroshima. The British were able to stem a Communist
insurgency in Malaysia; in the early 1960s thousands of American advisers arrived to quell a larger armed revolution in South Vietnam.

In November 1963 Kennedy was killed by a sniper armed with a tele-
scopic rifle. His successor, Lyndon Johnson, faced a world pockmarked
with Communist insurgencies as well as a massive flow of Soviet arms
into the Middle East. The following year China set off an atomic bomb,
joining its Soviet ally in the exclusive group of powers able to trigger mass

destruction. That year the black-comedic movie *Dr. Strangelove* was released, as bomb shelters were dug throughout the United States and schoolchildren indulged their elders by crouching with hands over their heads in air raid drills. Nikita Khrushchev, the man who had denounced Stalin, was overthrown in Russia and Western intelligence scrambled to interpret his downfall.

While America steeled itself to resist armed Communist aggression wherever it should appear, an amusing event with surprisingly profound consequences occurred in 1964 when the British "invaded" America, their spearhead a quartet of young men singing "Yeah, yeah, yeah." A shift in demographics toward young people, as a result of the post–World War II "baby boom," was due to drastically affect American culture in the latter half of the 1960s.

By the beginning of that year there were 21,000 U.S. servicemen in Vietnam: advisers and airmen as well as logistics personnel managing a steady stream of material aid. Nevertheless the South Vietnamese were losing the war. The military situation was disintegrating so rapidly that it would be mere months before Communist insurgents overran Saigon, relegating a small American ally to the ash heap of history. There was still time, during the first year of the Johnson administration, for the United States to back out of South Vietnam before it fell. To most Americans it was just one more messy trouble spot on the map, hard to find and even harder to assess. Johnson, however, was in no position to stand in the way of America's defense of a beleaguered friend. An accidental president, he would have been condemned by both the Republicans and the conservative wing of his own party; more important, he would have dismayed the nation by dropping the mantle of Kennedy, who had courageously stood up to Communism, committing America to fight for freedom around the globe.

It was also inconceivable that the pajama-clad South Vietnamese Communists, the Viet Cong, could defeat the U.S. armed forces. In August 1964 Congress overwhelmingly passed a resolution granting Johnson the power to defend South Vietnam with "all necessary measures." In November Johnson won a landslide victory in the presidential election. On an ominous note, however, at the end of that year the first North Vietnamese ground troops appeared in the South.

If the United States had learned anything in Korea, aside from maintaining preparedness for limited war, it was not to seek the destruction of Communist nations that bordered Red China. For this reason there was never a question of invading North Vietnamese territory, no matter how fervently the latter devoted themselves to invading the South. The Communists, on the other hand, had learned not to stage their people's revolutions in the form of massive blitzkriegs across the border, prompting a lethal reflex in kind from the American giant. The Vietnam War, as it was fought by non-South Vietnamese, thus began gradually, and took several years to become a major conflict with an out-of-control logic of its own. Shadowing the American war effort throughout was the suspicion that the conflict in Vietnam was more a civil war than a Communist aggression; however, the North's fierce commitment to international Marxism and its reliance on the enthusiastic support of the Soviet Union and China helped allay initial doubts about America's mission.

The trigger that propelled the war into high gear was pulled on February 7, 1965, when Viet Cong attacked the American air base at Pleiku, destroying a number of planes and inflicting 137 casualties. That afternoon 49 American jets soared above the Demilitarized Zone to hit North Vietnamese targets at the town of Dong Hoi. U.S. aircraft were already providing fire support for the Army of the Republic of South Vietnam (ARVN) in the South. Soviet Premier Alexei Kosygin was visiting Hanoi at the time and immediately vowed "all necessary assistance" to North Vietnam. On February 10, Viet Cong killed 23 more Americans in an attack on the town of Qui Nhon, prompting further U.S. air strikes. On March 2 the Johnson Administration dropped the idea of retaliatory strikes and launched a systematic bombing campaign against the North, dubbed "Rolling Thunder." On March 8, U.S. Marines came ashore at Da Nang.

The initial U.S. troop deployments were meant to protect bases and airfields but it soon became apparent that more than passive resistance was required in order to reverse the tide of the war. Already possessing the most mobile of armies since World War II, the Americans had added another level of transport in the helicopter, a machine that could deliver troops quickly anywhere within the battle zone. If the Communists could launch surprise attacks across the length of South Vietnam, so could the Americans. Combined with the world's most formidable air strength and unlimited artillery support, the United States offered an enormous military challenge to the Vietnamese Communists, dwarfing the effort of their previous foreign opponent, the French.

The first pitched battle between the United States Army and the North Vietnamese Army (NVA) took place in the Ia Drang Valley on November 14, 1965. A battalion of the 1st Cavalry Division (now helicopter borne, "airmobile") landed in the midst of two NVA regiments. Outnumbered, the Americans fought off waves of enemy attacks, though one platoon became isolated too far forward and was nearly wiped out. U.S. firepower was meanwhile brought to bear outside the landing zone perimeter and hundreds of NVA were killed. The second phase of the ground battle occurred days later when a marching battalion of Cavalry reinforcements collided with an NVA battalion in a stretch of high grass and trees. The fight turned into a melee of crossfire at close quarters, both sides intermixed, visibility limited to yards. The Americans suffered 151 dead but the North Vietnamese withdrew, followed by jet bombers and rocket-firing helicopters. When the final tally was compiled by the Americans who held the field, 1st Cavalry was found to have lost over 300 men while the NVA lost 3,000. The figures added up. The new U.S. strategy would become "search and destroy."

General William Westmoreland, commander of U.S. forces in South Vietnam, estimated that 1965 would be the year of buildup, 1966 the year of victory, and 1967 would be devoted to mopping up. By the end of 1967, in fact, when U.S. troop strength stood at nearly 400,000, the Communists were reeling, no longer within sight of forcing a positive decision in the South. The massive search and destroy missions—codenamed with innocuous American locales like Attaboro, Cedar Falls or Junction City—were accompanied by hundreds of smaller forays in which U.S. troops sought combat. Often a reconnaissance squad or platoon would be dropped by helicopter into enemy territory to locate Viet Cong or NVA. If resistance was encountered, reinforcements would fly in, or aircraft would be vectored to destroy the enemy positions with bombs, rockets or napalm. While the NVA operated from base camps, the indigenous Viet Cong were intertwined with the rural population in hamlets under their political control. During this period when the United States scoured the countryside, South Vietnam experienced a massive migration of population to its cities. The Americans proved ruthless with hamlets deemed ambiguous in rural areas, often setting homes to the torch to eliminate any possibility of resistance. In strategic border areas, U.S. soldiers fought pitched battles reminiscent of the Huertgen Forest or Saipan for small swathes of jungle territory against highly disciplined NVA regulars. On both sides, heroes were created by the score, most of them unrecognized by history because they had fallen to enemy fire.

By the end of 1967 the American public was encouraged to hear that the war was going well. According to official reports, every search and destroy mission had turned out to be a success and, aside from the occasionally well-placed Communist rocket or mortar round, every battle had been won, at least according to the body count. Body counts have been a standard feature of battlefield reporting since Cannae in 216 B.C., and the casualties in some battles, like Borodino, continue to be debated in current literature. In Vietnam, however, the system of determining victory in battle by body count alone caused faulty assessments of strategic, if not tactical, gains. Bodies were the Communists' primary resource, of which they had an abundant supply; imagine if Hanoi had determined its progress in the war by counting destroyed American machines. Further, once the Americans had determined the Vietnamese countryside to be a free-fire zone, and land therein became meaningless compared to attrition of the enemy, the mania for casualty statistics became subject to corruption. Officers intent on showing their units in a good light would demand ever higher body counts even if these were not warranted by the fighting. The mechanism of exaggeration by this measure is now well known: when counting bodies a lieutenant might include chickens, pigs or oxen, anything that might be dead. Others would include civilians. The number of cases where commanders simply reported "sixty" instead of "six" will never be ascertained.

Though many small-unit actions were fought in a fog of increasingly brutal ambiguity, South Vietnam had been retrieved from the brink of collapse. The full press of American military might was substantially more than Hanoi had bargained for, and by the end of 1967 the Communist military command knew that the prospect of victory was slipping away. Ho Chi Minh was near death and their forces in the South were on the defensive, more hiding from the Americans than effecting the revolution. The American public was gratified to learn that Westmoreland had seen "the light at the end of the tunnel." It was up to the Communists to prove that the United States was deluding itself and that the war was not even close to being over.

On January 30, 1968, the Communists launched their Tet Offensive, attacking hundreds of localities throughout South Vietnam, including 36 out of 44 provincial capitals. In the center of Saigon, Viet Cong sappers blew a hole through the wall of the U.S. Embassy and shot their way through the guards. It was the largest Communist offensive yet, aimed straight at the urban centers the Americans had thought were secure. Walter Cronkite, the respected American conveyor of news, opened his broadcast with the question, "What the hell is going on over there?" All the statistics, body counts, rosy reports and optimistic predictions of the Johnson administration were immediately seen to be lies. The Communists were everywhere, in strength, and on the attack. The light at the end of the tunnel had disappeared.

In military terms the Tet Offensive has been compared to the Ardennes Offensive, Hitler's last-ditch throw of the dice against the American Army in World War II. In that case, the U.S. public proved enthusiastic enough for the war effort to absorb a surprise enemy counterattack, even if that public, too, had been assured that the end was in sight. In Vietnam, on the other hand, the war had become too murky, government reports too unreliable. Statistics of favorable body counts paraded through the newspapers even as more American draftees continued to be fed into the conflagration. The only hard proof of events, to the average American, was the newsreel of Viet Cong in the center of Saigon or U.S. soldiers shouting for a medic behind a wall in Hue. Some television viewers saw the incredible sequence of a skinny Vietnamese civilian—a Viet Cong—with his hands tied behind his back suddenly shot in the head by a beefy South Vietnamese officer. What kind of war had the United States gotten into?

For Americans, Tet was the turning point of the war. Regardless of Communist casualties—North Vietnam's General Vo Nguyen Giap later admitted 40,000 dead out of 84,000 men committed in the assaults—the offensive had successfully struck straight at the heart of American support for the conflict. The overriding objective of United States political leaders henceforth became not how to win, but how to disengage.

Although it is a rare soul who doesn't believe that American involvement in Vietnam was ill-advised, debate continues over whether the United States, given a different strategy, could have achieved a lasting military victory over the Communists. Those who argue that America did not employ its full force, fighting with "one hand tied behind its back," generally refer to the use of airpower. It is indeed absurd that airstrike targets were selected by President Johnson and his advisers poring over maps in the White House, as opposed to commanders in the field. The three-and-a-half-year Rolling Thunder campaign was waged so incrementally as to forego any strategic effect, and was interrupted eight times by bombing halts in vain efforts to secure corresponding concessions from North Vietnam. Of course the United States, if it had chosen to, could have delivered such gigantic devastation from the sky that Hanoi would have ceased to function and not only NVA conscripts but their little brothers could have been killed before they even contemplated joining the assault on the U.S.-held South.

Johnson's problem, however, was that obliterating airpower such as the U.S. had wielded in its crusades against Germany and Japan was not available for use in Vietnam. If Hanoi had been made to resemble Tokyo in 1945 or Haiphong reduced to the same level of ruin as Hamburg in 1943, the U.S. would have opened the door to equally devastating escalations from North Vietnam's allies. Since North Vietnam was not supplied from its own industrial infrastructure, area bombing (or worse) could only have meant the targeting of civilians, and there was a clear line between interdicting military supplies and bombing North Vietnam "back to the stone age." If Red China had perceived the latter intent toward its ally, the war would have suddenly grown to frightful dimensions. The Soviet Union, for instance by enabling China to target U.S. aircraft carriers, would have helped in any way it could. In the West, moreover, public opinion had evolved against indiscriminate use of airpower in limited wars. By the time, in 1972, when America had achieved détente with the Communist superpowers, and was thus able to pound North Vietnam to its heart's content, the United States had already given up on the war, a fact that in Cold War terms was not a coincidence. The United States may have been allowed more latitude during its retreat than during the early years when it sought victory.

The larger question is whether airpower alone can ever effect a strategic decision in a conflict that involves opposing principles. The classic example is "what if" the Union had been able to bomb the Confederacy during the American Civil War. Would the South have given in sooner or would its entire population, rather than just soldiers, have been mobilized in resistance to a higher degree? Would its armies have been defeated or would they have switched to small-unit tactics to carry on the war in perpetuity? There's no doubt that air superiority can help win ground battles and that it can destroy an enemy's industry. However, Hiroshima aside, there is still no evidence that airpower alone can defeat a people or a cause in a major conflict.

In the Vietnamese ground war, although the Viet Cong and NVA were hopelessly overmatched by American firepower in the South, they had the advantage of not having to challenge it except when it suited them, and then only in hit-and-run attacks. While the U.S. did not have the option of retreating, for example, from any of South Vietnam's cities or district capitals, Communist withdrawals before American attacks, as long as the troops stayed in the field, served the gainful purpose of prolonging the war. The Vietnamese had unlimited time to achieve success, while the United States had a limited window of opportunity to win decisive results, after which public support for the effort would expire. The Communists also had the advantage of safe havens for their forces and supplies in Laos and Cambodia, where U.S. troops were forbidden to go.

An exception to Hanoi's adherence to Mao's dictum "When the enemy is strong, retreat; when the enemy is weak, attack" occurred at Khe Sanh in early 1968, when three NVA divisions laid siege to 6,000 Marines at an isolated camp near the Laotian border. In conjunction with the Tet Offensive undertaken by the Viet Cong, General Giap hoped to duplicate his success at Dien Bien Phu by destroying a large U.S. outpost with his NVA regulars. Although the siege was heavily publicized in the American press—by some as an example of Alamo-type bravery and by others as beleaguered U.S. troops facing doom—it ended anticlimactically. While the Viet Cong fell short of their Tet objectives, the Americans almost gleefully pounded the NVA troop concentrations around Khe Sanh with fighter bombers and B-52s. The U.S. Marines defended the ground with their customary grit and the NVA survivors soon drifted back into Laos. "But then," as historian James Harrison has remarked, "in the sort of bizarre irony that had become routine, two months later the U.S. command decided to abandon the Khe Sanh base, which in early July was promptly occupied by Communist troops." In North Vietnam's histories, Khe Sanh is viewed as a stellar, albeit hard-earned, victory.

The main reason the United States failed in Vietnam, and was doomed to fail, was the fanaticism of the Vietnamese Communists. While the premise of the Kennedy and Johnson administrations that the war in the South was a veiled aggression by the Soviet–Chinese axis was mistaken, so was the antiwar view that the conflict was purely nationalist in character. In fact, Vietnamese leaders such as Ho Chi Minh, Giap and Le Duan, as shown by extensive philosophical writings, had arrived at their own unique strain of Communism, parallel to but separate from those practiced by the Soviet Union and China. Combining principles of international Marxism, Confucianism and a touch of French Revolutionary ardor, and forged through successive wars against capitalist expeditionary forces, the moral determination of Ho Chi Minh's government, including its faith in ultimate success, was unshakable. Unlike its larger supporters, Hanoi was not concerned with geopolitics, personality cults, or even government, but with its struggle for a revolutionary ideal. Compared to the revolving door of ARVN generals seizing power in Saigon, vying with each other for American favor or for personal financial gain, the sacrificial nature of the true believers in Hanoi was the Communists' greatest advantage. Through 1968 the Communists had lost 500,000 men, compared to 36,000 American soldiers. Yet that year it was the United States which, quite sensibly, gave up.

The Tet Offensive was only the first shock of 1968, a year of turmoil throughout the world. On March 16 a company of the Americal Division marched into a hamlet north of Da Nang called My Lai and gunned down hundreds of Vietnamese noncombatants. The U.S. Army was able to keep the massacre out of the news for 18 months. President Johnson was nearly beaten in the New Hampshire primary election by antiwar Senator Eugene McCarthy, whose campaign staff consisted of college students. The following week the opponent Johnson most feared, Robert Kennedy, declared his candidacy. On March 31 President Johnson announced to the nation that he would not seek re-election, devoting himself instead to seeking peace in Vietnam. Four days later the black leader Martin Luther King, Jr. was killed by a sniper. Johnson had pushed through the most comprehensive civil rights legislation in U.S. history, yet black riots in northern cities had repeatedly forced him to rely on the National Guard to maintain order. King himself had stated, "The promises of the Great Society have been shot down on the battlefields of Vietnam." After King's death violent reaction burst out in 100 U.S. cities. In June, Robert Kennedy was assassinated at close range, while celebrating a primary victory in California, by a man named Sirhan Sirhan.

In May violent anti-government protests broke out in Paris and London. In both cases young people livid over American involvement in Vietnam demonstrated solidarity with the cause of people's revolution as well as intense animosity toward the individual materialism of their elders. In China, which had also experienced a baby boom, Mao Tse-tung was able to harness the country's youth, called Red Guards, to solidify his personal power in a Cultural Revolution. Teachers and intellectuals who demonstrated wayward thought were killed, beaten or put to work in the countryside where they could more closely identify with the masses. Future Chinese leader Teng Hsiao-p'ing, for one, was paraded through the streets of a rural village with a derogatory sign around his neck.

America, meanwhile, had not only experienced a violent surge of "black power," but a sexual revolution and burgeoning drug-use among its young. Stephen Holden of the *New York Times*, reflecting on recent neoconservative trends among young people, has commented:

> Post baby-boomers may find it hard to envision block after city block of shaggy-haired youth clad in exotic robes and finery milling in the streets. . . . But such scenes were commonplace in the late 60's. On Sunday afternoons . . . Central Park became a stoned soul picnic in which hundreds of artists, pot-smoking hippies, drug dealers and musicians congregated for fragrant, dreamy communal love-ins.

Each year millions of new 18-year-olds joined the "turn on, tune in, drop out" generation, while those of World War II and prior eras aged and their numbers diminished. Music became the driving force that defined "the generation gap," drawing a line between the hopeful idealism of young people on the one hand, and the murderous arm of capitalism, as represented by the U.S. Army in Vietnam, on the other. Politically, Eugene McCarthy could draw a crowd, and, until his death, Robert Kennedy, but neither could inspire the young masses to assemble with the same fervency as any number of rock bands. Still, while the rebellious youth of America lacked a Lenin or Robespierre, the one political idea they could rally behind en masse was opposition to the Vietnam War. In August 1968 Hubert Humphrey was nominated for president at the Democratic Convention in Chicago, while on the streets outside young antiwar protesters were clubbed and tear-gassed by the police. That month, too, the Soviet Union and its Warsaw Pact allies, together comprising half a million troops, invaded Czechoslovakia to quash a liberalization movement known to history as "Prague spring."

The North Vietnamese were aware of the antiwar turmoil in America and in both May and August 1968 launched mini-Tets in order to fan the flames. Both offensives were easily defeated by U.S. and ARVN forces. The American presidential election was meanwhile approaching its climax, Hubert Humphrey steadily gaining on his Republican opponent Richard Nixon. On the last day of October President Johnson announced a complete bombing halt of North Vietnam and maintained that serious peace talks would soon be in progress. Nixon, however, had anticipated an "October surprise" from the White House and passed word to South Vietnamese president Nguyen van Thieu that the South would gain better terms if the Republicans were in control. Thieu stalled the peace talks, thus halting a sudden upsurge in Humphrey's support. Nixon was elected president by less than a percentage point, only half a million votes.

The most difficult question of all regarding Vietnam is whether the United States would have been well advised to cut its losses and withdraw in the midst of the conflict rather than stay until Hanoi was compelled to sign a treaty that acknowledged the integrity of an independent South. One problem was that at the time there was no precedent in American history for admitting defeat in a major war. An evacuation contingency plan had been drawn up after Eighth Army's disaster in Korea in late 1950, but it was only to be used in the event of battlefield catastrophe, not a collapse of will. There was no prior case of the United States retreating before a foreign power because it had been defeated.

In the years since Vietnam, of course, the United States has been perfectly willing to retreat from foreign commitments, as from Beirut following the terrorist bombing of a Marine barracks and from Somalia after the Rangers' bloody firefight in Mogadishu. Such future embarrassments were precluded by the Clinton administration in 1998 by its resolve to employ ground troops only in "non-hostile environments." In 1968, to some, an open admission that American arms could fail against Communist forces during the Cold War implied frightening consequences. To others, pure pride was at stake, and there was the widespread additional fear that a sudden U.S. withdrawal would lay South Vietnam's non-Communists open to the type of horrendous massacre that would indeed occur later in Cambodia. Still, not only hippies and liberals but increasingly the country's intellectuals and its "East Coast establishment" became convinced after 1968 that America should withdraw from Vietnam as fast as possible.

Richard Nixon became president on January 20, 1969, and by the time he resigned five and a half years later had become the most vilified man in history. Although other world leaders, Hitler for example, have been judged more heinous over time, no other public figure endured a greater outpouring of unbridled hatred from his own people while in office. In many ways Nixon, an old-fashioned man from a small town and college, bereft of charm, good looks or physical grace, was the very antithesis of the "best and brightest" who were fast abandoning their support for the war. To the growing multitudes of young opposition activists, he became a lightning rod for frustration. Their animosity, in fact, was well founded, because despite millions of adherents, and an even greater volume of heartfelt passion, the antiwar movement failed. America remained in the conflict for four more years.

The problem with Lyndon Johnson's war effort was that it sought to subdue the deep, almost religious, conviction of Ho Chi Minh and his fellows in the destiny of their people's revolution by pure military force. The Vietnamese Communists were familiar with military force, could wield enough of their own, thanks to the Soviet Union and China, and could wait for the U.S. storm of firepower to recede. Nixon's problem was how to retreat while maintaining the United States' reputation as the world's preeminent military power, a status he deemed necessary if he were to initiate his broader plan: détente among the superpowers. Humiliation in Vietnam would not serve his larger purpose of tying the United States, the Soviet Union and Red China into a tacit alliance of self-interest designed to eliminate the threat of nuclear holocaust. The most psychoanalyzed, as well as hated, U.S. president, Nixon's conviction that U.S. initiatives to diffuse the Cold War could not be undertaken except from a "position of strength" may have been a symptom of a demented personality; then again, perhaps history will come to judge him more dispassionately than many U.S. citizens did at the time. In either case, the North Vietnamese had encountered a less "sensitive" American leader than the retired Lyndon Johnson, and the U.S. antiwar movement, to its horror, discovered that Nixon viewed persistence in Vietnam as but one vital aspect of his larger agenda.

On February 22, 1969, the North Vietnamese, as if to test the will of the new American president, launched another offensive throughout the South, resulting in over 1,100 U.S. dead. Nixon, hamstrung by public opinion against further ground escalation, responded by ordering U.S. aircraft to bomb North Vietnamese sanctuaries in Cambodia. These bombing missions were unpublicized so as not to compromise the neutrality of Cambodian leader Prince Sihanouk, who approved of them. The North Vietnamese also kept quiet since they were not supposed to be in Cambodia in the first place; only the American public remained uninformed of the strikes, made primarily by B-52s.

In 1969 the war entered a low-intensity phase while the Communists licked the wounds from all their failed offensives and attempted to halt the erosion of their infrastructure in the countryside. While prior to Tet the South Vietnamese government had claimed control of two-thirds of the country, it now claimed up to 90 percent, though the true loyalty of many docile hamlets was questionable. On May 10 the U.S. 101st Airborne Division decided to hunt NVA in the A Shau Valley and achieved a bloody success in a battle dubbed "Hamburger Hill." The American public recoiled at the fighting and word was passed to the new U.S. commander, General Creighton Abrams, to desist in such ventures. The days of "search and destroy" missions thus came to a close as the new American imperative became to hold down casualties. Losses still occurred on patrols, primarily from booby traps, mines, snipers and long-range weapons fire, but massive U.S. sweeps no longer sallied into Communist strongholds challenging the enemy to fight pitched battles.

In March the West was astonished to learn that fighting had broken out on the Soviet–Chinese border. In one clash 31 Soviets were killed, and each side reinforced their strength in the area to a million men. The Soviets supported their divisions with 120 ballistic missiles and Mao Tsetung ordered his faithful Red Guards to start building underground bunkers in China's cities. Unfortunately, while to Americans the crack in the Communist monolith seemed like positive news, it had an even better effect on North Vietnam. The Communist superpowers subsequently competed for influence with the fighting purists in Hanoi, who played their patrons off against each other while increasingly viewing both with contempt.

In the United States antiwar protests tailed off with the end of the spring 1969 college semester, but then almost half a million young people gathered in upstate New York for a musical festival-cum-political rally. The largest gathering of partisans since V-J Day, outnumbering, too, U.S. troop strength in Indochina, Woodstock marked the apex of 1960s countercultural idealism. Peace and love were reaffirmed as political principles as the youth sought to inaugurate a new era in world history: "The Age of Aquarius." However, "Woodstock Nation" was short-lived. Two months later the same promoters attempted to stage a similar event at the Altamont Speedway in California with the British group the Rolling Stones as featured attraction. A gang of Hell's Angels motorcyclists was hired to guard the stage and the bikers earned their pay, which was beer, by beating concertgoers and musicians with pool cues. The climax came after dark when a young black man with a gun rushed the stage and Hell's Angels stabbed him to death while the Rolling Stones sang one of their hits, "Sympathy for the Devil." Altamont, together with the Charles Manson murders that summer, and increasing evidence of serious drug abuse among teenagers, fueled a growing backlash against the American counterculture. "Flower power" had in any event been suffering a noticeable decline as chants of "Make love, not war" or "Peace now" had given way at many rallies to "Ho Ho Ho Chi Minh" or "Off the pigs."

One effect of Nixon's election was to liberate congressional Democrats from having to loyally stand behind Johnson's Vietnam policy. With the exception of some southern legislators, nearly the entire Democratic Party abandoned its support for the war. Constantly assailed by demonstrators and with scant institutional support in Washington, the Republican president found it necessary to go over the heads of Congress in national addresses to the public. On November 3 Nixon declared the existence of a "great silent majority" upon whom he counted for help in resolving America's dilemma in Vietnam. His tactic of marginalizing the antiwar movement—in effect, by aligning the Vietnam War with the American culture war into the same "us versus them"—earned a 77 per-

cent approval rating from the public. On the following April 30, his rat-
ings having slipped again, he beseeched the public with his "pitiful, helpless giant" speech, this time with maps, requesting support for a specific
military operation.

The following day, U.S. and ARVN troops invaded Cambodia. The
South Vietnamese attacked the "Parrot's Beak," a jagged peninsula of
Cambodian territory that pointed toward Saigon, while 31,000
Americans, primarily the 1st Cavalry Division and the 11th Armored
Cavalry Brigade, crossed the border at the "Fish Hook" to the north. The
American antiwar movement exploded upon news of the Cambodian
incursion and four students were tragically shot dead by National
Guardsmen at Kent State University in Ohio. "Soldiers are cutting us
down," sang the group Crosby, Stills, Nash and Young. Fortunately U.S.
troops in Cambodia had less cause to repeat that lament than they might
have expected, although fatalities exceeded 300. Simultaneously bombarded by aircraft and guns, and surprised by the largest offensive in two
years, the Communists abandoned their base camps and supplies. For two
months U.S. and ARVN forces trampled the former NVA sanctuaries,
blowing up bunkers and seizing stores, including almost 23,000 assault
rifles, thousands of other weapons and 7,000 tons of food. Even the
enormous will power of the North Vietnamese could not overcome such
material losses and no serious attacks were launched against South
Vietnam from that direction for the remainder of the year.

Nixon tried to ameliorate his hard-nosed pursuit of "peace with
honor," as opposed to simply peace, by pulling Americans out of South
Vietnam and by 1971 U.S. troop strength had fallen by over 200,000
from its high of 541,000 in April 1969. Not surprisingly, the North
Vietnamese felt no need to pursue a negotiated settlement, despite the
best efforts of Nixon's National Security Adviser, Henry Kissinger, since
America was withdrawing anyway. Their point was that the United States
should continue its retreat and then the Vietnamese themselves would
resolve their internal differences. Meantime the United States frantically
tried to build up the South Vietnamese Army, at the same time encouraging land reform and local government policies that might motivate the
South Vietnamese populace to defend the Saigon government.

While the Nixon administration remained determined to paper over
America's retreat by gaining a peace agreement with North Vietnam that

did not look too much like surrender, the morale of U.S. ground forces went into a tailspin. Drug use and insubordination among the troops reached alarming levels, as did the practice of "fragging"—assaults on undesirable officers, often with fragmentation grenades. The cultural turmoil of the late 1960s had produced crops of draftees alienated from the very institutions they were supposed to be fighting for. While conscripts arrived in 1965 to the tune of "I Want to Hold Your Hand," now they were arriving to the sound of "Purple Haze." Though evidence of army deterioration, through court martial and drug treatment data, was disturbing enough, General Phillip Davidson, the chief U.S. intelligence officer in Vietnam, commented:

> The most prevalent form of low morale and lack of discipline never appeared in the statistics—what the "grunts" called "search and evade." In its simplest form, this consisted of going on a patrol or search operation and intentionally not finding any enemy. Either the patrol just sat down shortly after leaving the patrol base, or it searched an area known to be free of the enemy. The patrol leader returned with a false report of his route and a negative report of enemy contact.

While the late 1960s was no period for a Western country to fight a war, and America fielded perhaps the most dispirited army in its history during the retreat stage of Vietnam, it also fielded volunteer elites—LRRP Rangers, Green Berets, SOG teams, Navy SEALs—of legendary prowess. In addition, the manning and maintenance of American airpower did not allow for deterioration in combat efficiency. The trick was to keep enough pressure on the Communists to buy time for the South Vietnamese to take over while the United States disengaged. In military terms it's impossible to argue that America was defeated in the war, but then again neither were the North Vietnamese. The Viet Cong had been the biggest losers, having shot their bolt during the Tet Offensive and thereafter having been stripped of most of their territory by pacification. The war was approaching its logical climax as a match between ARVN and NVA. In early 1971 it was decided by the Nixon administration, with the complicity of the Thieu administration in Saigon, to put ARVN to the test.

NVA strength in the South flowed from the Ho Chi Minh Trail, a network of roads that bypassed the Demilitarized Zone between North and South Vietnam through neighboring Laos and Cambodia. Although continually bombarded by U.S. aircraft throughout the war, a major land operation had never been launched into Laos despite the obvious fact that cutting the trail would deprive the NVA of supplies. John Prados, in his history *The Blood Road*, quotes two North Vietnamese generals of disparate opinions. General Le Trong Tan, when asked how the Americans could have won the war, replied:

> If they had been wise they should at a certain point in time have cut a specific section of The Trail and taken over that area. Then we would have been stuck. We would never have been able to fight and win as we did. If they had been brave enough to do so, they would at least have severely disrupted our strategic network.

General Nguyen Van Vinh, on the other hand, stated in a 1966 interview:

> They speak glibly of offensives through Laos. But where are they going to get the troops? What do they think we will be doing meanwhile? . . . How can they dream of pushing through a couple of hundred miles of jungle and mountains in southern Laos and occupying it? That might look very attractive on the maps they print in their newspapers. Perhaps it is of comfort to the U.S. public. But in fact they cannot do it.

In any event, once the United States realized it could not force a decision on South Vietnamese territory alone it had run out of time; Nixon had barely been able to launch the 1970 Cambodian incursion through a firestorm of public protest. Afterward, Congress had repealed the Tonkin Gulf Resolution, and the subsequent Cooper-Church Amendment had decreed that U.S. troops were forbidden to cross national borders without the permission of Congress. The most ambitious offensive operation of the war would have to be undertaken by the South Vietnamese.

U.S. troops paved the way for the attack by reopening Route 9, the east–west road that led to Laos through Khe Sanh. Engineers descended on the famous, abandoned firebase and constructed a new airstrip for helicopters while repairing the old strip for use by cargo-carrying C-130s. U.S. artillery deployed along the border to support the attack. In anticipation of media scrutiny, SOG and LRRP teams were pulled in from Laos and ARVN units stripped of their American advisers, though according to Cooper-Church U.S. airmen could still aid the offensive. The South Vietnamese attack resembled World War II's Operation Market Garden, with airborne troops delivered in advance of an armored thrust down the road. The geographic objective was the Laotian town of Tchepone, almost due west of the Vietnamese DMZ. The South Vietnamese intended to stay in Laos for three months, destroying NVA base areas and physically obstructing the Ho Chi Minh Trail until spring monsoons effected their own slowdown of enemy supplies. Code-named Lam Son 719, after a hero in Vietnamese history, the operation would forestall the Communists while buying time for America to finish handing over responsibility for the war to ARVN.

Lam Son 719 started out promisingly as the ARVN 1st Armored Brigade broke across the border and covered six miles the first day. Helicopters of the U.S. 101st Airborne Division deposited units of ARVN 1st Airborne and 1st Infantry plus Rangers and Marines at predesignated firebases north and south along the flanks of Route 9. On the second day the offensive was stalled by bad weather, but on the third, February 10, 1st Armored linked up with the ARVN 9th Airborne Battalion at the town of Ban Dong, halfway to Tchepone. Then, however, the attack halted in the face of growing NVA resistance.

The North Vietnamese, who had good intelligence in Saigon and within ARVN, apparently knew about Lam Son 719 before it began. The American preparation of reoccupying Khe Sanh and plowing up to the Laotian border during the prior week had also provided a clue about the route of attack. The NVA already had its 320th and 304th Divisions in the vicinity of Tchepone, but prior to the offensive its 308th Division, from North Vietnam, and 2nd Division, from the South, began heading for the point of incursion. The question for ARVN General Hoang Xuan Lam became whether to press the attack at full speed through what turned out to be NVA superiority, or to keep his head close to, but not inside, the tiger's mouth.

The Americans, for their part, were furious at the South Vietnamese for not advancing faster, especially when they learned President Thieu had secretly ordered Lam to abandon the offensive once he had suffered 3,000 casualties. General Davidson stated:

> Such an order stifles boldness, the one ingredient which might have successfully concluded the mission and have curtailed ARVN losses. Actually, Thieu's order guaranteed that ARVN would lose the initiative and take heavy casualties as the troops hunkered down in their fire bases to await the onslaught by the ever-increasing forces of the enemy.

On the other hand, the South Vietnamese might have been behaving rationally. On February 14 the first NVA tanks were encountered: light PT-76s from the direction of Tchepone. Then heavy T-54s were met, accompanying the Communist 308th Division dispatched from North Vietnam. The ARVN 39th Ranger Battalion was wiped out at its firebase north of Route 9. To its left, an entire airborne brigade was overrun, its commander and his staff captured. To the south of the road, ARVN 1st Infantry and 1st Marines were under equally fierce attack that included bombardments from 130mm guns. In the center of the wedge the South Vietnamese fought their first tank battles of the war with their cousins from the North, American-made M-41s proving inferior to the Soviet-built T-54s. All this time U.S. gunships, fighter bombers, rocket-firing helicopters and B-52s were laying as many explosives on the NVA as they could fly into Laos.

If the United States had suffered strategic disappointment in Vietnam, it was not about to endure the additional stigma of an actual battlefield defeat. ARVN had to get to Tchepone, whether they liked it or not, and the choppers of the 101st Airborne, reinforced by additional companies from the quiet sectors of South Vietnam, would get them there. Three firebases named for movie stars—Lolo (for Gina Lollobrigida), Liz and Sophia—were designated in a string toward the town from the southeast while a fourth, less jauntily named Hope, was assigned to the north above Route 9. At LZ Lolo the U.S. lost 11 helicopters, 40 others damaged, just trying to get the South Vietnamese troops in. At LZ Hope, the first CH-46 helicopter blew up after being hit by AA fire, and 13 other choppers were damaged, but two battalions of the ARVN 2nd Airborne Regiment were landed. The 2nd Battalion made its way through abandoned NVA supply dumps and enemy bodies left by air strikes to enter Tchepone on March 8. The airborne troops spent several hours "holding" the deserted town, whereupon the operation was called a success and the South Vietnamese turned to the problem of getting their forces out of Laos.

The initial offensive had pitted 17,000 ARVN troops against 22,000 defenders. By mid-March the NVA had concentrated 40,000 men against an ARVN force that had been whittled down to less than 10,000. Thieu, having committed his best units, refused to pour additional men into the operation.

The ARVN 1st Armored Brigade was ambushed during its retreat along Route 9, and learned from prisoners that more ambushes were in place farther down the road. Its remaining tanks set off cross-country for the border. On reaching the Sepon River, U.S. aircraft dropped bulldozers and bridge materials to help the the ARVN tanks and vehicles negotiate a ford. The airborne battalion that had reached Tchepone tried to make its way out through the jungle, at one point coming across a vast bunker complex. After numerous firefights with the NVA it was finally overwhelmed in an ambush on March 21. At the isolated firebases, nearly all of which were now surrounded by NVA, the South Vietnamese troops had to be pulled out by air. The task was complicated by 20 NVA antiaircraft battalions with guns ranging from 23mm to 100mm, and thousands of enemy troops with small-arms ringing the pickup zones. Tom Marshall, in his memoir *The Price of Exit*, captures the peril of the chopper crews:

> Haze from the bombings mixed with the dust and fires from heavy combat under way on the pickup zones. In addition, in an attempt to mask the approach of the Hueys, there had been extensive use of white phosphorous smoke. But the smoke only added to the confusion. . . . As the elements approached the pickup zone, the ships spread out into a loose trail formation to keep a one-minute separation between aircraft. They would not sit the aircraft on the ground, but come to a five-foot hover above the PZ. The crew chief and the door gunner would then pull aboard enough ARVNs for a safe full load. They knew that if they set down on the ground they would be swamped by panicked ARVNs and would not be able to take off.

Each helicopter could carry only 10 South Vietnamese soldiers and so a steady stream of U.S. aircraft was ordered to fly into intense fire. The ARVN troops were not protecting their own perimeters and U.S. pilots suspected the Communists were using frightened clusters of soldiers as lures in order to bag the greater prizes: U.S. choppers. Marshall recorded a sequence involving a lieutenant pilot of the 101st Airborne:

> Fischer continued his approach, taking his own hits as he crossed the perimeter of the pickup zone. Because he was coming hot and fast, he stood the bird on its tail and pulled in all the power he could to stop it. As he leveled the Huey to hover, his first sight was ARVN soldiers. Through his chin bubble, he saw them crouching in the middle of the pickup zone. It was a demoralizing, disgusting sight to Fischer: none had weapons.
>
> At the same time, all hell opened up inside the Huey as AK-47 rounds turned the Huey's underside into a magnesium Swiss cheese. . . . The crew chief and door gunner were shooting their M-60 machine guns while hollering to Fischer to "Keep moving! Keep moving! Still taking hits." By this time, North Vietnamese were waiting on the very edge of the PZ. His climb out was made to the east under continuous fire.

Later, after the best efforts of heavier U.S. helicopters to extract wrecked machines from the battle zone—upon which they could be declared "damaged" rather than destroyed—the Americans announced a total of 108 choppers lost in Laos. In addition to that number, over 600 others were shot up. To the Nixon administration, any chance of calling Lam Son 719 an ARVN victory was betrayed by video footage of terrified South Vietnamese soldiers clinging desperately to the skids of those U.S. choppers that made it out. The wisdom of launching ARVN, in its first major offensive operation, against a heavily defended section of the Ho Chi Minh Trail within a day's march of the North Vietnamese border can be questioned; on the other hand, the NVA did not have an easy time against the South Vietnamese, and an even worse one against the nearly 10,000 sorties delivering 52,000 tons of bombs by American aircraft during the six-week-long campaign.

On April 25, 125,000 antiwar demonstrators congregated in

Washington, and then a quarter of a million arrived on May 15 to "shut down the government." Ironically, while Saigon and Hanoi enjoyed relative calm, Washington was under siege. Helmeted Guardsmen with riot batons dragged 12,000 protestors into custody while the White House was cordoned off by armed troops, machine guns on its roof. By now the war had gone on so long that the peace movement had become ingrained in the culture of an entire generation, opposition to the continued killing not unlike a rite of passage for young people coming to political awareness. In Vietnam the year 1971 saw continued progress in pacification of the South's countryside, while the Communists, despite the successful defense of their main supply route, seemed unable to mount serious military operations. American troops continued to spill out of the country, only 133,000 left by the end of the year. The front-line fighting had been handed over to ARVN as U.S. soldiers devoted themselves to air support and logistics.

In February 1972 Richard Nixon made history by becoming the first U.S. President to visit China. The famous scenes of Nixon and Kissinger smiling and toasting Mao Tse-tung and Chou En-lai provided confidence in the West that the Cold War had been diffused. In Hanoi, however, the images from the Forbidden City, together with the news that Nixon and Kissinger would soon be visiting Moscow, created anxiety among the Communist world's longest-suffering true believers.

On March 30 the North Vietnamese launched their most powerful series of attacks yet, dubbed the "Easter Offensive." Having abandoned their attempt to foment a people's revolution in the South, they now tried conquest with regular troops, columns of tanks and battalions of heavy artillery. For the first time since 1968's Tet Offensive they succeeded in capturing a district capital, Quang Tri in the north. Four NVA divisions spearheaded by 200 tanks crashed across the DMZ. The ARVN 3rd Infantry Division collapsed before their onslaught but the 1st Infantry and 1st Marines were able to stand firm before Hue. In the Central Highlands, NVA attacking out of Laos and Cambodia besieged Kontum, and farther south the Viet Cong 5th, 7th and 9th Divisions (now manned with North Vietnamese regulars) surrounded An Loc. ARVN fought surprisingly well, rushing its airborne brigades and Ranger units to support infantry and territorials at each endangered spot. The key to defeating the offensive, however, still lay with the Americans.

Overriding objections from both the State Department and the Pentagon, Nixon ordered American air strength in the theater doubled. While U.S. aircraft flew barely 1,000 sorties in March, over 35,000 were launched the following two months by the Americans and South Vietnamese. B-52s from Guam, aimed at map coordinates and precisely timed by U.S. advisers on the ground, pulverized NVA attacks, at times only several hundred yards away from ARVN defense lines. The U.S. Navy also rushed to stations where it could pound enemy concentrations around the coastal cities in the north.

Not satisfied with tactical strikes in the South, at the height of the offensive Nixon ordered U.S. bombers to once again hit North Vietnam. These attacks were opposed by Creighton Abrams, who could see that ARVN was hanging on by its fingernails, but Nixon and Kissinger were determined to counter the North Vietnamese offensive with an escalation of their own. Further, the days of micro-managing the bombing campaign from Washington were over. "I don't want any more of this crap about the fact that we couldn't hit this target or that one," Nixon told the Chairman of the Joint Chiefs of Staff. "This is your chance to use military power effectively to win this war." Having played the "China card," and confident in his negotiating power with the Soviet Union, Nixon not only allowed U.S. aircraft to bomb Haiphong harbor but on May 8 ordered the port to be sealed off with mines. Several Soviet ships suffered casualties in the attacks but the Kremlin declined to call off the summit meeting they had scheduled with Nixon later that month.

The Easter Offensive lost its momentum, and in the weeks that followed the Communists suffered heavy casualties from ground fighting and air attacks. On June 26 they sent word to the Americans requesting a resumption of the Paris Peace Talks. ARVN spent the rest of the summer retaking Quang Tri and other locales while the bulk of the NVA once more retreated to the remote jungles of South Vietnam and to their bases in Laos and Cambodia. It had been a "near-run thing." For a few weeks South Vietnam's army had teetered on the brink of oblivion, and, more ominously, a tidal wave of refugees had begun to swell. Hanoi's problem was that by gambling on a quick victory with massed conventional forces it had provided U.S. aircraft their best targets of the war. Rather than saturating square miles of jungle to destroy a hidden enemy battalion, U.S. pilots had found columns of enemy tanks and supply convoys in the open. It was apparent, too, that in making the transition from guerrilla to conventional warfare the NVA still had much to learn about handling armor formations, coordinating arms, and large-scale supply.

In Moscow in late May, the Nixon–Kissinger team arrived at a Strategic Arms Limitation Treaty with the Soviet Union, the first step back from the nuclear arms race that had frighteningly accelerated for two decades. Unfortunately, although the Soviets had little interest in continuing the Vietnam War, the United States learned they were unable to desist supplying Hanoi as long as the North Vietnamese could gain arms from China. The Chinese had already informed Nixon that the people's revolution in Vietnam required their continuing support, particularly since such matters, and influence, could not be left to the Soviets. The crack in the Communist monolith had resulted in a triangle of superpowers within which no party was willing to quit. The only true fruition of American hopes would occur if the South Vietnamese proved capable of standing on their own.

A seemingly minor event occurred on June 17, 1972, when five men were caught trying to break into Democratic campaign headquarters at the Watergate Hotel in Washington. The Nixon administration denied any responsibility for the burglary.

On July 19 Henry Kissinger resumed discussions with his counterpart from North Vietnam, Le Duc Tho. The Paris Peace Talks, which had been going on since the end of the Johnson administration, began to bear fruit in the fall of 1972. Essentially, the Americans offered a withdrawal of their forces, including airpower, if the North Vietnamese would agree to stop trying to conquer the South. The United States agreed that all forces, other than their own, could be held in place in South Vietnam, pending a domestic political resolution. President Thieu suddenly became alarmed that America was planning to abandon his country, leaving a quarter of a million NVA troops at his throat. Kissinger flew to Saigon and bluntly warned Thieu that these were the best terms he would get; more influentially, Nixon wrote Thieu a private letter stating that if North Vietnam broke the terms of the agreement, it was his intention "to take swift and severe retaliatory action." On October 26, Kissinger announced, "Peace is at hand." Ten days later, Nixon won a landslide victory over his "dove" opponent, George McGovern, by taking 49 of 50 states, the only exception being the Kennedys' homeground, Massachusetts.

Newly empowered, Nixon lost patience when the North Vietnamese began to contest the peace agreement that had been all but signed. The Vietnam War had not only chased Lyndon Johnson from office but had lasted throughout Nixon's first term. It's impossible to gauge how tired the American public—both hawk and dove—had become of the affair and how much they wished the interminable conflict to end.

On December 19 Nixon ordered the "Christmas" bombings, in which United States airpower suddenly had free rein over North Vietnam such as it had never enjoyed during the crucial years of the war. For the first time, B-52s hit Hanoi and Haiphong. Soviet or Chinese intervention was no longer a possibility and American aircraft wreaked massive destruction across the North. Amazingly, North Vietnam claimed only 1,700 civilian fatalities after the gigantic blitz of explosives. At the same time, 15 B-52s were shot down and many others damaged by the torrent of SAMs that climbed into the sky from NVA batteries. Much of the world's press was aghast at what seemed like a homicidal tantrum on the part of the U.S. President. But the bombings achieved their purpose. Once the smoke had cleared the North Vietnamese promptly returned to the conference table, and on January 26, 1973, signed the agreement that allowed the United States to finish retreating from its Vietnam venture with what Nixon had craved: "honor." America remained the world's preeminent military power, although, as Nixon, Mao and Brezhnev all realized, heaven help the next nation that chose to project its force so far away and for so difficult a cause.

In 1973, the Thieu government held its own in the "war of the flags," when Southern hamlets would show their allegiance after being politically won over by either the Viet Cong or Saigon, or by whoever's forces happened to be nearby. In June a new U.S. ambassador, Graham Martin, arrived in the country. A former New Deal Democrat, Martin, perhaps due to the combat death of one of his sons who was a helicopter pilot, had hardened into a fierce anti-Communist. While many U.S. officials who dealt with the Thieu administration had grown cynical or fatalistic, the tall, patrician ambassador stood out in the American community with his unwavering—some would say eccentric—faith in South Vietnam's cause.

In October of that year the Soviet-supplied nations of Syria and Egypt launched a huge, two-pronged conventional assault against

American-supplied Israel. At the height of the crisis the Soviets threatened to unilaterally intervene and Nixon responded with a worldwide military alert while the U.S. Air Force conducted an emergency airlift of supplies. The Israelis prevailed in counterattacks on both fronts and the Arab states sued for peace.

Though no longer front page news in America, the ground fighting between Vietnamese intensified in 1974 with the North claiming, perhaps rhetorically, 31,000,000 violations of the Paris Peace Accords by Saigon. Meanwhile, the Ho Chi Minh Trail became a two-lane highway covered with asphalt and paralleled by a fuel pipeline. Without U.S. air interdiction, Soviet and Chinese supplies flowed south intact, providing the NVA with the greatest arms buildup in its history. That year South Vietnam suffered runaway inflation and widespread unemployment. American aid shrank due to anxiousness in the U.S. Congress simply to be done with the entire bottomless Indochina morass. Inflation in America, including a 400-percent increase in the price of oil following the Yom Kippur War, further devalued the appropriations meant for South Vietnam. The worst news for Saigon was that in August 1974 Richard Nixon was forced to resign his presidency in disgrace. His effort to withstand the Watergate scandal failed after prosecutors discovered a secret White House taping system that had recorded his private conversations with advisers. It turned out Nixon had indeed been informed of the 1972 burglary, perpetrated by his Committee to Re-elect the President, but had tried to cover up the crime.

In American cultural terms, Nixon's fall brought down the curtain on the Sixties, an era that had begun with the shocking death of John Kennedy. By this time the only Americans left in South Vietnam were aid workers, diplomats, businessmen and CIA operatives. The country had resumed its status as just one more messy troublespot on the map. U.S. college students and blacks were no longer rioting; even the rock music of the era slipped away, several of its icons having died of accidental drug overdoses or from inadvertent asphyxiation. The new musical idiom of the Seventies was disco, a form solely intended to inspire people to dance. As for Vietnam, Kissinger admitted to a journalist that all he had sought was a "decent interval" between American withdrawal and total defeat.

Interestingly, when the North Vietnamese Politburo planned its 1975 offensive, it anticipated major gains that year, but with final conquest of the South not expected until one last push in 1976. The CIA made a similar analysis. It may be assumed in retrospect that while South Vietnam was an artificial state, maintained in the face of a diehard opponent only by strenuous Western efforts, no one realized how fragile the South's structure really was. Further, in military terms the war had come full circle. Having begun with conventional Western arms attempting to combat a grassroots insurrection and guerrilla tactics, the war now featured a conventional Communist army set to assault a South Vietnam intent on pacification and economic reforms. The South had a policy of organizing ARVN units on a territorial basis, stationing them close to their homes on the theory that they would fight harder to protect their families than if they were among strangers. Although on paper ARVN had enough manpower, its mobile reserve was limited to the 1st Airborne Division, 1st Marines and a number of independent Ranger groups. While the NVA was on the rise in weapons and tactics, ARVN, formed and trained in the profligate American mold, was in material decline due to shortages. U.S. aid, over $2 billion in 1973, had been cut by over 50 percent in 1974 and then by another 30 percent for 1975. Without maintenance, much of the high-tech equipment bequeathed by the Americans was useless against the strictly practical armaments driven south by the twentieth-century Spartans of North Vietnam.

After the United States accomplished its withdrawal of troops in January 1973, President Thieu decreed a policy of "no retreat" from any government-held territory. This forward defense concept was punctured on January 6, 1975, when the NVA finished overrunning Phuoc Long province, just 80 miles north of Saigon. Already stretched thin, ARVN couldn't retake the place and it was seen that the NVA was now heavier than ever with tanks, long-range artillery, antiaircraft weapons and motor transport. Although Hanoi only meant the attack as a preliminary thrust, the fall of Phuoc Long set important precedents: first, the Communists had never before been able to conquer and then hold onto an entire province; second, NVA main force regiments proved superior to their counterparts in mobile firepower—the days of hit-and-run had ended; third, though both sides looked over their shoulder throughout the battle for American aircraft to appear, for the first time in a decade these were not forthcoming. Hanoi continued its buildup with new confidence. In

Saigon the frightening reality set in that ARVN was finally on its own, and the intruder had crashed through the door.

Debate raged within North Vietnam's high command about where to launch the real thrust in 1975. Since the northern provinces—including Quang Tri City, Hue and Da Nang—were defended by ARVN's elite troops, and Saigon was strongly held, all agreed that the Central Highlands should continue to be the main point of impact. However, within that region some argued that the key ARVN bases of Pleiku and Kontum should be the next target. It was a southern-based general, Tran Van Tra, who argued that the initial objective should be the city of Ban Me Thuot. After securing the support of Kissinger's old negotiating partner, Le Duc Tho, now the second most important man in North Vietnam, Tra's views were adopted. The NVA would strike not at ARVN strength but at a point in the center of South Vietnam where the enemy was weakest. Still, no one guessed that the fall of Ban Me Thuot would carry such extreme consequences. A sledgehammer can sooner or later crush a stone, but sometimes all it takes is one blow at a crucial point.

On the ARVN side, General Pham Van Phu, commanding II Corps, anticipated an offensive against Pleiku and Kontum. His intelligence officer urgently tried to convince him to reinforce Ban Me Thuot but Phu was reluctant to weaken the defenses of Pleiku on the very eve of an enemy attack. His corps strength consisted of the 22nd and 23rd Infantry Divisions (the former deployed near the coast) along with 7 regiments of Rangers, one armored brigade and a number of territorial battalions. Facing him were five NVA divisions and fifteen independent armor, artillery and AA regiments.

On March 1 the Communist offensive in the center of South Vietnam began with small attacks around Pleiku. Mobile units simultaneously slipped from their jungle base camps to cut the roads that connected Pleiku, Kontum and Ban Me Thuot with the coast. Skirmishes flared throughout the Highlands except at Ban Me Thuot, where the bulk of NVA strength was stealthily moving into position. An NVA prisoner revealed the enemy plan, but Phu still refused to reinforce the single regiment of 23rd Division that held the town.

At two in the morning on March 10, the Communists launched a set-piece attack on Ban Me Thuot with their 10th, 316th and 320th Divisions. A storm of artillery fire shattered ARVN strongpoints as NVA

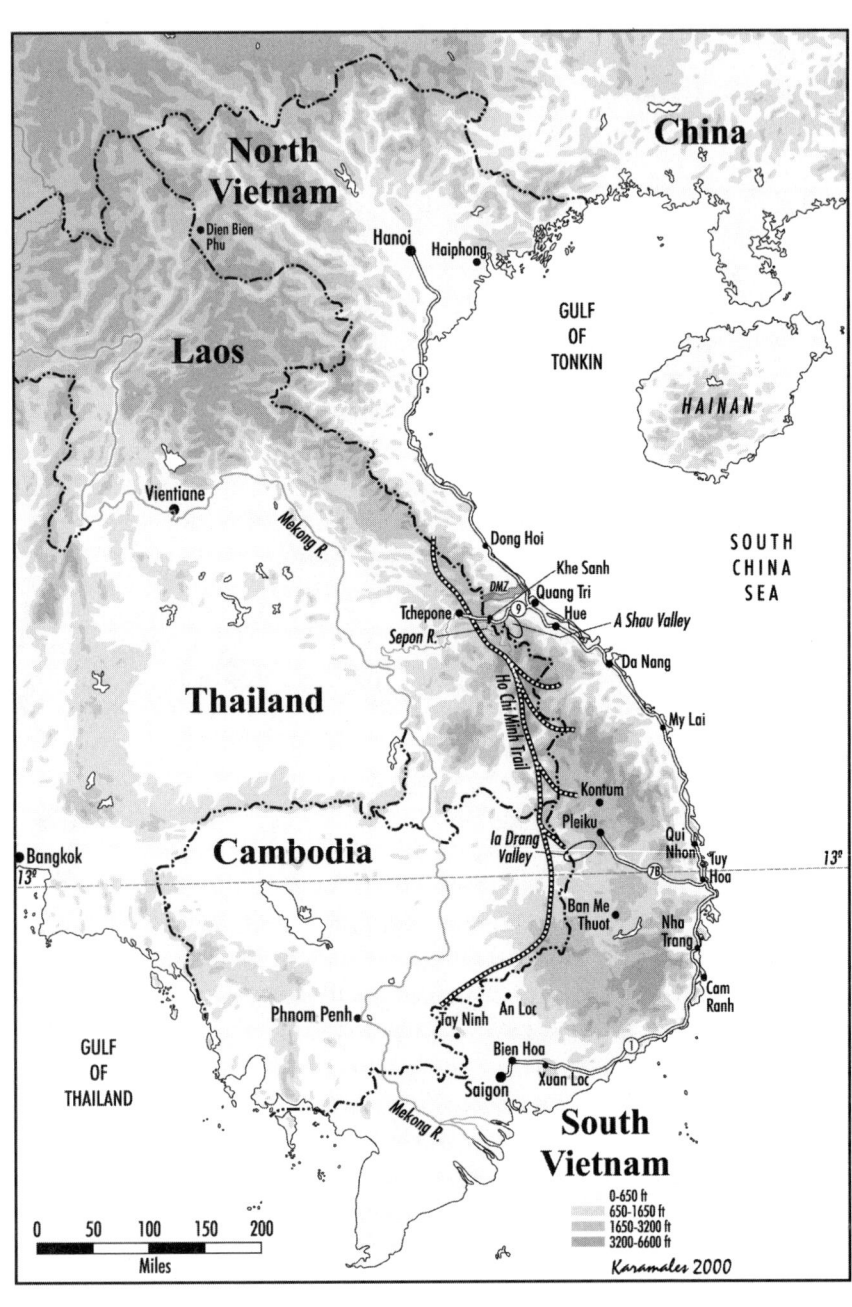

Karamales 2000

infantry overran the outer defenses, only to be met by counterattacks in the town center. At daylight NVA tanks rumbled in from the north and west; ARVN 23rd Division's headquarters was in plain sight as NVA guns zeroed in. The South Vietnamese fought back desperately but by night-fall Ban Me Thuot had fallen. Fighting continued on the outskirts of town and at the airfield. On March 12 General Phu ordered a counterat-tack with 23rd Division's remaining two regiments from Pleiku. Since the roads were cut, these had to be airlifted into position, to landing zones where NVA AA battalions were already lying in wait. Enemy tanks and artillery broke up the survivors before they could get into formation. Many soldiers deserted to join their families, now streaming out of the area.

For President Thieu, who had hoped until the last moment that an avenging U.S. air force would suddenly re-emerge in the skies to devas-tate the Communists, the overwhelming assault on Ban Me Thuot prompted a drastic revision in strategy. His "no retreat" edict had been a bluff to stiffen morale and convince the Americans that South Vietnam was determined to defend itself. Now he called a conference of his mili-tary advisers to suggest a more realistic alternative: abandon the Central Highlands and the northern provinces. Hold on to the coast if possible, but if need be withdraw all forces to a southern enclave below the 13th parallel. Just days into the Communist offensive Saigon gave up on half the country. In base camps, fortifications and staging areas throughout the northern provinces, South Vietnamese soldiers who had prepared to resist the NVA onslaught found that the government had no faith in their ability to hold. A massive retreat was underway, in many cases before the defenders had even seen the enemy.

North Vietnamese commander Van Tien Dung, who had set up his advance headquarters in the Central Highlands, seemed almost bemused by the effect of his initial thrust: "The battle of Ban Me Thuot, followed by the attack that wiped out reinforcements from the 23rd Division . . . threw the enemy into panic and confusion, not only at the division and army corps level, but all the way to the top."

Thieu conferred with General Phu of II Corps, informing him to abandon Pleiku and Kontum in the Highlands. The decision was made to evacuate not via the major thoroughfares to the coast, which the North Vietnamese had cut, but along an old French road, Route 7B. This road, which had long been abandoned and almost reclaimed by the jungle, featured numerous destroyed bridges and a final stretch mined by the South Korean expeditionary force that had fought alongside its American allies until 1972. By sneaking out through Route 7B the ARVN units from the Central Highlands might evade their NVA pursuers and reassemble on the coast for a counterattack. Thieu insisted that only regular ARVN units should evacuate, in secrecy, leaving territorials and civilians behind. The column would be led by engineers, with Rangers as rearguard. In theory it might have worked.

The "Convoy of Sorrow," as the retreat was termed by a South Vietnamese journalist, met disaster. In Pleiku and Kontum there was no way to disguise that the army was in retreat, and on which road. Accurate numbers were never determined, but up to 200,000 civilians trailed along, obstructing the roads, consuming supplies and obscuring the difference between a military operation and a desperate mass exodus. Remaining territorial battalions—mostly Montagnard tribesmen—dissolved into anarchy and began to plunder the wreckage of Saigon's authority. Worse, the NVA soon realized what was going on. General Dung had repeatedly inquired about Route 7B, only to be assured it was unusable. Now he found the enemy stretched out across that road in ragged formation. The NVA 320th Division was ordered to race after the fugitives on a parallel course; the 968th Division was ordered to jump through Pleiku in direct pursuit.

Though they had stolen two days' march on the NVA, the huge column of South Vietnamese became clogged in the jungle. Food and water were a problem; old people and children were the first to drop out, dying along the way. Then Communist mortars and artillery found the road and bodies piled up, often ground into the mud by drivers of vehicles frantically trying to get through. NVA infantry set up ambushes and roadblocks. ARVN units lost their cohesion as soldiers, mixed with their families, refused to obey orders. The South Vietnamese Air Force, flying at high altitude to avoid AA fire, attempted to hold off the NVA but its bombs fell indiscriminately, killing as many friendlies as enemy. On nearing the coast the column had to detour because the South Korean mines couldn't be removed. After days of suffering across 150 miles, the column came to a succession of NVA fireblocks. Two battalions of Rangers backed by tanks chewed their way through, at one point beating off an attack by an NVA regiment. On March 25 the head of the column reached Tuy Hoa on the coast. About 20,000 out of 60,000 soldiers and less than a third of the refugees got through. The survivors were useless for any grand counterattack plans that Thieu might have imagined.

In the north the Communist offensive began with modest mobile penetrations, but after Ban Me Thuot, followed by ARVN's startlingly sudden evacuation of the Central Highlands, the North Vietnamese switched to full-blooded assaults. By March 20 they had seized Quang Tri near the DMZ and had gotten in between Hue and Da Nang. Thieu went on television to announce that Hue would be defended, but privately he authorized General Ngo Quang Truong of I Corps to abandon the ancient city. He also ordered ARVN's elite 1st Airborne Division to move by sea from Da Nang to Saigon; at first the Marines moved to replace the Airborne but within days they too would be ordered south. The effect of ARVN's elite units vacating their positions was to unhinge the civilian population, which began flowing toward the coast in a huge tidal wave as refugees. During 1968's Tet Offensive the one city the Communists had been able to briefly hold was Hue; after it was retaken thousands of civilians were found buried in mass graves, either executed by Communists or obliterated by U.S.–ARVN retaliatory firepower. Now hundreds of thousands of people streamed toward safe haven at Da Nang.

At Hue, ARVN 1st Division was ordered to evacuate by sea but NVA artillery spotted the embarkation points and reduced the division to a shambles. The survivors who arrived at Da Nang were a panicked mob, without weapons. Advance NVA units also reached the coast south of Da Nang, prompting another flow of refugees toward the port. All of I Corps, including much of the civilian population, was falling back on the city, South Vietnam's second largest, and the mass of humanity in the city was slipping out of control. ARVN 3rd Division was barely holding off NVA thrusts from the west; 2nd Division on the outer defenses of Da Nang in the south was disintegrating, even though the enemy hadn't yet arrived in strength; soldiers deserted to find their loved ones or simply ran to escape the obvious debacle.

It would be convenient to attribute the collapse of I Corps to enemy pressure but the NVA, despite its new weapons and large-unit tactics, was not wholly responsible. Once the Central Highlands collapsed, a form of psychosis took hold in the northern provinces after they had been geographically separated from Saigon. There was little fighting, and though the NVA lapped at the edges of the panicked masses the impetus to flee stemmed more from fear than from actual bloodshed. One problem on the Communist side was that the South Vietnamese were collapsing faster than they could pursue. Another was that the North Vietnamese Army, which had become set in its ways due to interminable incremental warfare, had never seen this kind of success before and was unprepared. Hard-driving General Dung complained that the men were barely rising to the task:

> In these new, more demanding conditions, we were still caught up in the old style with its many long, drawn-out meetings. No one had yet stood up to make decisions and organize action rapidly. In some situations the infantry were out of hand. They had field radios, but instead of using them went rattling around stringing telephone wires. They had vehicles taken from the enemy and POWs who knew how to drive, but didn't dare use them to drive the troops, and instead just kept on walking. It was not as if the enemy were organized, waiting for us with defenses prepared; they were disorganized, falling apart, yet when we atack we still demand full discussion. . . . The enemy air attacks were limited and at high altitude, and when they dropped their bombs their aim was not precise, yet we still did not send our troops into battle during the day—waiting late, wasting time. We had to cover these problems time and time again . . .

To empathize with the NVA, it took time to sink in among the Communist rank and file that the Americans were no longer in the war. The vicious blows of aircraft catching their troops in the open or the vibrations of rotary-winged craft growing louder, signaling the sudden attack of an airmobile unit in their rear, did not occur. Meanwhile, the evacuation of Da Nang turned ugly. In a city swollen with half a million refugees and thousands of leaderless ARVN troops, the air and sea evacuation became anarchic. Soldiers pillaged and raped helpless civilians; aircraft were mobbed; men were crushed trying to hang on to landing gear or fell to their deaths from precarious fingerholds after takeoff. In bitter frustration soldiers fired at helicopters carrying press photographers or that swept in to take away VIPs. Aboard the jammed ships pushing off from the harbor refugees suffered indignities at the hands of armed thugs. The Marines, considered to be the toughest troops in ARVN, proved the most brutal during the chaos. I Corps commander General Truong had to swim to a lifeboat to escape, his command shattered. The North Vietnamese entered Da Nang with four divisions on March 30, restoring order, and perhaps wondering how the war had suddenly become so easy.

In Saigon, U.S. Ambassador Graham Martin recognized that, beginning with Pleiku, all eyes had been on the reactions of the remaining Americans. Once U.S. personnel scrambled for the airfields, everyone knew that the Communists were coming in, no matter what exhortations came over Saigon radio. Martin had no taste for the frenzy that had seized Da Nang and resolved to hold his Saigon Embassy staff in place—an island of stability in a sea of fear. Nevertheless, other U.S. agencies and private organizations made their own decisions. Transports and airliners began streaming out of Tan Son Nhut Airport west of Saigon, packed with U.S. businessmen, aid workers, foreign nationals and Vietnamese families. After March 31, on orders from Washington, the U.S. Defense Attaché's Office began evacuating nonessential personnel.

On the coast, ARVN 22nd Division held its own in battles against the NVA 3rd Division, but then other enemy units arrived from the west, followed by still others coming down from the north. On April 1 the remnants of 22nd Division, some 2,000 men, were pulled off by sea. Farther south, NVA 320th Division emerged from the jungle after wreaking havoc on the pitiful mob from the Central Highlands, and easily captured the terminus of that retreat at Tuy Hoa. ARVN's 3rd Airborne Brigade attempted to hold off the NVA 10th Division above the important coastal city of Nha Trang, but by April 2 had been reduced to 300 fleeing survivors. General Phu had already deserted his corps headquarters at Nha Trang to hospitalize himself in Saigon (he would later commit suicide). On April 3 the ARVN command staff at Cam Ranh Bay evacuated. City after city along the coast was swept by panic, in the words of correspondent Richard Butler, "as if Da Nang were a contagion carried by the demobbed, bitter Marines and foot soldiers who made it out of the northern provinces." In some cases the towns fell to drunken looters days before NVA soldiers appeared.

The North Vietnamese, of course, had found their strategic plan disrupted. They simply hadn't considered that South Vietnam would fall in one big rush and now they needed to reassess their plans. After an urgent meeting of the Politburo, Le Duc Tho traveled south to confer with the army's high command. This was, literally, a once-in-a-lifetime opportunity for Hanoi's leaders; the mood was euphoric yet the discussions grave. The NVA had few difficulties with combat strength and logistics because resistance had been so feeble; the main problem was time. President Thieu's plan to form a hard enclave in the south might provide ARVN a greater density of troops, less territory to defend and nowhere else to retreat. If the South held out until the rainy season, only weeks away, the North would lose its momentum. The biggest worry was that if the South managed to stabilize its front the B-52s might return. Even with Nixon ousted and discredited, who could be sure that American airpower would continue to sit idle if the South demonstrated that it was willing to bravely hold on? The answer to the North's dilemma was clear: just like ARVN, abandon the northern and central provinces. All units were to head immediately for Saigon.

Logistics were rerouted and divisions received new orders; only the most essential mopping-up operations were allowed in the conquered provinces. Endless convoys streamed south, not just along the Ho Chi Minh Trail and other concealed roads, but on South Vietnam's highways. NVA commander Dung, who had fought for 30 years, almost always at a materiel disadvantage, waxed ecstatic:

> From the North all manner of vehicles rushed day and night, nose to tail, jumping past the way stations to carry men and supplies to the front . . . seeing our large artillery, our antiaircraft guns and missiles, our units of tanks and armored cars, and our amphibious vehicles, the newly liberated compatriots . . . could not conceal their surprise. Because of the distortion and deceit of the United States and Saigon propaganda which they had heard for so long, they had never before been able to imagine that Uncle Ho's soldiers would have so many vehicles and artillery pieces, or that they would be so young and handsome, so kind and cheerful.

In Saigon the depressing reality set in that from all of I and II Corps only one weakened division—the 22nd—was able to reform intact for defense of the capital. Survivors of airborne, marine and ranger units had been air- or sealifted back to III Corps from their various catastrophes, but the bulk of ARVN units in the northern half of the country had disappeared. III Corps had its own 25th, 5th and 18th Divisions, arrayed in an arc from west to north and east of Saigon, and an armored brigade and territorial militias that were still in place. During the first week of April there was little fighting except west of Saigon, where ARVN 25th Division grappled with the Viet Cong 5th Division that had attacked from Cambodia. Meanwhile, seventeen other Communist divisions were on the way.

Evacuating the 6,000 Americans still in South Vietnam was not an overwhelming problem. Even under normal circumstances Tan Son Nhut was one of the world's busiest airports and a steady stream of outgoing passengers caused hardly a stir. The vast majority of fugitives were Vietnamese resigned to beginning a new life abroad. Filling in the cracks of the evacuation was a company called Air America, whose CIA pilots dashed back and forth across the country retrieving American personnel or their affiliates, sometimes behind enemy lines. With the battle for Saigon approaching, the U.S. Embassy remained calm, Ambassador Martin grimly certain that the second his staff made for the exit the South Vietnamese would realize all was lost.

Tragedy struck on April 4 when the first plane of Operation Babylift—a huge C-5A Galaxy transport—crashed in a rice paddy north of Tan Son Nhut. The plane had been packed with 243 orphans, strapped two to a seat, but the cargo doors blew out at 23,000 feet and the pilots lost control. Seventy-eight of the small children died along with 57 adult nurses and crew. The Communists thought the whole idea of Americans shipping orphans out of the country was sinister; however in this case they encountered implacable opposition. Female aid workers who had cared for the abandoned children were firmly resolved not to leave their charges to the mercies of the NVA. Adoption agencies in the U.S. had already arranged homes for most of the infants, and Operation Babylift continued.

On April 8 a South Vietnamese F-5 fighter bomber peeled off from its formation and attacked the Presidential Palace in Saigon, hitting it with two bombs. The pilot then landed at an airfield near NVA-held Da Nang. Hanoi said the pilot had been one of theirs all along.

By April 9 the NVA had assembled enough strength in the south to mount attacks designed to strangle the capital. While direct assaults were launched against ARVN 5th and 25th Divisions from the north and northwest, other NVA divisions attacked from the southwest to cut Saigon's connection to the Mekong Delta. The hardest battle, however, took place east of Saigon where ARVN 18th Division held Xuan Loc, a vital position on Route 1 that protected Bien Hoa air base. Three NVA divisions, the 6th, 7th and 341st, unleashed a huge artillery bombardment followed by massed infantry attacks. For almost the first time in the campaign the ARVN troops not only held firm under intense fire but counterattacked, pushing the NVA back. Six times in four days the Communists launched massive assaults against Xuan Loc and each time were repulsed.

In Saigon a glimmer of optimism shone through the fog of impending doom. With its back to the wall ARVN was finally fighting. 18th Division's wounded were flown out in a relay of Saigon's last remaining helicopters, which simultaneously brought in ammunition and reinforcements. Correspondents were allowed to view the carnage and Richard Butler quoted 18th Division commander Le Minh Dao claiming, "I don't care how many divisions the other side sends against me. I will knock them down." The NVA had to pull its armor out of the battle to regroup and resupply.

Elsewhere the Communists chopped away at the outer defenses of the capital. General Dung commented, "Northwest of Saigon, the enemy's 25th Division still held fast to Tay Ninh. Indeed they were in a position where if they held it they would die, and if they abandoned it they would die, too." Assaulted by a corps comprised of the NVA 70th, 316th, 320th and 968th Divisions, the ARVN troops found themselves bypassed by enemy units, their supply dumps destroyed. Within days 25th Division would break up, its units falling back through NVA ambushes.

At Xuan Loc, the 18th Division also became gradually surrounded as enemy units from II Corps advanced in its rear from the southeast. NVA IV Corps to its front was reinforced, and head-on attacks came in again. Thirty-seven enemy tanks were knocked out in front of the South Vietnamese lines. This epic battle featured ARVN use of a "daisy cutter" bomb, a ghastly weapon that the Americans had used to clear landing zones in the jungle. The explosion of ignited gases and shrapnel covering nearly half a mile wiped out the headquarters of NVA 341st Division. The Communists protested the use of mass destruction weapons, even as they poured 20,000 artillery and rocket rounds into Xuan Loc. On April 22nd the 18th Division was forced to retreat along Route 1, fighting its way through enemy attacks on either flank.

The fall of Xuan Loc marked the end of hopes that the South's army could hold off the North's. Bien Hoa had come within range of mortars as well as artillery. A total of eight divisions were approaching from the east. In the west the NVA's only problem was crossing rivers with their heavy equipment against scattered resistance. The Communists below Saigon made just enough progress against 22nd Division to sever communications with the Delta. The only remaining option was a political solution, and toward this end, on April 21, President Thieu resigned.

It may seem curious that the South Vietnamese, even at this late stage, thought they could gain a negotiated settlement with North Vietnam, but their belief stemmed from decades of listening to the Communists themselves. The Viet Cong and NVA had always claimed to be fighting a revolution with popular support. Their official animosity had been aimed at the French and Americans, not fellow Vietnamese, and in any event the North had always sworn that it was not the aggressor. It was thought in Saigon that Hanoi would be embarrassed if, at the climax of its proletarian triumph, the "masses" were seen to be regiments of Soviet-built T-54s.

The South Vietnamese general Duong Van ("Big") Minh had stayed at the fringe of Saigon politics throughout the war, often under a dark cloud of suspicion that he had secret connections with Hanoi. Now he was welcomed as president in the hope that he really did have those connections. The Communists, however, refused to deal with him. They had no interest in negotiations with Saigon politicians; yet, during the last week of April a lull once again fell over the battlefield. The new factor was alerts at U.S. air bases in Thailand, Guam and the Philippines, and a steady buildup of the U.S. 7th Fleet off the South Vietnamese coast. The Americans had returned; this time to get their people out.

The day Thieu resigned, President Gerald Ford ordered a full-scale evacuation of American nationals. Americans had been drifting out of the country for over a month, but those who chose to stay until the very end,

including the U.S. Ambassador, now had a way to disguise their procras-
tination: a multitude of South Vietnamese were also trying to get away. A
stream of airliners and cargo planes packed with refugees flowed out of
Tan Son Nhut while thousands of people waited their turn at the gates.
From the coast hundreds of boats overflowing with refugees set off to
reach the American fleet. Graham Martin informed Washington that the
evacuation was indeed underway—only he and his staff weren't joining.

One thing that gave the Americans confidence was the seamless evacuation they had made of several hundred people from Pnomh Penh on
April 17. In that case the Cambodian population had not been seized by
refugee fever; their war had not been as long and bitter as the one in
Vietnam, and most citizens calmly resigned themselves to accommodate
the Communist victors, Pol Pot's Khmer Rouge.

In South Vietnam there was far greater fear. Eight Americans captured at Ban Me Thuot, six of them missionaries, had disappeared;
reports of executions, even beheadings, came in from Da Nang and other
conquered cities. In Washington the imperative was to get the U.S. personnel away; in Saigon the urgency was to evacuate as many Vietnamese
friends, employees and endangered officials as possible. A problem was
that more and more Vietnamese kept showing up. A unit of ARVN
Marines asked to get their families out; requests came in from Washington to evacuate the Vietnamese employees of IBM, Chase Manhattan
and Mobil Oil. Groups would arrive in convoys of buses, all with special
priorities and high-level patrons. Ken Moorefield, attached to the U.S.
Embassy, who was helping to supervise the stream of refugees out of Tan
Son Nhut, wrote:

> Then there were hundreds of unmarried Vietnamese women with
> Amerasian children. Some carried wrinkled letters from the
> American fathers that they'd been treasuring for several years. . . .
> And I had no authority, mandate, or instructions how to cope
> with this. I concluded that I'd better take initiative on my own
> because we were running out of time. God, the first day I must've
> signed hundreds and hundreds of documents.

The North Vietnamese Army, which fidgeted outside Saigon for days,
witnessed the progression of giant aircraft coming in and out of Tan Son

Nhut. Fully armed with SA-7 missiles and a panoply of high-caliber AA guns, they could have closed the air corridor, but were reluctant to do so. For one thing, why not let the Americans bring out their puppets? Traitors and lackeys would only take up space in the re-education camps to come. Another reason for the NVA to hold its fire was that a sudden splash of dead Americans would only earn savage retaliation. So far the American escort gunships and jets, too, were holding their fire.

The U.S. Embassy's position was delicate. As the outer defenses of Saigon crumbled it became clear the NVA could rush the city at any time. Just as worrisome were the mobs of surly ARVN troops roaming the streets and milling ominously around the airport. On the afternoon of April 28, five A-37 jets came screaming in from the north to bomb Tan Son Nhut, destroying eight aircraft including a big C-130. At first people assumed it was ARVN pilots running amok but in fact it was the Communists' answer to peace overtures made by Big Minh. The attack was led by the defector who had bombed Thieu's Presidential Palace weeks before, with North Vietnamese pilots he had trained to fly the captured American aircraft.

At 4:00 on the morning of April 29 the NVA opened fire on the airport with artillery and rockets. Over 300 rounds came in, destroying aircraft, fuel tanks, the control tower and barracks. Two U.S. Marines on the perimeter suffered a direct hit, the last Americans of the war to die on the ground. A lone ARVN C-119 gunship took off to blast the NVA batteries while two smaller craft went searching for advancing enemy infantry. The C-119 flew back to restock with ammunition and went out again, only to be shot out of the sky by an SA-7. As the day wore on, the artillery became more accurate, concentrated on the runways, as NVA observers near the airport spotted the fire. Informed that fixed-wing aircraft could no longer fly out of Tan Son Nhut, Ambassador Martin drove to the airport to see for himself. Reports came in from the outskirts of Saigon that the North Vietnamese had begun to move. Ken Moorefield said, "The panic really started to set in at this point. People were ricocheting around the city frantically looking for a way out." With time running down, the Americans had only one remaining option: evacuation by helicopter.

Operation Frequent Wind began with U.S. Marines airlifted in from the 7th Fleet to protect the chopper landing zones at Tan Son Nhut. Unfortunately, even though the NVA hadn't yet arrived, the Marines were forced to exchange fire with groups of disgruntled ARVN soldiers. The South Vietnamese Air Force was ordered to evacuate its planes, and scores of F-5s and A-37s dodged the incoming fire and craters on the ground to climb into the sky. Once aloft, a number of pilots unloaded their ordnance on enemy lines before making their way to Thailand. From the U.S. 7th Fleet, the first wave of 36 Chinook helicopters, flanked by Cobra attack choppers, flew in from the coast to the airport around 3:00. The pilots knew that NVA antiaircraft guns were locked on but the Communists declined to fire. Offshore the fleet was buzzed with ARVN helicopters carrying officers and their families. At 1:00 Air Marshall Nguyen Cao Ky, in his private copter, arrived aboard the carrier *Midway*. Most of the South Vietnamese helicopters were pushed overboard from the ships; some pilots were asked to ditch their craft into the sea, whereupon they were retrieved by lifeboats.

The operation at Tan Son Nhut ended shortly before midnight. At first there were 3,000 refugees waiting evacuation, but more arrived throughout the evening. Finally everyone had been put aboard Chinooks. The occasional artillery round that came in from the NVA helped keep the crowd down. The U.S. Marines were the last to leave, taking care to blow up the airport's radar, computers and communications equipment before heading back to the fleet.

In Saigon, 13 rooftops had been designated as landing zones where Hueys and other small craft could take people out. The center of the vortex, however, turned out to be the U.S. Embassy. CH-46s could land on the roof while even larger CH-53s could land on the grounds. Outside the Embassy's gates were thousands of Vietnamese waiting, sometimes struggling, to get in. Men would make it over the wall only to become more desperate at finding themselves separated from their families on the other side. During the afternoon the 44 Marine guards at the Embassy were reinforced with an additional platoon from Tan Son Nhut. South Vietnamese police, on the promise of evacuation for them and their families, helped maintain order.

About 400 Americans were left in Saigon, and Ambassador Martin was ordered by National Security Adviser Brent Scowcroft, then Henry Kissinger, to get them out. Martin, however, realized that as soon as the Americans were gone the airlift would end. He responded by taking a leisurely walk to his home to retrieve some things and to destroy important papers. Across the city Americans were attempting to make sure their Vietnamese contacts, employees and friends were safely evacuated. Ironically, one of the biggest problems became a shortage of buses—people couldn't make it to the departure points. Foreign Service Officer Shep Lowman found a crowd of Vietnamese at his home. When he realized he couldn't guarantee their evacuation through the Embassy, he implored them to head for the docks where barges were pushing off half-full. Without ground transport he was never sure how many of the people who had trusted him made it to safety. Lowman himself had only to show up at the Embassy to be promptly escorted through the crowd. Afterward he was philosophic to journalist Harry Mauer about what the evacuation had accomplished:

> When you get on that chopper and you fly away and leave behind hundreds of friends and acquaintances to what you know is a very bad fate, it's a hard time. It takes awhile to get over that. And I didn't think the evacuation had been mishandled, actually. I thought it was handled fairly well. There are a lot of complaints about it, but most of them come from unrealistic expectations about what an evacuation can do. We took out a tremendous number of people. The performance of the pilots and helicopters was remarkable.

For the Vietnamese at the Embassy it was first come, first served; maids and cooks with their families intermingled with senators and businessmen. The French Embassy adjoined the American one and a hole was knocked in the wall so that U.S. personnel wouldn't have to struggle through the crowd. The French, who had 15,000 nationals in the country, weren't evacuating. During the afternoon a helicopter with a burst fuel line crashed in the ocean killing the pilot and a crewman—the very last U.S. casualties of the war.

As April 30 began the sounds of fighting grew louder on the periphery of Saigon as the NVA closed in. Scattered ARVN units were still resisting. Alan Carter of the U.S. Information Service, another American wracked with guilt over Vietnamese friends left behind, recalled:

> During the last hours in the embassy there was a sense almost of

orderliness. People were doing what they had to do—blowing safes or destroying files, or sitting and waiting to be taken out. . . . As we got up in the air a little bit, it was like watching a Roman carnival. Two ammo dumps were going off—one just north of the city, and one at the opposite end. So a lot of stuff was flying around in the air, and it looked like fireworks.

The evacuation was being monitored closely in Washington, where frustration with Graham Martin was growing. Chinooks had been shuttling from the fleet all afternoon and evening, yet whenever the Ambassador was asked for a count of how many people were remaining his estimate remained unchanged. In effect, Martin was holding himself hostage until the helicopters finished evacuating all the Vietnamese who wanted to leave. Just after midnight Martin was asked for a hard count, and, encountering the crowd at a low ebb, replied that there were 726 people left. At 3:15 in the morning of April 30 a helicopter pilot handed him a message from the U.S. commander-in-chief Pacific, relayed from President Ford: "On the basis of the reported total of 726 evacuees, CINCPAC is authorized to send 19 helicopters and no more. The President expects Ambassador Martin to be on the last helicopter." The words "no more" were underlined.

By the time the last choppers arrived, the crowd at the Embassy had grown again to 1,100, more people continuing to arrive. At 4:40 a helicopter named *Lady Ace 09* arrived at the Embassy and its pilot informed Martin that he was to get on it; again, an order from the President. Arriving on the carrier *Okinawa* and asked for a statement from journalists, Martin replied, "I'm hungry." At 8:00 in the morning, full daylight, one last CH-46, escorted by six Cobra gunships, flew into Saigon. Its pilot, Chris Woods, landed on the roof of the Embassy and took on 11 fellow Marines, the last American soldiers out. Hundreds of despairing South Vietnamese were left on the grounds and outside the gate, thousands more across the city. On other rooftops throughout Saigon, Air America choppers continued bringing out civilians until the last possible minute. However, by mid-morning, the evacuation was over.

Big Minh, who to his credit remained in Saigon, announced at 10:30 that his government was anxious to effect an "orderly transfer of power" to the Provisional Revolutionary Government—the Viet Cong. North

Vietnamese General Dung, however, would have none of that "sleight of hand." At noon NVA tanks broke through the gates of Independence Palace in the center of Saigon. Minh and his staff were waiting inside. Minutes later the country of South Vietnam ceased to exist.

It was a consolation to guilt-ridden American evacuees that there was no sudden bloodbath of revenge against the people they had left behind. In fact, the immediate mood was euphoria on the part of the conquerors and curiosity on the part of Southerners about their country cousins whom they were seeing up close for the first time. The NVA was extremely well disciplined and its soldiers amazingly young. There was some sentiment that Saigon had fallen to rubes. Eventually, however, the Communists needed to place over a million South Vietnamese in "re-education camps" in order to bring them into line with the new order; over the next eight years 65,000 were executed because they were incorrigible or guilty of crimes. Another year would pass before the "boat people" phenomenon began in earnest.

In America there was some celebration among former antiwar protestors of the Communist victory, but for most people the end of the decent interval prompted—or aggravated—a sense of loss. It was not exactly like losing a loved one but perhaps more like the feeling of running over someone's dog; profound regret mixed with a certain helplessness that there was nothing one could do. With the fall of South Vietnam, the slogan of the last four years of the war, "peace with honor," was stripped of its premise and revealed in its true light: "retreat." The Americans did accomplish their long retreat with honor, at least on the battlefield, but after South Vietnam disappeared the painful question remained: had America staged an excruciatingly gradual withdrawal from 1968 to 1973 only to disguise the genuine result of its efforts, a strategic defeat? It was small comfort to the American public that when NVA tanks parked in front of the U.S. Embassy in Saigon on the morning of April 30, 1975, the war had at long last reached an unambiguous conclusion.

After its failure to resist the Communist tide in Vietnam, United States troops fought no more "hot" battles in the Cold War. In fact, they didn't have to. North Vietnam, having gained a huge, finely honed military machine and a wealth of captured equipment, invaded Cambodia in late 1978, in part to stop the madness of the "killing fields" perpetrated by their former partners, the Khmer Rouge. Two months later Red China invaded Vietnam, only to find their former clients superior to Chinese soldiers on the battlefield. The Soviet Union, meanwhile, set up a deep-water naval base at Cam Ranh Bay and parked long-range bombers at Da Nang. On Christmas Day, 1979, the Soviets invaded their neighbor Afghanistan, embarking on their own "Vietnam" and, although United States troops were not involved, the Afghani resistance found itself eagerly supplied from a bottomless well of American weaponry.

In 1989, of course, the Soviet Union began to fall apart at the seams and its essential component, Russia, became a parliamentary democracy. China was rocked by massive demonstrations in Tiananmen Square that included students parading replicas of the American Statue of Liberty. The Communist world that had seemed so fearsome throughout 40 years of Cold War thus rapidly evaporated, leaving behind only a few scattered puddles. At this writing, Russia and China are engaged in a difficult struggle to join the global economy led by America. The United States has enjoyed an unprecedented stretch of prosperity, although its political and military leaders have been on constant alert not to repeat their previous mistake of overreaching. In Asia, only North Korea and the reunited nation of Vietnam continue to fiercely adhere to Marxist principles—both nations ever-committed to struggle, and determined to go it alone.

Afterword

Taking risks in warfare, as in everyday life, will result in either failure or success. It can be inspirational to learn of bold, aggressive ventures—the Normandy invasion, for example—that achieve triumph against dangerous odds. Strangely enough, utter catastrophes such as the Alamo or Ishandlwana, drenched in noble heroism as they are, stir the inspirational juices as well. Perhaps more relevant to the human experience, however, are cases where things did not work out quite as planned and a middle road between triumph and disaster must be taken: retreat. Retreats, as we have seen, occur in a variety of forms, sometimes accompanied by new opportunities for success; at other times simply as an extended prelude to disaster.

George Washington's army and Britain's army of 1940 were able to recover from their early travails, the survivors able, after several more years, to enjoy the sweet smell of total victory. The Americans in Korea similarly recovered from a terrible retreat, although the ultimate solution to that conflict affirms that success, as well as failure, more usually appears in various degrees. To this day, patriots in both America and China can claim victory in Korea by virtue of having not given in to immense enemy power. Both sides are, of course, correct. Napoleon's experience in Russia provides the best example of seeing no light at the end of the tunnel; no choice other than to admit disaster. But then, that giant of European history took on the most formidable foe of all: the weather. The Nez Perce were defeated excruciatingly short of their goal; however, even if they had been able to elude the U.S. Cavalry, one wonders whether they could have prevailed over their next opponent: the twentieth century.

The Nazi war against the Soviet Union is described here during a middle stage, when there was one more surge of the seesaw to come; within that giant conflict, both sides were able to recover from several crippling retreats. It is not entirely surprising that after saving, at great cost, the West from one of its own—a capitalist state, albeit a despicable aberration—the Soviet Union entered the Cold War with an enhanced moral foundation underlying its revolutionary ideology and its new status as a superpower. Around the world European supremacy was in decline and materialistic values had gained a bad reputation. In Vietnam, moral conviction, not greater force, was the primary factor that the United States failed to bring strongly enough to bear. It may seem odd that America retreated before an army that it could have, and repeatedly did, defeat on the battlefield; then again, subsequent events proved that the Cold War, in its later stages, was better fought without guns.

Retreats can thus head in a number of directions. At times the route of withdrawal is filled with those who have lost heart; at other times with courageous fighters determined to remain in the field. History has a habit of assigning clear winners or losers to battles of the past; however, those who retreat, depending upon their motivation, determination and the direction they take, can, in fact, be either.

Source Notes

For full titles and other bibliographical information on books or documents cited, the reader is referred to the bibliography.

WASHINGTON IN NEW YORK

Page 9. "I only wish . . .": quoted in Bobrick, 143; "Cover Americans . . .": quoted in Fleming, 141.

Page 11. "When I can be convinced . . .": quoted in Irving, 156; "wretchedly profane": Ketchum, 200.

Page 12. "I dare say . . .": quoted in Hibbert, 68; "such a dirty . . .": quoted in Irving, 203; "every day . . .": quoted in Commager, 153.; "neither directly . . .": quoted in Irving, 213; "One must not be . . .": Ibid., 214; "I have the greatest reason . . .": quoted in Freeman, 4.

Page 13. "If she ships of war . . .": quoted in Irving, 222.

Page 14. "under intense fire . . .": quoted in Scheer and Rankin, 154.

Page 15. "key to the whole continent": quoted in Gallagher, 14; "I thought all London . . .": quoted in Ketchum, 106.

Page 16: "an advantageous field of battle . . . possession of it.": quoted in Freeman, 87.

Page 18. "a kind of fire . . .": quoted in Gallagher, 93.

Page 21. "Lord Stirling gave battle . . .": quoted in Gallagher, 101; "The Hessians and our brave Highlanders . . .": quoted in Commager, 443.

Page 22. "We were greatly shocked . . .": quoted in Gallagher, 119.

Page 23. "I had at the moment . . .": Clinton, 43–44; "6,000 to defend": Ibid.

Page 24. "The having to deal with . . .": quoted in Scheer and Rankin, 195; "Howe is either our friend . . .": quoted in Bobrick, 215.

Page 25. "I suppose you'll endeavor . . .": quoted in Pearson, 179.

Page 26. "Good God . . .": quoted in Hibbert, 128. (Irving [280] and others quote Washington inquiring instead: "Are these the men with whom I am to defend America?") "Nothing could equal . . .": quoted in Scheer and Rankin, 207; "the enemy appeared . . .": quoted in Commager, 468.

Page 27. "it has given spirits . . .": quoted in Commager, 469; "The prisoners we took . . .": Ibid., 470; "On my going to the foreposts . . .": Clinton, 47.

Page 29. "Providence, or some good honest fellow . . .": quoted in Freeman, 205; "we have reason to suspect . . .": quoted in Commager, 474–75; "There is a radical evil . . .": Ibid., 479; "just dragged from the tender scenes . . .": Ibid., 482.

Page 31. "If only General Lee . . .": quoted in Freeman, 221; "Yonder is the ground . . .": quoted in Scheer and Rankin, 220.

Page 32. "If the battle should be lost . . .": Clinton, 51.

Page 33. "The left of the regiment . . .": quoted in Commager, 490.

Page 35. "creep along up the rocks . . .": Ibid., 494.

Page 36. "Oh General, why would you . . .": quoted in Freeman, 253.

Page 38. "Our people, instead of behaving . . .": quoted in Freeman, 279; "His Excellency thinks . . .": Ibid., 260; "I am of opinion . . .": quoted in Irving, 300.

Page 39. "I do not mean to flatter . . .": quoted in Irving, 300–01; "I received your most obliging . . .": Ibid., 307.

Page 40. "Good God, have I come . . .": quoted in Freeman, 274; "Affairs appear in so important a crisis . . .": quoted in Irving, 303.

Page 41. "If I was not taught to think . . .": quoted in Freeman, 282–83; "I am in hopes here to reconquer . . .": quoted in Irving, 318.

Page 42. "The force I have is weak . . .": quoted in Freeman, 284.; "The militia in this part . . .": Ibid., 286; "I am much surprised . . .": Ibid. "butchered more men . . .": Horace Walpole quoted in Hibbert, 264.

Page 43. "The dread of instant death . . .": quoted in Commager, 501; "Here, sir, are the . . . see to the guard.": quoted in Irving, 320; "The sentries were struck with a . . .": quoted in Commager, 502; "could not be approached . . .": quoted in Wheeler, 168.

Page 44. "most active and most enterprising . . .": quoted in Scheer and Rankin, 235; "The ingenious maneuver . . .": quoted in Bobrick, 225.

NAPOLEON IN RUSSIA

Page 47. "The French people watched . . .": quoted in Ros, 81.

Page 60. "You are compromising . . .": quoted in Palmer, 54.

Page 61. "placed it in the same category . . .": Clausewitz, 133.

Page 62. "The saddest thing of all . . .": Uxkull, in *Arms and the Woman*, 1966.

Page 63. "He who has Smolensk . . .": quoted in Riehn, 217.

Page 65. "This prince, herculean in . . .": quoted in Brett-James, 100; "The King of Naples . . .": Ibid., 101.

Page 66. "Kutusov's arrival . . .": Clausewitz, 140.

Page 70. "The King of Naples charged . . .": quoted in Cate, 239; "At 8:30 we were put on a hill . . .": Uxkull, in *Arms and the Woman*, 1966..

Page 71. "So it's your turn . . . Guard becoming involved": quoted in Brett-James, 130; "gradually took on the character . . .": Clausewitz, 155.

Page 72. "When I said that all important . . .": quoted in Brett-James, 133; "I must be able to see the chessboard . . .": quoted in Cate, 243.

Page 73. "A knight in the strictest sense . . .": quoted in Brett-James, 160.

Page 74. "A demon inspires these people . . .": quoted in Palmer, 148.

Page 75. "How much longer . . . we who started it": quoted in Cate, 316; "The Don regiments continue . . .": quoted in Brett-James, 207.

Page 76. "Supply officials and actors . . .": Ibid.; "Everywhere you looked . . .": from *Voina I Mir*, 1868. Tr. by Stefan Korshak.

Page 77. "We have never understood . . .": Clausewitz, 173.

Page 79. "The naked masses . . .": quoted in Brett-James, 222; "sit and moan beside . . .": Ibid., 226; "singing in chorus . . .": Ibid.

Page 80. "We heard them shouting . . .": Bourgogne, 80–81.

Page 81. "A Marshal of France . . .": quoted in Palmer, 228.

Page 82. "Never did a battle won . . .": quoted in Cate, 364; "The bravest of the brave": quoted in Palmer, 229. "On this day . . .": quoted in Brett-James, 240; "One cannot imagine . . .": Ibid., 226; "I noticed that people with dark hair . . .": Ibid., 237.

Page 83. "In every bivouac . . .": Walter, 72.

Page 84. "his reputation as a soldier . . .": Clausewitz, 179.

Page 86. "Every moment I found . . .": Austin, 300; "One had to beware . . .": Brett-James, 259; "a beautiful lady . . .": Austin, 304.

Page 87. "To be on foot . . . one step further forward": Walter, 86; "everyone was screaming . . .": Ibid., 85; "One night we bivouacked . . .": from *Dvenatsaty God v Vosponinaniakh Sovremennikov*, 1912. Tr. by Stefan Korshak.

Page 88. "The roads were like . . .": Bourgogne, 214; "Murat was huddled . . .": quoted in Brett-James, 278; "I saw terrifying corpses . . .": Ibid., 281.

Page 89. "We had just been served . . .": Ibid., 288.

THE NEZ PERCE IN MONTANA

Page 93. "the people": Hampton, 20.

Page 94. "much more cleanly . . .": quoted in Josephy, 14; "the most friendly . . .": quoted in Hampton, 23.

Page 95. "drunkenness, gambling, crime . . .": quoted in Hampton, 30; "It can hardly be anticipated . . .": quoted in Brown, 33.

Page 96. "It is a great pity . . .": Ibid., 38.

Page 97. "The region of country . . .": quoted in Brown, 41.

Page 98. "I have heard the officers discuss him . . .": Ibid., 97; "This Joseph will admit . . .": Ibid., 73.

Page 99. "General Howard is promenading . . .": Ibid., 84; "who had a heavy gutteral . . .": quoted in Hampton, 51; "My conduct was somewhat . . .": quoted in Howard, 127; "We have put all non-treaty . . .": quoted in Hampton, 55; "Yesterday [the Nez Perce] had a grand. . .": quoted in Howard, 152.

Page 101. "Oh shoot me!": quoted in Hampton, 9.

Page 102. "Don't shoot . . .": Ibid., 12; "We had to run her down . . .": Ibid., 13.

Page 103. "One thing is certain . . .": Ibid., 60.

Page 107. "Leave us alone . . .": Ibid., 93.

Page 111. "Although we outnumbered . . .": quoted in Brown, 191.

Page 112. "stealthily crawling through the grass . . .": Ibid., 194; "The hills on the opposite side . . .": Ibid.

Page 113. "victory barren of results": quoted in Hampton, 118.

Page 114. "Not only was Joseph . . .": Ibid., 121.

Page 115. "narrow defile . . .": Brown, 203.

Page 118. "In a fight between an officer. . .": Ibid., 254; "Many women and children . . .": Ibid.

Page 119. "Few of us will soon forget . . .": Ibid., 258; "If the worst comes . . .": quoted in Howard, 252.

Page 121. "They take no guns . . .": quoted in Brown, 276.

Page 123. "The Chief sat by the fire . . .": quoted in Howard, 280.

Page 124. "I ran about 50 feet . . .": quoted in Brown, 329.

Page 127. "an Indian mounted each horse . . .": Ibid., 358; Only one warrior, Teeto . . .": quoted in Beal, 195–96.

Page 128. "It was the most horrible . . .": quoted in Beal, 197; "I find it impossible . . .": Ibid., 198.

Page 129. "One young buck . . .": quoted in Hampton, 280.

Page 130. "This gallop forward . . .": quoted in Howard, 316.

Page 131. "Enemies right on us!": quoted in Hampton, 290; "A strange Indian chief . . .": quoted in McWhorter, 482; "My father told me to run . . .": Ibid., 483–84.

Page 132. "I dashed unarmed . . .": quoted in Beal, 216; "I thought it singular . . .": quoted in Brown, 395.

Page 133. "I am the only damned man . . .": quoted in Beal, 215; "Ten or twelve soldiers . . .": Ibid., 216.

Page 135. "Where the sun now stands . . .": quoted in Josephy, 630; "Thus has terminated . . .": Report of Secretary of War, 1877.

Page 136. "I am very sure . . .": Wood, 6.

THE ALLIES AT DUNKIRK

Page 142. "senseless": Guderian, 90; "I don't think . . .": Ibid., 92; "the hard task ahead . . .": Ibid., 91.

Page 149. "Suddenly there was . . .": Guderian, 100.

Page 150. "Klotzen, nicht kleckern": Ibid., 105.

Page 151. "We have been defeated . . . the battle": Churchill, 42.

Page 154. "Fighters alone . . .": Terraine, 130.

Page 155. "The rest of the division . . .": Chapman, 158.

Page 156. "It was intensely interesting . . .": quoted in Hamilton, 368; "Utter dejection . . . strategic reserve?": Churchill, 47; "I admit this was . . .": Ibid.

Page 157. "Presently I asked . . .": Ibid., 49.

Page 158. "And then what are you . . . ask me this question": Guderian, 92; "Corps headquarters remained . . .": Ibid., 110.

Page 159. "The enemy platoon commanders . . .": quoted in Horne, 428.

Page 160. "There had been no miracle. . . .": Chapman, 178.

Page 161. "All in a state . . . thought of plan": Ibid., 184; "God help the BEF": quoted in Horne, 498.

Page 162. "I brought every available gun . . .": Ibid., 505.

Page 165. "The attack on the town . . .": Guderian, 116.

Page 167. "We have reason to . . .": Churchill, 71; "Only a miracle . . .": quoted in Gelb, 124.

Page 168. "make use of the period . . .": Guderian, 117.

Page 171. "We took 20,000 . . .": Ibid., 118.

Page 174. "The continued existence . . .": quoted in Terraine, 76; "It would be agreeable . . .": Ibid., 155.

Page 175. "The panzer generals . . .": Taylor, 267.

Page 176. "The effect on Barker . . .": quoted in Hamilton, 388.

Page 177. "His Majesty's Government . . .": Churchill, 110.

Page 178. "We must be very careful . . .": Ibid., 115; "History is full . . .": Lord, 274.

Page 179. "While to uninformed Continentals . . .": Churchill, 578.

FIRST PANZER ARMY IN THE CAUCASUS

Page 183. "Kick in the door . . .": quoted in Seaton, *The Battle for Moscow.*

Page 187. "Are we supposed to . . . ": quoted in Hingley, 337.

Page 188. "Even if the coming . . .": Gehlen, 55.

Page 190. "The defenders would not . . .": quoted in Carell, 472.

Page 192. "Due to our forced . . .": Zhukov, 125; "could have taken Stalingrad. . .": quoted in Mellinthin, 192.

Page 193. "It will, in addition . . . fast formations": quoted in Carell, 477.

Page 194. "If I do not get . . .": quoted in Goralski, 174; "His persistent under estimation . . .": quoted in Shirer, 917; "It was to lay . . .": Manstein, 260.

Page 197. "Tanks! Tanks!": quoted in Tieke, 64.

Page 198. "The Soviet armies had learned . . .": Glantz (Symposium), 38.

Page 201. "One should never . . .": Warlimont, 257.

Page 202. "You have to spare the nerves . . .": quoted in Irving, 425.

Page 203. "The entire army . . .": quoted in Tieke, 172; "A great drama . . .": Ibid; "Earlier than anticipated . . .": Ibid., 175.

Page 206. "There was fierce . . .": Keitel, 185.

Page 210. "Manstein's coming": quoted in Beevor, 291.

Page 211. "Hitler was still not . . .": Manstein, 392.

Page 214. "A tank battalion . . .": quoted in Tieke, 264; "We had to drive . . .": Ibid., 269.

Page 215. "If Hitler thought . . .": Manstein, 386; "For the first time . . .": Werth, 531.

Page 216. "Darkness fell . . .": quoted in Tieke, 277.

Page 217. "During the day . . .": Ibid., 295–96.

Page 218. "Soviet intelligence . . .": Glantz (Symposium), 69.

EIGHTH ARMY IN KOREA

Page 226. "By 1949, we were . . .": Ridgeway, 11.

Page 229. "retreating, panting . . .": quoted in Toland, 61.

Page 235. "There will be no Dunkirk . . .": quoted in Blair, 167; "Glad you British have . . .": Hastings, 89.

Page 237. "The enemy, although we . . .": Ridgeway, 30.

Page 239. "We want you to feel . . .": quoted in Alexander, 236.

Page 240. "We are no longer . . .": quoted in Toland, 241.

Page 245. "one of the most offensive . . .": quoted in Spurr, 162.

Page 246. "After all, there's a lot . . .": quoted in Hastings, 130.

Page 247. "The coordinated action . . .": quoted in Appleman, 102; "abandon all their . . .; Ibid., 104.

Page 251. "All up and down the . . .": quoted in Knox, 625.

Page 254. "How many hordes . . .": quoted in Alexander, 311.

Page 255. "The tank drivers . . .": Appleman, 197.

Page 256. "The developments resulting . . .": quoted in Blair, 464.

Page 261. "The tanks in the division . . .": Appleman, 266.

Page 263. "The Chinese hit us . . .": quoted in Knox, 647.

Page 264. "For the next 500 yards . . .": quoted in Appleman, 264.

Page 265. "It was not what was desired . . .": Ibid., 269.

Page 266. "Get the hell out of . . .": Ibid., 271.

Page 268. "Geysers of flame . . .": Paik, 110.

Page 271. "The men I met . . .": Ridgeway, 86; "It has always been tempting. . .": Ibid., 76.

Page 272. "It is not a question . . .": Ibid., 264.

THE FALL OF SOUTH VIETNAM

Page 279. "What the hell is . . .": quoted in Davidson, 483.

Page 281. "But then, in the sort of bizarre . . .": Harrison, 270.

Page 283. "The promises of the Great . . .": Isaacs & Downing; "Post babyboomers . . .": *New York Times,* April 18, 1999.

Page 289. "The most prevalent form . . .": Davidson, 618.

Page 290. "If they had been wise . . .": quoted in Prados (*The Blood Road*), 325; "The speak glibly of . . .": Ibid., 360.

Page 292. "Such an order . . .": Davidson, 646.

Page 293. "Haze from the bombings . . .": Marshall, 257.

Page 294. "Fischer continued his . . .": Ibid., 259.

Page 296. "I don't want any more . . .": quoted in Morrison, 545.

Page 303. "The battle of Ban Me Thuot . . .": Dung, 99.

Page 306. "In these new, more demanding . . .": Ibid., 87.

Page 307. "as if Da Nang were . . .": Butler, 214.

Page 308. "From the North all manner . . .": Dung, 134–35.

Page 310. "I don't care how many . . .": quoted in Butler, 247; "Northwest of Saigon . . .": Dung, 170.

Page 312. "Then there were hundreds . . .": quoted in Santori, 232.

Page 313. "The panic really started . . .": Ibid., 234.

Page 315. "When you get on that . . .": quoted in Mauer, 612–13; "During the last hours . . .": Ibid., 605.

Page 316. "On the basis of the reported . . .": quoted in Todd, 366.

Select Bibliography

WASHINGTON IN NEW YORK

Bobrick, Benson. *Angel in the Whirlwind: The Triumph of the American Revolution*. New York: Penguin Books, 1997.

Clinton, Sir Henry. *The American Rebellion*. New Haven, CT: Yale University Press, 1954.

Commager, Henry Steele, and Richard B. Morris. *The Spirit of 'Seventy-Six: The Story of the American Revolution as Told by Participants*, Vol. I. Indianapolis/New York: The Bobbs-Merrill Co., 1958.

Fleming, Thomas. "George Washington, General," in Robert Cowley (ed.): *Experience of War*. New York: W.W. Norton, 1992.

Freeman, Douglas Southall. *George Washington: A Biography*, Vol. IV. New York: Charles Scribner's Sons, 1951.

Gallagher, John J. *The Battle of Brooklyn, 1776*. New York: Sarpedon, 1995.

Hibbert, Christopher. *Redcoats and Rebels. The American Revolution Through British Eyes*. New York: W.W. Norton, 1990.

Irving, Washington. *George Washington, A Biography*. (Charles Neider, ed.) Garden City, NY: Doubleday & Co., 1976.

Ketchum, Richard M. *The Winter Soldiers*. Garden City, NY: Doubleday & Co., 1973.

Leckie, Robert. *George Washington's War: The Saga of the American Revolution*. New York: HarperCollins, 1992.

Moore, Frank (comp.) and John Anthony Scott (ed.). *The Diary of the American Revolution*. New York: Washington Square Press, 1967.

Pearson, Michael. *Those Damned Rebels: The American Revolution As Seen Through British Eyes*. New York: G.P. Putnam & Sons, 1972.

Scheer, George F., and Hugh F. Rankin. *Rebels & Redcoats*. New York: World Publishing Co., 1957.

Ward, Christopher. *The War of the Revolution*, Vol. I. New York Macmillan, 1952.

Wheeler, Richard. *Voices of 1776*. New York: Thomas Y. Crowell Co., 1972.

NAPOLEON IN RUSSIA

Austin, Paul Britten. *1812: The Great Retreat*. London: Greenhill Books, 1996.

Bourgogne, François. *Memoirs of Sergeant Bourgogne, 1812–1813*. London: Constable, 1996.

Brett-James, Antony (trans. & ed.). *1812: Eyewitness Accounts of Napoleon's Defeat in Russia*. London: Macmillan and Co., 1967.

Cate, Curtis. *The War of the Two Emperors: The Duel Between Napoleon and Alexander: Russia, 1812*. New York: Random House, 1985.

Chandler, David G. *The Military Maxims of Napoleon*. London: Greenhill Books, 1987.

Clausewitz, Carl von. (Peter Paret and Daniel Moran, ed. & trans.) *Historical and Political Writings*. Princeton, NJ: Princeton University Press, 1992.

Delderfield, R.F. *Napoleon's Marshals*. New York: Stein and Day, 1962.

Fregosi, Paul. *Dreams of Empire: Napoleon and the First World War, 1792–1815*. New York: Birch Lane Press, 1990.

Haythornthwaite, Philip J. *Napoleon's Military Machine*. New York: Sarpedon, 1995.

Horne, Alistair. *How Far from Austerlitz? Napoleon 1805–1815*. London: Macmillan, 1996.

Kaiser, David. *Politics & War: European Conflict from Philip II to Hitler*. Cambridge, MA: Harvard University Press, 1990.

Nafziger, George F. *Napoleon's Invasion of Russia*. Novato, CA: Presidio Press, 1988.

Nosworthy, Brent. *With Musket, Cannon and Sword: Battle Tactics of Napoleon and His Enemies*. New York: Sarpedon, 1996.

Palmer, Alan. *Napoleon in Russia: The 1812 Campaign.* New York: Simon and Schuster, 1967.

Riehn, Richard K. *1812: Napoleon's Russian Campaign.* New York: McGraw-Hill, 1991.

Smith, Digby. *The Greenhill Napoleonic Wars Data Book.* London: Greenhill Books, 1998.

Villahermosa, Gilberto, and Matt DeLaMater. "The Battle of Borodino: Revisiting Napoleon's Bloodiest Day," in *Napoleon Journal,* No. 14, Fall 1999.

Walter, Jakob. (Marc Raeff, ed.) *The Diary of a Napoleonic Foot Soldier.* New York: Doubleday, 1991.

Napoleon: His Life, His Wars, His World, No. 6, Nov. 1996.

THE NEZ PERCE IN MONTANA

Beal, Merrill D. *I Will Fight No More Forever: Chief Joseph and the Nez Perce War.* Seattle: University of Washington Press, 1963.

Brown, Mark H. *The Flight of the Nez Perce: A History of the Nez Perce War.* New York: G.P. Putnam's Sons, 1967.

Dillon, Richard H. *North American Indian Wars.* New York: Gallery Books, 1983.

Hampton, Bruce. *Children of Grace: The Nez Perce War of 1877.* New York: Henry Holt & Co., 1994.

Howard, Helen Addison. *Saga of Chief Joseph.* Caldwell, ID: The Caxton Printers, Ltd., 1941.

Josephy, Alvin M., Jr. *The Nez Perce Indians and the Opening of the Northwest.* Boston: Houghton Mifflin, 1997.

McWhorter, L.V. *Hear Me, My Chiefs: Nez Perce History and Legend.* Caldwell, ID: The Caxton Printers, Ltd., 1992.

Morgan, Ted. *A Shovel of Stars: The Making of the American West 1800 to the Present.* New York: Simon & Schuster, 1995.

Taylor, Marian W. *Chief Joseph: Nez Perce Leader.* Philadelphia: Chelsea House Publishers, 1993.

Wood, Charles Erskine Scott. "The Pursuit and Capture of Chief Joseph." Reprinted in *Archives of the West 1874–1877* by The West Film Project and WETA, 1996.

THE ALLIES AT DUNKIRK

Bailey, Anthony. "Bloody Marvellous," in *Military History Quarterly*, Vol. 3, No. 2, Winter 1991.

Carse, Robert. *Dunkirk 1940*. Englewood Cliffs, NJ: Prenctice-Hall, 1970.

Chapman, Guy. *Why France Fell: The Defeat of the French Army in 1940*. New York: Holt Rinehart and Winston, 1968.

Churchill, Winston S. *Their Finest Hour*. Cambridge, MA: Houghton Mifflin Co., 1949.

Collier, Richard. *1940: The Avalanche*. New York: The Dial Press, 1979.

Gelb, Norman. *Dunkirk: The Complete Story of the First Step in the Defeat of Hitler*. New York: William Morrow, 1989.

Guderian, Heinz. *Panzer Leader*. New York: E.P. Dutton, 1952.

Hamilton, Nigel. *Monty: The Making of a General*. New York: McGraw-Hill, 1981.

Hart, B.H. Liddell. *The German Generals Talk*. Orig. 1948. New York: Quill, 1975.

Horne, Alistair. *To Lose a Battle: France, 1940*. Boston: Little Brown and Co., 1969.

Irving, David. *Hitler's War*. New York: Viking, 1990.

Keegan, John. *The Second World War*. New York: Viking, 1989.

Killen, John. *A History of the Luftwaffe, 1915–1945*. New York: Doubleday, 1967.

Lord, Walter. *The Miracle of Dunkirk*. New York: Viking, 1982.

Lukacs, John. *The Duel: 10 May–July 1940: The Eighty-Day Struggle Between Churchill and Hitler*. New York: Ticknor & Fields, 1991.

Meyer, Kurt. *Grenadiers*. Winnipeg: J.J. Fedorowicz Publishing, 1994.

Shirer, William. *The Rise and Fall of the Third Reich*. New York: Simon & Schuster, 1960.

Stein, George H. *The Waffen SS, 1939–45*. Ithaca, NY: Cornell University Press, 1966.

Taylor, Telford. *The March of Conquest: The German Victories in Western Europe, 1940*. New York: Simon and Schuster, 1958.

Terraine, John. *A Time for Courage: The Royal Air Force in the European War, 1939–1945*. New York: Macmillan Publishing Co., 1985.

Thompson, Laurence. *1940*. New York: William Morrow, 1966.

FIRST PANZER ARMY IN THE CAUCASUS

Beevor, Antony. *Stalingrad: The Fateful Siege: 1942–1943.* New York: Viking, 1998.

Carell, Paul. *Hitler Moves East, 1941–1943.* Winnipeg: J.J. Fedorowicz Publishing, 1991.

Gehlen, Reinhard (David Irving, tr.). *The Service: The Memoirs of General Reinhard Gehlen.* New York: World Publishing, 1972.

Glantz, David (ed.). *Art of War Symposium: From the Don to the Dnepr: Soviet Offensive Operations—December 1942–August 1943.* Center for Land Warfare: U.S. Army War College, 1984.

———. *The Role of Intelligence in Soviet Military Strategy in World War II.* Novato, CA: Presidio Press, 1990.

———. *Kharkov 1942: Anatomy of a Military Disaster.* Rockville Centre, NY: Sarpedon, 1998.

Goralski, Robert, and Russell W. Freeburg. *Oil & War: How the Deadly Struggle for Fuel in WWII Meant Victory or Defeat.* New York: William Morrow & Co., 1987.

Hingley, Ronald. *Joseph Stalin: Man and Legend.* New York: McGraw-Hill, 1974.

Keitel, Wilhelm. *The Memoirs of Field Marshal Keitel.* Walter Gorlitz, ed. New York: Stein and Day, 1966.

Lucas, James. *Alpine Elite: German Mountain Troops of World War II.* London: Jane's Publishing Co., 1980.

Manstein, Erich von. *Lost Victories.* Novato, CA: Presidio Press, 1982.

Mellenthin, F.W. von. *Panzer Battles.* Tulsa: University of Oklahoma Press, 1956.

Tieke, Wilhelm. *The Caucasus and the Oil: The German-Soviet War in the Caucasus, 1942/43.* Winnipeg: J.J. Fedorowicz Publishing, 1995.

Warlimont, Walter. *Inside Hitler's Headquarters 1939–45.* Novato, CA: Presidio Press, 1990.

Werth, Alexander. *Russia At War, 1941–1945.* New York: Carroll & Graf Publishers, 1984.

Zaloga, Steven. "Technological Surprise and the Initial Period of War," in *The Journal of Slavic Military Studies*, Vol. 6, No. 4, Dec. 1993.

Zhukov, Georgi K. (Harrison Salisbury, ed.). *Marshal Zhukov's Greatest Battles.* New York: Harper & Row, 1969.

EIGHTH ARMY IN KOREA

Alexander, Bevin. *Korea: The First War We Lost.* New York: Hippocrene Books, 1986.

Appleman, Roy E. *Escaping the Trap: The US Army X Corps in Northeast Korea, 1950.* College Station: Texas A&M University Press, 1990.

———. *Disaster in Korea: The Chinese Confront MacArthur.* College Station: Texas A&M University Press, 1989.

Blair, Clay. *The Forgotten War: America in Korea, 1950–1953.* New York: Times Books, 1987.

Boettcher, Thomas D. *First Call: The Making of the Modern U.S. Military, 1945–1953.* Boston: Little, Brown and Co., 1992.

Cohen, Eliot A., and John Gooch. *Military Misfortunes: The Anatomy of Failure in War.* New York: The Free Press, 1990.

Hallahan, Robert F. *All Good Men: The 52nd Field Artillery Battalion in Korea, 1950–1953.* n.p., 1999.

Hastings, Max. *The Korean War.* New York: Simon & Schuster, 1987.

Hopkins, William B. *One Bugle No Drums: The Marines at Chosin Reservoir.* Chapel Hill, NC: Algonquin Books of Chapel Hill, 1986.

Hoyt, Edwin P. *The Day the Chinese Attacked: Korea 1950.* New York: McGraw-Hill, 1990.

Knox, Donald. *The Korean War, An Oral History: Pusan to Chosin.* New York: Harcourt Brace Jovanovich, 1985.

———. *The Korean War: Uncertain Victory.* New York: Harcourt Brace Jovanovich, 1988.

Marshall, S.L.A. *The River and the Gauntlet: Defeat of the Eighth Army by the Chinese Communist Forces, November 1950, in the Battle of the Chongchon River.* New York: William Morrow & Co., 1953.

Paik, Sun Yup. *From Pusan to Panmunjom.* McLean, VA: Brassey's, Inc., 1992.

Ridgeway, Matthew B. *The Korean War.* Garden City, NY: Doubleday & Co., 1967.

Spurr, Russell. *Enter the Dragon: China's Undeclared War Against the U.S. in Korea, 1950–51.* New York: Newmarket Press, 1998.

Toland, John. *In Mortal Combat: Korea, 1950–1953.* New York: William Morrow & Co., 1991.

Wilkinson, Allen Byron. *Up Front Korea.* New York: Pilot Books, 1974.

THE FALL OF SOUTH VIETNAM

Butler, David. *The Fall of Saigon.* New York: Simon & Schuster, 1985.

Colby, William. *Lost Victory.* Chicago: Contemporary Books, 1989.

Davidson, Phillip B. *Vietnam at War: The History, 1946–1975.* Novato, CA: Presidio Press, 1988.

Dung, Van Tien. *Our Great Spring Victory.* New York: Monthly Review Press, 1977.

Giap, Vo Nguyen: *How We Won the War.* Philadelphia: Recon Publications, 1976.

Harrison, James P. *The Endless War: Vietnam's Struggle for Independence.* New York: The Free Press, 1982.

Isaacs, Jeremy, and Taylor Downing. *Cold War: An Illustrated History, 1945–1991.* Boston: Little, Brown & Co., 1998.

Kalb, Marvin, and Bernard Kalb. *Kissinger.* Boston, Little Brown, 1974.

Marshall, Tom. *The Price of Exit.* New York: Ivy Books, 1998.

Maurer, Harry. *Strange Ground: Americans in Vietnan, 1945–1975: An Oral History.* New York: Henry Holt, 1989.

Morrison, Wilbur H. *The Elephant and the Tiger: The Full Story of the Vietnam War.* New York: Hippocrene Books, 1990.

Olsen, James S. and Randy Roberts. *Where the Domino Fell: America and Vietnam 1945 to 1990.* New York: St. Martin's Press, 1991.

Prados, John. *The Blood Road: The Ho Chi Minh Trail and the Vietnam War.* New York: John Wiley & Sons, 1999.

———. *The Hidden History of the Vietnam War.* Chicago: Ivan R. Dee, 1995.

Pratt, John Clark (ed.). *Vietnam Voices: Perspectives on the War Years, 1941–1982.* New York: Penguin Books, 1984.

Santoli, Al. *To Bear Any Burden: The Vietnam War and Its Aftermath in the Words of Americans and Southeast Asians.* New York: E.P. Dutton, 1985.

Schulzinger, Robert D. *A Time for War: The United States and Vietnam, 1941–1975.* New York: Oxford University Press, 1997.

Todd, Oliver. *Cruel April: The Fall of Saigon.* New York: W.W. Norton, 1990.

Young, Marilyn. *The Vietnam Wars, 1945–90.* New York: HarperCollins, 1991.

Index